Riti Thapar Kapoor, Mohd Rafatullah and Aseem Vashisht (Eds.)
Electronic Wastes

Also of interest

Electronic Wastes.
Pollution, Management Strategies, Policy Frameworks
Edited by Riti Thapar Kapoor, Mohd Rafatullah and Aseem Vashisht, 2026
ISBN 978-3-11-224806-5, e-ISBN (PDF) 978-3-11-224807-2

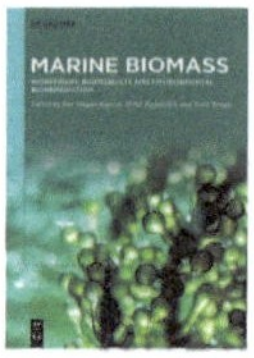

Marine Biomass.
Biorefinery, Bioproducts and Environmental Bioremediation
Edited by Riti Thapar Kapoor, Mohd Rafatullah, Norli Ismail, 2024
ISBN 978-3-11-135358-6, e-ISBN (PDF) 978-3-11-135395-1

Bioremediation Technologies.
For Wastewater and Sustainable Circular Bioeconomy
Edited by Riti Thapar Kapoor, Mohd Rafatullah, 2023
ISBN 978-3-11-101665-8, e-ISBN (PDF) 978-3-11-101682-5

BioChar.
Applications for Bioremediation of Contaminated Systems
Edited by Riti Thapar Kapoor, Maulin P. Shah, 2022
ISBN 978-3-11-073858-2, e-ISBN (PDF) 978-3-11-073400-3

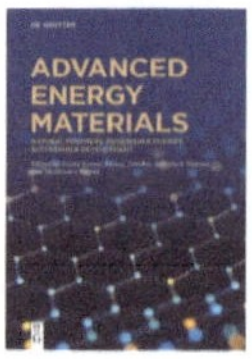

Advanced Energy Materials.
Natural Polymers, Renewable Energy, Sustainable Development
Edited by Sushil Kumar Verma, Sonika Gupta, Abhishek Sharma, Shailendra Rajput, 2025
ISBN 978-3-11-128890-1, e-ISBN (PDF) 978-3-11-128907-6

Biomedical Waste Management.
Bioremediation and Recycling
Edited by Neha Suyal, Prasenjit Debbarma, Manali Singh, Anita Mishra, Roop S. Bora, 2024
ISBN 978-3-11-130316-1, e-ISBN (PDF) 978-3-11-130528-8

Electronic Wastes

Biological Treatments, Valorization, Circular Economy

Edited by
Riti Thapar Kapoor, Mohd Rafatullah and Aseem Vashisht

DE GRUYTER

Editors
Riti Thapar Kapoor
Professor
Centre for Plant and Environmental Biotechnology,
Amity Institute of Biotechnology,
Amity University Uttar Pradesh,
Noida - 201 313
Uttar Pradesh, India

Mohd Rafatullah
Associate Professor
Department of Mathematics and Natural Sciences,
Prince Mohammad Bin Fahd University,
P.O. Box 1664,
Al-Khobar 31952,
Saudi Arabia

Aseem Vashisht
Associate Professor
University Centre for Research and Development,
Chandigarh University,
Mohali, Punjab, 140413,
India

ISBN 978-3-11-222549-3
ISBN 978-3-11-222550-9 (PDF)
ISBN 978-3-11-222551-6 (EPUB)
DOI https://doi.org/10.1515/9783112225509

Library of Congress Control Number: 2026934614

Bibliographic information published by the Deutsche Nationalbibliothek
The Deutsche Nationalbibliothek lists this publication in the Deutsche Nationalbibliografie; detailed bibliographic data are available on the Internet at http://dnb.dnb.de.

De Gruyter and Walter de Gruyter GmbH are part of De Gruyter Brill.
www.degruyterbrill.com

Questions about General Product Safety Regulation:
productsafety@degruyterbrill.com

Cover illustration: photoguns/iStock/Getty Images Plus
Printing: CPI books GmbH, Leck

MIX
Papier | Fördert gute Waldnutzung
FSC® C083411
FSC www.fsc.org

Preface

Rapid economic development, urbanization, global demand, expansion of the digital infrastructure and technical developments have invented different electronic devices that have transformed living standards and overall well-being. Electronic wastes (e-wastes) often known as electronic trash, a type of garbage generated by the industrial world, are one of the most difficult and rapidly expanding issue. Electronic wastes are electronic or electrical equipments that are wastes or not functional and reached the end of life in their service. Approximately nine hundred types of electronic equipments are available in the market. Electronic waste is heterogenous in nature, and consists of heavy metals, dye, plastics, xenobiotic substances like brominated flame retardants, polycyclic aromatic hydrocarbons, polychlorinated biphenyls that are non-biodegradable which pose environment and human health risks. As per the report of global e-waste monitor, in year 2022, around 62 million tonnes of e-wastes have been generated which may rise to 82 million tonnes by the year 2030. Most of the electronic devices are designed with short lifecycle and companies frequently release new models, thus making fast disposal of electronic products. With continous increase in e-wastes, safe disposal of e-wastes has become a global problem. Very less electronic goods under working conditions are used for second hand, as 20% e-wastes are recycled, and around 80% are discarded or recycled by unskilled workers in informal settings, who handle e-wastes manually and pose environment risks. The indiscrimate disposal of e-wastes in landfills or open burning allow toxic components to enter in the groundwater, soil and air, thus adversely affecting ecosystem and human health.

For addressing e-wastes challenges, multi-faceted strategies are needed. Microbes release enzymes such as catalase, reductase, laccase, dioxygenase which can convert toxic components of e-wastes into less toxic forms. Microbes can remediate different toxic constituents such as polychlorinated biphenyls, polyaromatic hydrocarbons in sustainable and eco-friendly way. By leaching process, some microbes assists in extraction of metals from e-wastes. The existing socio-economic system follows a linear approach of take – make – dispose. Electronic wastes can be reduced by applying 3Rs principles such as reduce, reuse and recycle. The e-wastes can be used for the development of value-added products such as construction materials etc. There is a need to develop recycling infrastructure for designing electronic devices which can be easily recycled and resued. Application of single strategy cannot resolve the problem, combined efforts which include technological advancements, policy reforms, global collaborations are required for effective management. Amid these challenges, circular economy emphasizes resource efficiency, material recovery, and sustainable design, thus by increasing the life span of electronic devices by reducing the consumption of raw materials. Therefore, circular practices not only conserves natural resources but also achieves the objectives of sustainable development goals.

 | https://doi.org/10.1515/9783112225509-202

This book highlights biological treatment methods for management of electronic wastes, and circular strategies for valorization of electronic wastes for environmental clean-up. By integrating theory with real-world applications, this book aims to provide readers with a comprehensive understanding of both the challenges and opportunities inherent in this field. The readers of this book includes researchers, environmentalists, engineers, industry persons and policy makers who want to utilize green technologies for removal of e-wastes in feasible way. The editors gratefully acknowledge the authors for their contributions in form of book chapters and learned reviewers for their insightful suggestions, which have enhanced the overall quality of the book. We would also like to express our sincere thanks to De Gruyter Brill Editorial Office, especially Ms Helene Chavaroche for bringing us this opportunity and guiding us throughout the book publication process. Our sincere thanks to Ms Ute Skambraks, Ms Adelinde Luyken, Ms Sathiyavathi Pajaniradj and other publication team members for their continuous guidance and support.

Riti Thapar Kapoor
Professor
Centre for Plant and Environmental Biotechnology
Amity Institute of Biotechnology
Amity University Uttar Pradesh
Noida – 201 313, Uttar Pradesh, India

Mohd Rafatullah
Associate Professor
Department of Mathematics and Natural Sciences
Prince Mohammad Bin Fahd University
P.O. Box 1664, Al-Khobar 31952, Saudi Arabia

Aseem Vashisht
Associate Professor
University Centre for Research and Development,
Chandigarh University, Mohali, Punjab
140413, India

Contents

Deeksha Ranjan, Laxmi Singh, Richa Choudhary*

Electronic wastes: sources, composition, types, and toxicity

Abstract: With the rapid advancement of science and technology, electrical and electronic equipment have become indispensable in modern life, serving critical roles in communication, healthcare, education, entertainment, and industry. However, once these devices reach the end of their functional life, they are discarded as waste electrical and electronic equipment, contributing to the growing global challenge of electronic waste (e-waste). The increasing production, consumption, and obsolescence of electronic and electrical equipment have accelerated the accumulation of e-waste, now recognized as one of the fastest-growing solid waste streams worldwide. E-waste is a complex mixture, containing economically valuable materials such as copper, aluminum, silver, and gold, alongside hazardous substances, including lead (Pb), mercury (Hg), cadmium (Cd), hexavalent chromium [Cr(VI)], and brominated flame retardants (BFRs). Improper disposal or informal recycling practices release these toxicants into soil, water, and air, leading to ecosystem contamination, bioaccumulation, and significant risks to human health, including neurotoxicity, endocrine disruption, and carcinogenesis. The present chapter provides a comprehensive overview of the sources, classification, composition, and toxicity of e-waste, highlighting its dual role as both a secondary resource reservoir and an environmental hazard.

Keywords: Electronic waste, electrical equipment, composition, toxicity, sustainable management

1 Introduction

The upregulated use of the electronics sector over the recent decades has completely transformed our day-to-day life. Devices ranging from smartphones, computers, and medical devices to home appliances, electronic vehicles, and other electronic equipment, have become necessary for communication, healthcare, education, and industry.

***Corresponding author: Richa Choudhary**, Department of Life Sciences, School of Biosciences and Technology, Galgotias University, Greater Noida, Uttar Pradesh, India, e-mail: sinharicha11@gmail.com
Deeksha Ranjan, Department of Applied Sciences and Humanities, Faculty of Engineering and Technology, Rama University, Kanpur, Uttar Pradesh, India
Laxmi Singh, Department of Biotechnology, Indian Institute of Technology, Madras, Chennai 600036, India

 | https://doi.org/10.1515/9783112225509-001

But this digitally driven revolution has also led to unforeseen and growing environmental issues with the production of e-waste or electronic waste [1]. E-waste includes all the discarded electrical and electronic equipment (EEE) and parts thereof that have reached the end of their operating life or due to becoming obsolete, as a result of rapid technological change and product turnover. It is one of the most rapidly expanding waste streams globally, with far-reaching implications for sustainability, human health, and resource management. As per the Global E-waste Monitor 2024, approximately 62 billion kg of e-waste were generated globally in 2022, marking an increase of around 25% over the past 5 years [2]. About 22.3% of this waste was collected and recycled, but the major part was improperly disposed of in open dumps and landfills. The growth of consumer electronics, along with decreased product life, has started “Planned obsolescence,” where products are deliberately engineered for quick replacements [3]. This has increased the trade of electronic devices, even in slow-developing economies that had been slow in embracing digital technologies. E-waste is a heterogeneous mixture of valuable metals like gold, silver, palladium, and copper, as well as toxic substances like lead, mercury (Hg), cadmium (Cd), chromium (Cr), and brominated flame [4]. However, the recovery of precious metals provides a profitable prospect for secondary resource production, uncontrolled treatment, and recycling procedures, particularly in open dumps and landfills that release hazardous substances into the environment. Open burning, acid leaching, and dismantling in uncontrolled facilities expose the workers and the surrounding communities to a mixture of harmful substances [5]. Contamination of air, soil, and water systems results in bioaccumulation of heavy metals at different trophic levels of the food chain and serious health effects, including neurotoxicity, carcinogenicity, reproductive abnormalities, and endocrine disorders [6].

The composition and toxicity of e-waste depend on several factors, like age, type, and manufacturing substances [7]. Older cathode-ray tube (CRT) displays, for example, have high volumes of leaded glass, while contemporary printed circuit boards are infused with intricate polymers, microchips, and rare earth metals. The widespread application of nanomaterials and lithium-ion batteries in contemporary electronics introduces new contaminants, whose long-term toxicity and environmental behavior are not well understood [8]. Therefore, e-waste composition analysis offers important information regarding possible exposure routes and risk assessment models required for waste management sustainability. Other than environmental pollution, e-waste poses a contradictory problem to the world’s circular economy. Discarded electronics can be used as an urban mine because of the recoverable metals, reducing reliance on mining and other environmental damage [9]. In many underdeveloped countries, safe recovery procedures are not followed because of a lack of advanced recycling technologies, efficient collection mechanisms, and regulatory enforcement. Informal recyclers often use improper methods to recover copper and gold, like burning cables in an open area or using acid baths, which leads to the release of dioxins and furans and other persistent organic pollutants (POPs) in the atmosphere [10]. All these emissions influence the air quality and contribute to global chemical pollution.

Policy-wise, e-waste management is still not aligned properly. The Basel Convention controls the transboundary movement of hazardous waste, including e-waste, but illegal shipments from developed to developing countries are still underway in the name of "Used electronics" [11]. This trade offloads the environmental cost to places with no ability to recycle or dispose of hazardous parts safely. In addition, although the European Union's Waste Electrical and Electronic Equipment (WEEE) Directive and India's E-Waste Management Rules (2022) have laid down guidelines for producer responsibility and formal recycling, worldwide enforcement is uneven. Numerous low- and middle-income nations are still lacking strong monitoring mechanisms, resulting in both the coexistence of formal and informal sectors, with the latter prevailing because of economic benefits.

Another aspect of e-waste relates to socioeconomic forces driving its spread. Urbanization, growth in income, and the desire for a digital lifestyle have raised per capita electronic use, even in rural areas. The faster rate of innovation, like upgradation in mobile phones, smart devices, and home appliances, has rendered repair and reuse less financially appealing. To combat this, scientific studies over the last 10 years have more and more stressed the toxicological complexity of e-waste. Experiments have shown synergistic effects of mixed metal and organic contamination, where combined pollutants are more toxic than individual substances [12]. Interactions among metals like cadmium (Cd) and lead (Pb) with organic flame retardants enhance oxidative stress and interfere with plant, animal, and human metabolic pathways [13]. In addition, e-waste disposal in landfills is the major cause of leachate contamination, which poses long-term threats to groundwater and agricultural systems [14]. It is hence imperative to know the composition and transformation of e-waste-derived contaminants to make predictive models of environmental behavior and risk.

2 Sources of e-waste

The generation of electronic waste is massively attributed to the wide range of sources that go down to the broad use of EEE in the modern world [15]. As a result of accelerated technological innovation, the products are being replaced at quite short intervals, thus causing the discarded material flow to be increasingly complex and diverse [16]. The sources of e-waste may be classified as those of households, commercial, industrial, institutional, and infrastructural areas, each having its own characteristics, compositions, and management challenges (Figure 1). Understanding the sources is very important for the implementation of effective collection, recycling, and policy strategies that intend to lessen the risks to the environment and human health.

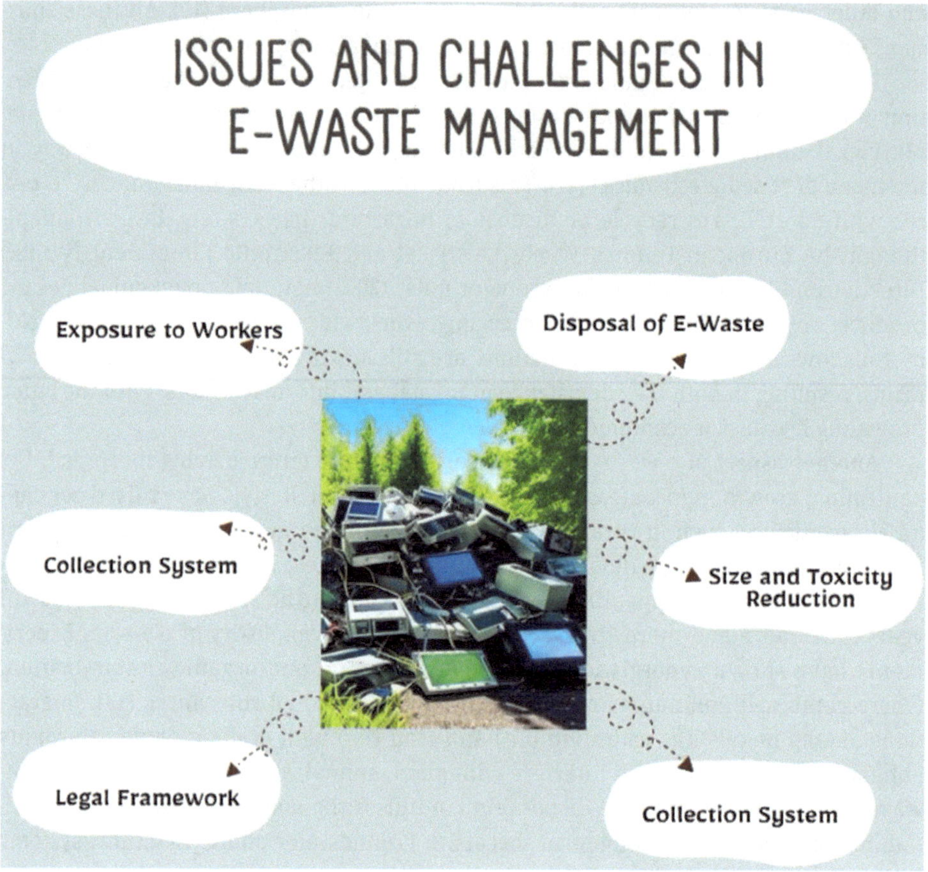

Figure 1: Major issues and challenges faced in the management of e-waste.

2.1 Household sources

Home appliances and gadgets are among the major sources of e-waste worldwide. The common devices that mainly fall under this category are mobile phones, televisions, laptops, tablets, refrigerators, washing machines, air conditioners, and microwave ovens, but it is not limited to these. The sector has, due to the ongoing and rapid technological advances, caused the average lifespan of consumer electronics to be drastically shortened. For instance, the average time interval for changing a smartphone is less than 3 years, which is mainly influenced by software updates, performance upgrades, and consumers' tendency to purchase the latest model [17]. Similarly, the transition from CRT televisions to LED and smart TVs has resulted in an increase in the number of obsolete devices that have reached landfills, each having substantial amounts of leaded glass and plastic housing [18]. In developing countries, a significant amount of e-waste is generated from second-hand and imported electronic devices that become obsolete not long after resale. In the absence of structured degrading systems in most regions, such appliances are primarily disposed of through household garbage or other informal collection means. Then, there are batteries, CFLs, and small accessories of IT, like chargers, routers, and keyboards; although each item may appear insignificant, their cumulative impact due to high consumption volume represents a major environmental concern [19].

2.2 Commercial and office equipment

The commercial sector, including offices, banks, educational institutions, and service industries, generates a considerable share of e-waste. Equipment such as desktop computers, printers, scanners, photocopiers, telecommunication devices, servers, and networking hardware forms a substantial component of this category [19]. One of the major reasons for the frequent obsolescence of electronic hardware is the regular upgrading cycles, corporate technology refresh programs, and the transition to cloud computing. To keep up with efficiency and cybersecurity standards, businesses are replacing electronic systems every 3–5 years [20]. Consequently, there is a huge volume of redundant but still operable equipment that is being piled up. Though some devices are resold or donated to small organizations, a large number of them are discharged into waste streams because of incompatibility and performance degradation. Besides this, printer cartridges, toners, and storage drives with sensitive data are giving additional problems in the matter of disposal; thus, it is often necessary to have secure data destruction before recycling [21]. Despite the potential for reuse, inadequate awareness and lack of incentives hinder systematic collection from commercial entities in most regions.

2.3 Industrial and manufacturing sources

Industrial sources are more technologically complicated sources of electronic waste. The factories, production plants, and manufacturing facilities use a variety of electronic systems, including but not limited to automation controls, programmable logic controllers, sensors, measuring instruments, and robotic components; these eventually become obsolete as technology upgrades, due to calibration issues, or when mechanical failures occur [22]. Besides that, pre-consumer electronic waste is generated by the manufacturing process itself in the form of defective chips, circuit boards, wires, soldered assemblies, and testing units. Such material streams often possess a high concentration of precious metals, thus having a high value for recovery [23]. However, due to improper management, contamination by solvents, heavy metals, and other hazardous substances used in production commonly occurs [24]. In addition to electric/electronic devices, industrial e-waste includes large-scale generators, transformers, and control panels used in energy and utility installations. Disposal is further involved by the toxic composition of capacitors and insulating oils that contain polychlorinated biphenyls (PCBs) [25]. This is a type of POP that requires focused treatment to avoid its release into the environment. Thus, besides the size, industrial wastes are also chemically complicated and need specialized methods of recycling.

2.4 Institutional and healthcare sources

Large-scale institutions, like hospitals, research laboratories, and universities, are major generators of e-waste due to their high reliance on advanced diagnostic and analytical equipment. Electronic instruments within the healthcare industry, particularly X-ray machines, ECGs, MRI machines, infusion pumps, and laboratory analyzers, do reach a stage where they get outdated or cease to function. The high cost and technical specificity of medical electronics usually restrict their potential for reuse. Besides, electronic waste from medical devices has dual risks – chemical and biological – since many of these discarded devices may contain radioactive isotopes, mercury-based switches, or even residual biomedical contaminants. If not disposed of properly, such equipment is very likely to cause environmental contamination and occupation exposure risks [26]. On the other hand, at universities and research institutions, there are outdated laboratory instrumentation, computers, spectrophotometers, and electronic teaching aids that generate e-waste [10]. Due to the frequent changeover to newer technologies by such organizations, in search of precision and efficiency, outdated equipment piles up at an alarming rate. Proper segregation and recycling of institutional e-waste hold the key to avoid hazardous exposure and data security breaches associated with laboratory and academic devices.

2.5 Information and communication (ICT) technology infrastructure

The boom of digital infrastructure and migration toward 5G has added a new dimension to the sources of e-waste. Telecommunication towers, data centers, and network servers have to be constantly upgraded to meet the growing demands for bandwidth and connectivity. As older hardware and communication modules are retired, they make significant additions to the e-waste stream [27]. Servers, routers, modems, power supplies, and cooling systems from data centers are replaced regularly, with the goal of energy efficiency and an increased processing capacity. These components consist of metals, plastics, and rare earth elements; therefore, they offer a good recycling perspective, but often contain hazardous elements like cadmium and lead solder [28]. The concept of the Internet of Things further increases this problem because billions of connected sensors and devices that are being used, become obsolete in due course, establishing a source of e-waste that is dispersed and hard to collect with conventional systems.

2.6 Agricultural and transportation electronics

In the last decade, the emerging sectors of agriculture and transportation have become increasingly important sources of e-waste, besides being unconventional. Agriculture, on one hand, has incorporated electronic systems in irrigation controls, GPS-guided tractors, drones, and soil sensors [29]. These improve productivity and have indeed introduced new waste streams. Some devices have short functional lives and are difficult to repair in rural areas, leading to disposal without proper recycling. On the other hand, modern transportation vehicles incorporate a number of electronic subsystems like engine control units, navigation modules, entertainment panels, and EV batteries [30]. The strong movement toward hybrid and fully electric vehicles has started increasing the numbers of lithium-ion batteries that need to be managed at their end of life. Improper disposal of such components can release heavy metals and electrolytes, which are highly toxic to the environment. Thus, both agriculture and transportation are growing sectors contributing to the global burden of e-waste [1].

2.7 Military and aerospace equipment

Though often overlooked in civilian discussions, defense and aerospace sectors produce high-value and technologically advanced e-waste. The obsolescence of radar systems, navigation units, communication devices, satellites, and weapon control systems due to rapid advancements in the field results in the frequent generation of e-waste [31]. Many of these components contain rare earth elements, high-performance alloys, and polymers. Because confidentiality and security are a concern, open recycling of much of this waste does not take place, with management being carried out through specialized government facilities. The sheer material intensity and hazardous nature of such e-waste, however, makes it an important category for study.

3 Composition of e-waste

The composition of electronic waste reflects the material complexity of modern EEE. On the contrary, e-waste is characterized by a heterogeneous mixture of metals, plastics, glass, ceramics, and various chemical compounds that together form intricate electronic assemblies, unlike traditional waste streams [32]. The different recyclable materials, including valuable precious metals as well as dangerous toxicants, set out both the environmental risk and the economic potential of the recycling process [33]. Information on the makeup of e-waste is therefore crucial for the implementation of sustainable management strategies and to the upgrading of resource recovery and reduction of the ecological and human health impacts.

3.1 Various metallic components of e-waste

Metal fractions represent one of the most critical and valuable parts of all e-waste, making up about 40–60% of its total weight. The metallic fraction can be broadly categorized into base metals, precious metals, and specialty or rare earth elements. Base metals such as iron (Fe), copper (Cu), aluminum (Al), tin (Sn), and zinc (Zn) have widespread applications in structural components, wiring, and circuit connections [34]. Copper is among the most abundant economically recoverable metals present in printed circuit boards, cables, and connectors [35]. Its electrical conductivity is supreme, and hence it is irreplaceable for signal transmission and thermal management. Iron and steel occur in frames, casings, and mechanical parts of larger appliances such as refrigerators and washing machines, while aluminum is used in heat sinks, casings, and lightweight components in laptops and mobile devices. Recovery of base metals results in a good amount of money but if the recycling is done in an improper way, especially when cables are burned in the open to extract copper, then toxic fumes are released and the environment is polluted. Environment-friendly hydrometallurgical and bioleaching methods have been developed as viable solutions to the problem of pollution, resulting from smelting for the recovery of base metals [36] (Table 1). Electronic waste turns out to be a very rich secondary source of precious metals such as gold (Au), silver (Ag), platinum (Pt), and palladium (Pd). These metals are utilized in microelectronic parts due to their good conductivity, corrosion resistance, and catalytic properties [37]. As an example, gold is used in connector pins, bonding wires, and microprocessors; silver is found in solder and contact switches; while palladium and platinum are used in multilayer ceramic capacitors and hard disk drives [38, 39]. Even though their percentage by weight is very small (usually less than 0.1%), the economic value of these metals is very high. As per the estimates, a ton of thrown-away mobile phones can provide more gold than a ton of high-grade ore [40]. But it is not easy to get these metals because the extraction methods need to be very advanced, and they have to go through selective leaching and solvent extraction. If the metals are not handled properly, it can lead to the wastage of precious metals and the release of pollutants like cyanide or aqua regia, which are used in recovery operations that are done in an unregulated manner, thereby polluting the environment [41]. A number of rare earth elements, such as neodymium (Nd), dysprosium (Dy), europium (Eu), terbium (Tb), and yttrium (Y), are used for magnets, phosphors, and displays in electronics [42]. The main uses of neodymium and dysprosium are in high-performance magnets, which get to be the core components of hard drives, speakers, and wind turbine generators, while europium and terbium are used in fluorescent displays and LEDs [43, 44]. Besides that, indium (In), gallium (Ga), and tantalum (Ta) are great contributors in semiconductors, touchscreens, and capacitors [45]. The limited and geopolitically concentrated supply of REE mining makes e-waste an increasingly significant alternative source of materials. Nevertheless, the very low concentrations and the complicated association of these elements in the devices make recycling extremely difficult. Their recovery is still a research frontier, and progress in green chemistry and selective bioleaching is being considered as a possible facilitator.

3.2 Nonmetallic components of electronic waste

The nonmetallic fraction, which includes plastics, glass, and ceramics, makes up 40–60% of total e-waste, by mass. Although these materials provide important insulation, protection, and structure, they also present significant waste management problems due to their persistence and potential toxicity [19]. Plastics are found in every electronic product, from casings and insulation to circuit board components. Common types include PVC, ABS, PC, and PE. Most of these materials are durable, flame-resistant, and lightweight. Their composition includes, but is not limited to, BFRs like TBBPA and PBDEs that prevent ignition while releasing toxic brominated dioxins during incineration [46]. The other challenge involves mixed plastic wastes, which are generated by the dismantling of electronics and cannot easily be recycled due to multiple polymer types and metal fragments embedded within the wastes [47]. Several studies are underway to explore mechanical separation and feedstock recycling or chemical depolymerization, in hopes of recovering useful polymer material. However, economic viability remains low. Nevertheless, developing bioplastic-based electronic components could reduce this challenge in the future. Glass plays an integral role in display units, CRTs, and optical components. CRT glass, once the dominant fraction of e-waste, contains as much as 25% lead oxide, with additions of barium oxide to provide radiation shielding [48]. When the glass is landfilled or crushed, the lead is leached into the surrounding soil and groundwater. Modern flat panel displays are based on aluminosilicate and borosilicate glasses, which are far less toxic, but their recyclability is still rendered difficult by ITO coatings and layer structures [49]. Ceramics add a further complicating dimension, as they are generally stable and nonbiodegradable, complicating waste treatment [50]. However, given their inert nature, ceramic components have relatively low direct toxicity, compared with plastics or metals, though they do contribute to landfill volume.

3.3 Hazardous components and toxic substances

One of the main characteristics of e-waste is that, along with valuable materials, it also has hazardous substances. In many electronic components, chemicals are used that are classified as toxic, carcinogenic, or environmentally persistent [12]. As a result, the highly problematic elements in e-waste are lead (Pb), cadmium (Cd), mercury (Hg), arsenic (As), and chromium (Cr), among others [51]. Lead is in widespread use in solder, glass, and cables, while cadmium appears in rechargeable batteries (Ni–Cd), chip resistors, and semiconductors, and mercury is used in switches, thermostats, and fluorescent backlights [52, 53]. These elements are capable of bioaccumulation and may cause neurological, renal, and developmental toxicity in humans. Chromium, especially in its hexavalent form (Cr^{6+}), is used in corrosion-resistant coatings, with carcinogenic risks [54]. Poor informal recycling practices, such as manual dismantling and open burning, release these metals into the air, soil, and water, resulting in long-term contamination. For example, studies in e-waste recycling hubs, including Guiyu in China and Agbogbloshie in Ghana, have shown soil and river contamination levels of lead and cadmium that are several orders of magnitude above safe limits [55]. In addition to heavy metals, electronic waste comprises a substantial amount of organic pollutants such as PCBs, brominated flame retardants (BFRs), and phthalates [12]. These chemicals are widely recognized as POPs due to their characteristics of bioaccumulation and interference with the endocrine system. Incomplete combustion of plastics and flame retardants during open burning and smelting generates dioxins and furans, two of the most toxic environmental contaminants known. PCBs are usually made of epoxy resins and phenolic compounds that release hazardous fumes in case of inappropriate processing [56]. Simultaneous exposure to heavy metals and organics is linked to synergistic toxicity effects, presenting a severe threat to human health and ecological balance [57]. Contemporary electronic devices increasingly employ nanomaterials and lithium-ion batteries, thus introducing new classes of contaminants. Nanoparticles of Ag, TiO_2, and carbon nanotubes are used for electric conductivity and durability [58]. However, because of their nanoscale dimensions and high reactivity, they may easily disperse into the environment, where their long-term ecotoxicological effects are still under question. The same can be said about lithium-ion batteries, containing electrolytes based on lithium salts and organic solvents, aside from cobalt and nickel oxides in cathodes. Such batteries lead to fire hazards due to mismanagement. In the last few years, increased adoption of electric vehicles and portable electronics has dramatically increased this emerging waste stream, which demands advanced recycling and recovery technologies.

4 Types of electronic waste

Classification of e-waste is of importance from various aspects, as it gives a clear overview of the management, treatment, and recycling potential of electronic waste. Various classification schemes have been proposed, based on functionality, usage patterns, material complexity, and regulatory frameworks, due to the wide diversity in EEE in use worldwide. Economic, technological, and regional differences in production and consumption of electronic appliances also lead to a different categorization of e-waste. E-waste is systematically classified in various domains, ranging from a traditional "white–brown–grey" goods model to functional grouping under the WEEE directive, and newer classification methods in modern waste management systems (Table 2).

4.1 Traditional classification: white, brown, and gray goods

One of the simplest and most often-used ways to classify e-waste is by dividing it into three fundamental categories [59]. The term "white goods" is a common usage used to refer to major home appliances that enable housework activities such as cooking, cleaning, and refrigerating. Examples of items in this category include refrigerators, washing machines, air conditioners, dishwashers, ovens, and other bulky electronic devices. They are termed "white" because they often come with a typical enamel coating with a white finish. White goods generate among the highest volumes of e-waste due to their size, high metal content, and relatively short service lifespan in developing countries [1]. From a recycling point of view, they are metal-rich, essentially composed of substantial amounts of steel, aluminum, and copper, which makes them suitable for mechanical shredding and material recovery processes. Brown goods include entertainment devices such as television, radio, video cameras, and music systems. Over history, these devices were contained in brown wooden or plastic casings; hence, they came to be referred to as "brown goods." Brown goods are comprised of complex electronic circuitry, cathode ray tubes in the older models, and printed circuit boards, very often with embedded lead, heavy metals, and BFRs in them [1]. Their recycling requires fairly advanced dismantling and chemical processing technologies if environmental hazards are not to be created thereby [60]. Gray goods include computers, laptops, printers, scanners, and other peripheral accessories of information and communication technology equipment. This category has rapidly grown with the spread of electronic devices and information systems worldwide. It consists of high amounts of plastics and silicon, integrated circuits, lithium-ion batteries, and precious metals, including gold, palladium, and silver used in microprocessors and connectors. The complexity and diversity of materials make them both valued yet difficult to recycle efficiently. The rapid obsolescence of ICT equipment, too, has made gray goods a large component of contemporary e-waste streams. While the tripartite categorization provides an initial understanding of e-waste sources, it is not adequate to define either regulatory or industrial management. More specific frameworks, including those outlined under the EU WEEE Directive, have thus been developed.

4.2 Classification under the European Union WEEE Directive

The European Union Directive on WEEE categorizes e-waste into ten major groups to standardize the management of e-waste across member states. This is one of the most widely adopted regulatory models in the world and has thus formed the basis for legislation in many countries. The European Union classifies electronic waste, under its WEEE Directive, into ten categories: large household appliances, small household appliances, IT and telecommunications equipment, consumer electronics, lighting equipment, electrical and electronic tools, toys and leisure equipment, medical devices, monitoring and control instruments, and automatic dispensers. This complex classification encompasses both household and special-category equipment, such as medical and industrial devices, to ensure that each group receives the appropriate emphasis in the recycling and regulatory processes.

Other established international classifications, for example, by the United Nations University and the Partnership on Measuring ICT for Development, define types of e-waste based on product function, material composition, and usage patterns. For example, they enumerate six broad categories: temperature exchange equipment, screens and monitors, lamps, large equipment, small equipment, and small IT and telecommunication equipment. The above frameworks help in standardizing the management of e-waste between jurisdictions.

5 Toxicity and environmental impacts of electronic waste

The main source of toxicity for electronic waste is the mixture of its different parts that include complicated organic compounds, heavy metals, flame retardants, and persistent pollutants. When waste is not properly disposed of and recycled, these materials are released into the environment and thereby become a great risk to both ecosystems and human health. The toxic effects of e-waste depend on the type of chemical contaminants that are involved, their bioavailability, and the routes of exposure [61]. Parts of electronic devices that are thrown away or burned can release toxins into the soil, water, and air, thus they cause pollution to local dumping sites that can spread through long-range atmospheric and hydrological transport.

5.1 Toxicity pathways and mechanisms

Main pathways of electronic waste toxicity include leaching, volatilization, dust dispersion, and combustion [62]. Combustion of cables in open burnings to retrieve copper or acid leaching to recover gold and silver are just some of the informal recycling activities, releasing a wide range of toxicants, including heavy metals like Pb, Cd, Hg, Cr, As, and organic pollutants such as polychlorinated biphenyls, dioxins, furans, and polybrominated diphenyl ethers [63]. These toxic substances may reach the food chain via multiple routes, including soil uptake by plants and bioaccumulation in aquatic organisms or by direct inhalation of contaminated particulates. Leaching is one of the most serious processes. Under acidic or oxidative conditions, heavy metals are solubilized from e-waste matrices to contaminate the soil and groundwater. For example, lead and cadmium, which easily leach out from solder, batteries, and plastics, bind to organic matter or mineral colloids in soils [64]. Once mobilized, they may be taken up through plants or infiltrate drinking water systems, leading to chronic exposure in nearby communities. Conversely, the volatilization of mercury, brominated compounds, and dioxins involves either the incineration or high-temperature exposure of electronic waste materials, which culminates in atmospheric release [4]. Such gaseous pollutants may undergo long-range transport and deposition, leading to global dispersal of contamination. Another important method of exposure involves dust dispersion. Finely distributed dust particles, produced during various crushing and shredding processes, carry a very high content of heavy metals and POPs [65]. These respirable particles (<10 μm) are readily inhaled and deposited deep in the pulmonary alveoli, where respiratory toxicity ensues with systemic distribution through the bloodstream. The overall toxic burden on humans and the environment is increased as a result of the additive impact of multiple exposure pathways.

5.2 Human health impacts

Human health effects related to e-waste exposure comprise a wide spectrum and are often cumulative in nature, involving various organ systems (Figure 2). For example, people who live close to recycling plants, mainly in developing countries, are exposed for long periods to intricate mixtures of harmful substances. Among the most studied heavy metals for their neurotoxicity and nephrotoxicity are lead (Pb) and cadmium (Cd). Pb interferes with calcium-dependent neurotransmission and impairs cognitive development of children, causing irreversible neurological damage [66]. Cadmium accumulates in kidneys and bones, leading to renal dysfunction, osteoporosis, and impaired reproductive health [67]. Hg, especially in methylated form, disrupts the functions of the central nervous system and causes motor and cognitive impairments [68]. Organic pollutants such as PBDEs and dioxins exhibit endocrine-disruptive properties. Generally, their mechanism of action is to mimic or block the activity of hormones and thus negatively affect the reproductive and developmental aspects. Because they are flame retardants of plastics and circuit boards, PBDEs have the tendency to bioaccumulate in the adipose tissue and breast milk, and represent an important route of exposure for infants [69]. Dioxins and furans, formed during the open burning of chlorinated plastics, are potent carcinogens that have been reported to cause immune suppression and teratogenicity [70]. Chromium compounds in metal plating and nickel from electronic components can induce allergic reactions, respiratory problems, and even carcinogenesis [9]. Occupational exposure is especially worrisome among informal recyclers. Workers dismantling devices without personal protection gear come into daily contact with acids, fumes, and contaminated dust. Epidemiological studies among e-waste workers have documented increases in skin lesions, respiratory distress, and blood parameter changes [71]. Cumulative exposure has made children involved in recycling suffer from stunted growth, reduced IQ, and impaired immune response [72]. Exposure of pregnant women to environmental contaminants from e-waste has resulted in a significantly increased frequency of miscarriages, preterm birth, and birth of low-weight infants, which is a clear indication of the passage of toxicants across the placenta [73].

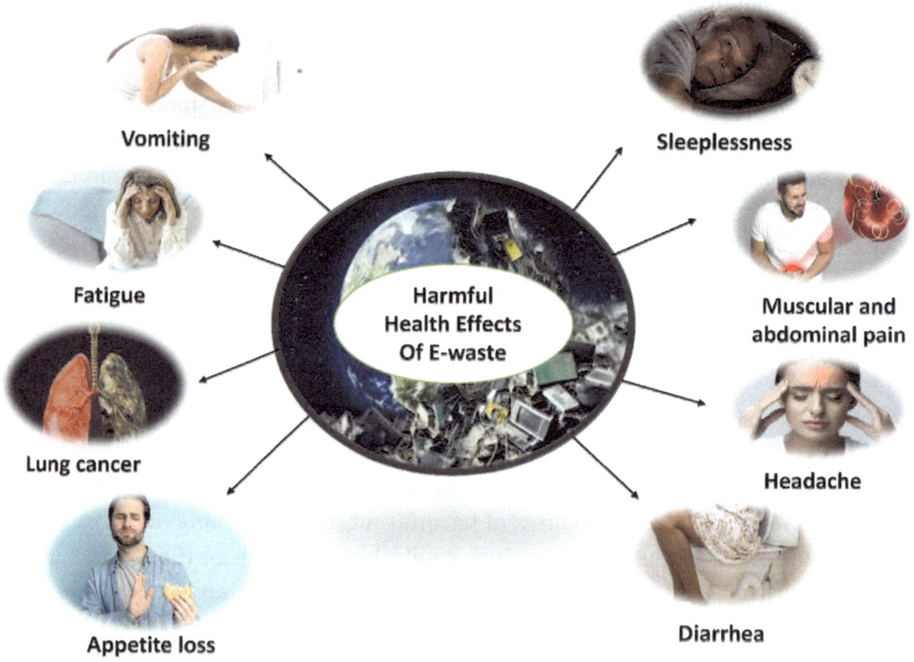

Figure 2: Various health problems associated with E-waste.

5.3 Environmental contamination

E-waste pollution has diverse effects that reach all layers of the environment, including soil, water, and the atmosphere. Soil pollution is caused by the deposition of heavy metals and organohalogens that leach from waste sites and the soil itself [74]. The soils within the e-waste recycling sites contain heavy metals that are several hundred times beyond the permissible standards. Of these heavy metals, the accumulation within the upper soil horizon depicts lead, copper, and zinc, which influence soil microbes and the enzymatic processes within the soil. Water pollution could be through runoff and the infiltration of leachates that contain dissolved metals and acids [75]. The leachates pose a threat to the water environment, influencing eutrophication and toxicity within the water bodies. The biomagnification of mercury and cadmium found within fish gets biomagnified through the food chain and eventually ends up within the nutritional levels of the human diets [76]. The impacts that have been observed and recorded concerning the highly polluted water bodies include reproductive impairments, skeletal abnormalities, and oxidative stress reaction among fish (Figure 3). Also, the issue of groundwater pollution, particularly within the developing world, is a concern since e-waste dumping sites are always situated near populated areas, and thus there is a drastic concern over the safety of drinking water. Concerning the emission of pollution within the atmosphere, the primary source within the e-waste environment is open burning and smelting [62]. Burning of plastics that are halogenated produces very dangerous gases such as hydrogen chloride, dioxin, and furans [70]. The effect of the gases produced poses severe respiratory irritation that not only brings immediate injury but also brings perpetual pollution of the climate and the atmosphere. Fine particulate matter, PM2.5 and PM10, generated through the burning of e-waste works toward the enhancement of the capacity of metals and organic chemicals within the atmosphere and thus increases the inhalability of the particles. Moreover, black carbon generated through e-waste further increases regional warming and the level of poor air quality [77].

5.4 Bioaccumulation and ecotoxicological effects

E-waste toxins are capable of biomagnification and bioaccumulation. Heavy metals such as mercury, cadmium, and lead, and hydrophobic organic compounds such as PBDEs, tend to biomagnify and bioaccumulate within biological systems. Phytotoxicity tests indicate that e-waste leachates reduce the germination of seeds, root growth, and chlorophyll concentration of plants [78]. The presence of copper and cadmium inhibits the action of the photosynthetic apparatus and triggers the production of ROS and disruptions of metabolism [79]. Marine ecosystems are home to organisms such as mollusks, crustaceans, and fish that tend to biomagnify large quantities of metals, thus inhibiting growth, reproduction, and mortality [80]. The mechanisms of biomag-

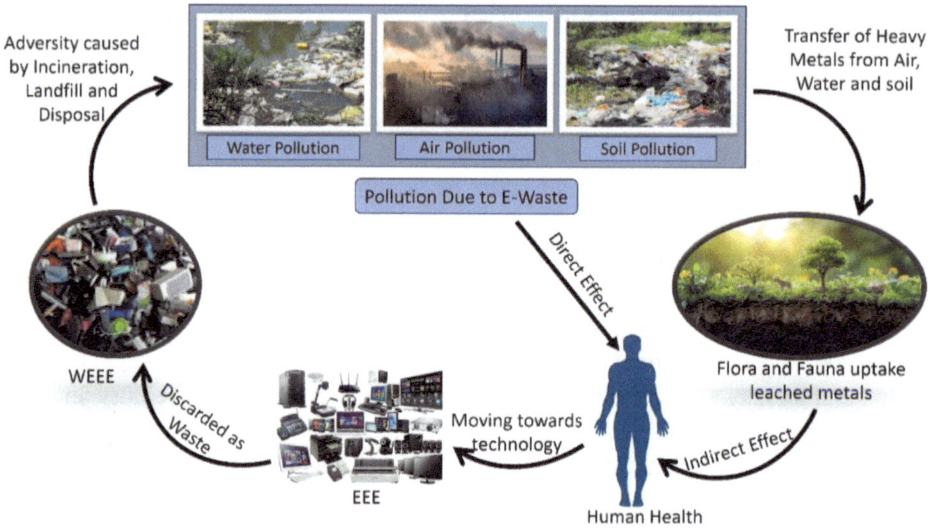

Figure 3: Schematic representation of e-waste cycling effect, illustrating disposal pathways and their direct and indirect impacts on the ecosystem.

nification include the interaction of metals with the SH groups of proteins and the replacement of essential metals such as calcium and zinc within enzymes [81]. This inhibits normal metabolism, thus leading to ROS and apoptosis. Due to hydrophobicity, persistent organic toxins readily biomagnify and are not readily metabolized [82]. The apex predators, prey of birds, fish-eating mammals, and humans are the most susceptible groups exposed to biomagnification.

5.5 Global and ecological implications

The environmental impact of e-waste does not end at the dump site but poses a global environmental crisis. Moreover, even the toxins released through the informally recycled e-waste sites of Asia and Africa might reach other ecosystems through atmospheric and oceanic currents and might harm other ecosystems located outside. The polar region has been found to have emissions of PBDEs and heavy metals through e-waste, over a distance of thousands of kilometers away [83]. The toxicity of e-waste is multifaceted and poses a threat to the environment itself since e-waste is a crisis of environmental pollution, and also since the crisis of e-waste poses a risk of public health emergencies. This poses a long-term effect of the emission of toxins and heavy metals that lead to the destruction of the soil and the resultant imbalance of ecosystems [84]. The issue of e-waste could be addressed well through the concept of a circular economy and a stringent control over the informally recycled sites and also a global approach toward curtailing the dangerous emission of e-waste and supplementing the technologies developed for e-waste management.

6 Conclusion

The increased volume and variety of electronic waste pose a severe threat to environmental sustainability and human health. E-waste comes from the continuously increasing number of sources – households, businesses, and industrial sectors – generating a wide array of devices, from large and small appliances to complex information technology equipment. Recycling through categories such as white goods, brown goods, and gray goods, or through internationally recognized directives, would facilitate better focused management and appropriate recycling. Electronic waste contains both valuable recoverable metals and hazardous constituents such as heavy metals and toxic chemicals. Unless properly disposed of, these materials tend to create a holistic threat to soil, water, and air quality, increasing ecological and human health risks. Major stakeholders in the management of e-waste include the government, producers, recyclers, and consumers. Effective policies regarding extended producer responsibility and structured collection and recycling programs offer paths toward mitigation of environmental harm and the recovery of resources. Public awareness of the dangers that result from poor disposal, initiatives in the development of formal recycling infrastructure, and strategic investment in safe and innovative technology are, therefore, very important steps toward containing the toxicity and volume of electronic waste. As the world increasingly depends on electronics, the shift toward sustainable handling of electrical and electronic waste would mark the defining ingredient in the face of health, resource security, and ecological well-being in the coming decades.

Table 1: A comparative study of new technologies for recycling e-waste.

Technology	Core principle	Key materials recovered	Advantages	Limitations/ challenges
Biohydrometallurgy	Microbial or fungal leaching using organic acids and redox reactions	Cu, Zn, Ni, Au, and Pd	Eco-friendly, low energy use, and minimal toxic by-products	Slow kinetics and limited scalability
Phytomining	Metal uptake by hyperaccumulator plants	Ni, Cd, and Au	Low-cost, suitable for contaminated soils, and produces biomass for valorization	Seasonal dependence and limited to low-metal concentrations

Table 1 (continued)

Technology	Core principle	Key materials recovered	Advantages	Limitations/ challenges
Advanced pyrometallurgy	High-temperature recovery using plasma arc or vacuum smelting	Cu, Fe, Al, and Au	High recovery efficiency and rapid processing	High energy demand and risk of toxic emissions
Hydrometallurgy (using DES/ILs)	Selective dissolution using green solvents	Au, Pd, and Cu	Mild conditions and solvent recyclability	High solvent cost and process optimization required
Nanotechnology-based recovery	Use of nanomaterials for adsorption, catalysis, or metal conversion	Pb, Hg, Cd, and Cu	High specificity and enables upcycling into value-added materials	Nanoparticle toxicity and cost of synthesis
AI and IoT-based systems	Data-driven collection, sorting, and tracking	All categories (traceability focus)	Enhances logistics and policy compliance	Requires digital infrastructure and interoperability

Table 2: Comparative classification frameworks for electronic waste.

Classification framework	Category	Representative devices	Key features/recycling implications	References
Traditional classification	White goods	Refrigerators, washing machines, dishwashers, ovens, and air conditioners	Large household appliances, high metal content, and easily recyclable through mechanical recovery	[1, 85]
	Brown goods	Televisions, radios, audio systems, and video cameras	Complex circuitry, CRTs, and hazardous components; and require advanced dismantling	[86]
	Gray goods	Computers, laptops, printers, scanners, and mobile phones	High content of plastics, rare metals, and semiconductors; major source of valuable recoverables	[85]

Table 2 (continued)

Classification framework	Category	Representative devices	Key features/recycling implications	References
EU-WEEE Directive classification	Large household appliances	Refrigerators, freezers, washing machines, and dryers	Bulk e-waste stream; mostly ferrous metals, plastic, and foam insulation	[87]
	Small household appliances	Irons, toasters, hairdryers, and coffee machines	Numerous small devices; major collection challenge	[88]
	IT and telecommunication equipment	Computers, routers, servers, and telephones	Contain precious and rare earth metals; growing global stream	[87]
	Consumer equipment	TVs, cameras, and musical instruments	Often contain toxic components (lead and brominated compounds)	[87, 89]
	Lighting equipment	Fluorescent lamps, LEDs, and mercury lamps	Contain mercury and phosphors; require safe disposal	[90]
	Electrical and electronic tools	Power drills, saws, and lawnmowers	Mixed materials; complex recycling logistics	[91]
	Toys, leisure, and sports equipment	Game consoles, treadmills, and electric toys	Increasing e-waste stream due to digitization of leisure devices	[92]
	Medical devices	Diagnostic devices, X-ray machines, and pumps	Need decontamination before recycling; often contain lead shielding	[93]
	Monitoring and control instruments	Smoke detectors, thermostats, and alarms	Contain sensors and PCB boards; moderate recyclable value	[2]
	Automatic dispensers	Vending machines and ATMs	Mechanically complex; mixed metal-plastic composition	[87]
Functional/ operational classification	Consumer electronics	Smartphones, tablets, and entertainment systems	Fastest-growing category; high obsolescence rate	[90]
	Industrial and commercial electronics	Servers, data systems, and control units	High metal concentration; longer life cycles	[90]
	Infrastructural or utility electronics	Smart meters, transformers, and network equipment	Require specialized handling; embedded systems	[90]

References

[1] Ghulam ST, Abushammala H. Challenges and opportunities in the management of electronic waste and its impact on human health and environment. Sustainability. 2023 Jan 18;15(3):1837.

[2] Serpe A, Purchase D, Bisschop L, Chatterjee D, De Gioannis G, Garelick H, Kumar A, Peijnenburg WJ, Piro VM, Cera M, Shevah Y. 2002–2022: 20 years of e-waste regulation in the European Union and the worldwide trends in legislation and innovation technologies for a circular economy. RSC Sustainability. 2025;3(3):1039–1083.

[3] Bisschop L, Hendlin Y, Jaspers J. Designed to break: planned obsolescence as corporate environmental crime. Crime, Law and Social Change. 2022 Oct;78(3):271–293.

[4] Malhotra P, Jain A. Chemical methods for the treatment of e-waste. In Waste Management and Resource Recycling in the Developing World 2023 Jan 1 (pp. 181–204). Elsevier.

[5] Kesar B, Joseph S. Impact of E-Waste Processing on Health and Environment. In Integrated Approaches for Sustainable E-Waste Management 2025 (pp. 283–308). IGI Global Scientific Publishing.

[6] Angon PB, Islam MS, Das A, Anjum N, Poudel A, Suchi SA. Sources, effects and present perspectives of heavy metals contamination: soil, plants and human food chain. Heliyon. 2024 Apr 15;10(7).

[7] Hossain R, Sahajwalla V. Current recycling innovations to utilize e-waste in sustainable green metal manufacturing. Philosophical Transactions of the Royal Society A: Mathematical, Physical and Engineering Sciences. 2024 Nov 4;382(2284):20230239.

[8] Dev RK, Yadav SN, Magar N, Ghimire S, Koirala M, Giri R, Das AK, Shah SK, Gardas RL, Bhattarai A. Recovery of rare earth elements (REEs) from different sources of e-waste and their potential applications: a focused review. Geological Journal. 2025.

[9] Murthy MK, Khandayataray P, Padhiary S, Samal D. A review on chromium health hazards and molecular mechanism of chromium bioremediation. Reviews on environmental health. 2023 Sep 1;38(3):461–478.

[10] Bhattacharjee J, De A, Kamila B, Mandal A. Recent scenario of e-waste recycling: chemical engineering. International Journal of Chemical Reactor Engineering. 2025 Jul 10;23(6):675–700.

[11] Sarkhoshkalat MM, Afkham A, Bonyadi Manesh M, Sarkhosh M. Circular economy and the recycling of e-waste. In New technologies for energy transition based on sustainable development goals: Factors contributing to global warming 2024 Aug 3 (pp. 319–354). Singapore: Springer Nature Singapore.

[12] Alabi OA, Adeoluwa YM, Huo X, Xu X, Bakare AA. Environmental contamination and public health effects of electronic waste: an overview. Journal of Environmental Health Science and Engineering. 2021 Jun;19(1):1209–1227.

[13] Bąkowski M, Winiarska-Mieczan A, Kwiecień M, Krusiński R, Jachimowicz-Rogowska K, Twarużek M, Grobelska D. Effect of cadmium and lead on living organisms–transport, accumulation, metabolism. Journal of Elementology. 2024;29(1):311–328.

[14] Adu JT, Aneke FI. Evaluation of heavy metal contamination in landfills from e-waste disposal and its potential as a pollution source for surface water bodies. Results in Engineering. 2025 Mar 1; 25:104431.

[15] Lamah L, Degny BF, Yable DD, Haba CT. Critical analysis of waste electrical and electronic equipment (WEEE) management in Côte d'Ivoire: challenges and prospects. Journal of Materials and Environmental Science. 2024;15(3):427. 440.

[16] Ilankoon IM, Dilshan RA, Dushyantha N. Co-processing of e-waste with natural resources and their products to diversify critical metal supply chains. Minerals Engineering. 2024 Jul 1; 211:108706.

[17] Oraee A, Pohl L, Geurts D, Reichel M. Overcoming premature smartphone obsolescence amongst young adults. Cleaner and Responsible Consumption. 2024 Mar 1; 12:100174.

[18] Kulkarni S. E-Waste Management and Its Impact on the Environment, Soil, and Human Health. In Green Sustainable Agriculture and Farming Practices 2025 Oct 14 (pp. 207–235). Apple Academic Press.
[19] Ajekwene K, Aigbokhan E, Akindele O, Yibowei ME, Momoh FP, Ugonna UK. Electronic waste (e-waste): sources, proliferation, effects & management in developing nations. IOSR Journal of Engineering. 2022; 12:12–27.
[20] Saeed S, Altamimi SA, Alkayyal NA, Alshehri E, Alabbad DA. Digital transformation and cybersecurity challenges for businesses resilience: issues and recommendations. Sensors. 2023 Jul 25;23(15):6666.
[21] Victor A, Arunkumar G, Rajkumar S, Selvanambi R. Survey on effective disposal of e-waste to prevent data leakage. Computer Assisted Methods in Engineering and Science. 2024;31(2):187–212.
[22] Rene ER, Sethurajan M, Ponnusamy VK, Kumar G, Dung TN, Brindhadevi K, Pugazhendhi A. Electronic waste generation, recycling and resource recovery: technological perspectives and trends. Journal of hazardous materials. 2021 Aug 15; 416:125664.
[23] Gulliani S, Volpe M, Messineo A, Volpe R. Recovery of metals and valuable chemicals from waste electric and electronic materials: a critical review of existing technologies. RSC Sustainability. 2023;1(5):1085–1108.
[24] Hassan AI, Saleh HM. Toxicity and hazardous waste regulations. In Hazardous Waste Management 2022 Jan 1 (pp. 165–182). Elsevier.
[25] Palaniandy P, Yusoff MS, Wang LK, Wang MH. Electronic and electrical equipment waste disposal. In Solid Waste Engineering and Management: Volume 3 2022 Jun 25 (pp. 115–161). Cham: Springer International Publishing.
[26] Kheirabadi S, Sheikhi A. Recent advances and challenges in recycling and reusing biomedical materials. Current Opinion in Green and Sustainable Chemistry. 2022 Dec 1; 38:100695.
[27] Ganguly R, Ganguly P, Matharasi T. Sustainable Mobile and Wireless Computing Solutions. In Energy Efficient Algorithms and Green Data Centers for Sustainable Computing 2025 (pp. 209–248). IGI Global Scientific Publishing.
[28] Quinto S, Law N, Fletcher C, Le J, Antony Jose S, Menezes PL. Exploring the e-waste crisis: strategies for sustainable recycling and circular economy integration. Recycling. 2025 Apr 14;10(2):72.
[29] Sarker MR, Abdolrasol MG, Mohamad Hanif Md S, Kadir RA, Ahmad MN, Olazagoitia JL. Advancing agriculture automation systems: technological innovations, possible applications, challenges, and recommendations. Advances in Agriculture. 2025;2025(1):5518653.
[30] Singh AK, Kumar R, Kumar B, Chaturvedi DK. Fundamentals of modern transportation systems. In Intelligent Control for Modern Transportation Systems 2023 Oct 16 (pp. 1–20). CRC Press.
[31] Kumar P, Khanna MK, Sharma N. E-waste in Space: Is Space Becoming a New and Limitless Dumping Ground?. In Integrated Waste Management: A Sustainable Approach from Waste to Wealth 2024 Apr 15 (pp. 323–347). Singapore: Springer Nature Singapore.
[32] Nazari L, Xu C, Ray MB. Recovery of metals from electronic waste. In Advanced and Emerging Technologies for Resource Recovery from Wastes 2021 Apr 2 (pp. 127–156). Singapore: Springer Singapore.
[33] Seif R, Salem FZ, Allam NK. E-waste recycled materials as efficient catalysts for renewable energy technologies and better environmental sustainability. Environment, development and sustainability. 2023 Jan; 18:1.
[34] Tunali M, Tunali MM, Yenigun O. Characterization of different types of electronic waste: heavy metal, precious metal and rare earth element content by comparing different digestion methods. Journal of Material Cycles and Waste Management. 2021 Jan;23(1):149–157.
[35] Fazari J, Hossain MZ, Charpentier P. A review on metal extraction from waste printed circuit boards (wPCBs). Journal of Materials Science. 2024 Jul;59(27):12257–12284.

[36] Gunarathne V, Rajapaksha AU, Vithanage M, Alessi DS, Selvasembian R, Naushad M, You S, Oleszczuk P, Ok YS. Hydrometallurgical processes for heavy metals recovery from industrial sludges. Critical Reviews in Environmental Science and Technology. 2022 Mar 19;52(6):1022–1062.
[37] Awan TI, Afsheen S, Mushtaq A. Overview of Noble Metals. In Influence of Noble Metal Nanoparticles in Sustainable Energy Technologies 2025 Jan 30 (pp. 49–78). Cham: Springer Nature Switzerland.
[38] Lanin VL, Emel'yanov VA, Petuhov IB. Chapter 1. Assembly and mounting of electronic devices: advancements in technology and equipment. Surface Engineering and Applied Electrochemistry. 2024 Jun;60(3):269–288.
[39] Behrens V. Electrical Contacts and Hazardous Substances-Technical Backgrounds, Legal Restrictions, and Substitutions. In 2022 IEEE 67th Holm Conference on Electrical Contacts (HLM) 2022 Oct 23 (pp. 1–9). IEEE.
[40] Manikkampatti Palanisamy M, Myneni VR, Gudeta B, Komarabathina S. Toxic metal recovery from waste printed circuit boards: a review of advanced approaches for sustainable treatment methodology. Advances in Materials Science and Engineering. 2022;2022(1):6550089.
[41] Chen Y, Qiao Q, Cao J, Li H, Bian Z. Precious metal recovery. Joule. 2021 Dec 15;5(12):3097–3115.
[42] Baranovskaya VB, Karpov YA, Petrova KV, Korotkova NA. Actual trends in the application of rare-earth metals and their compounds in the production of magnetic and luminescent materials: a review. Russian Journal of Non-Ferrous Metals. 2021 Jan;62(1):10–31.
[43] Heim JW, Vander Wal RL. NdFeB permanent magnet uses, projected growth rates and Nd plus Dy demands across end-use sectors through 2050: a review. Minerals. 2023 Sep 29;13(10):1274.
[44] Zhang Y, Liu J, Li P, Wu Y, Sun K, Fang Q, Zhu W. Characteristics and tunable luminescence of a phosphate glass doped with terbium and europium for W-LEDs. Ceramics International. 2025 Mar 8.
[45] Dasgupta A, Sen S, Singh P, Raman A. A comprehensive overview of the foundations of semiconductor materials. Semiconductor Nanoscale Devices: Materials and Design Challenges. 2025 Mar; 14:80–109.
[46] Farag S, Farag Y, Attia M. Recycling technologies for brominated flame-retardant plastics in e-waste. Cleaner Waste Systems. 2025 Oct; 1:100418.
[47] Lange JP. Managing plastic waste— sorting, recycling, disposal, and product redesign. ACS Sustainable Chemistry & Engineering. 2021 Nov 12;9(47):15722–15738.
[48] Cenci MP, Munchen DD, Mengue Model JC, Veit HM. Metal resources in electronics: trends, opportunities and challenges. Management of Electronic Waste: Resource Recovery, Technology and Regulation. 2024 Jan; 24:114–151.
[49] Fontana D, Forte F, Pietrantonio M, Pucciarmati S. Recent developments on recycling end-of-life flat panel displays: a comprehensive review focused on indium. Critical reviews in environmental science and technology. 2021 Mar 4;51(5):429–456.
[50] Bajpai A, Biswas K, Meshram A. Beneficiation and Management of Technological Waste: Opportunities and Challenges. CRC Press; 2025 May 22.
[51] Abubakar A, Zangina AS, Maigari AI, Badamasi MM, Ishak MY, Abdullahi AS, Haruna JA. Pollution of heavy metal threat posed by e-waste burning and its assessment of human health risk. Environmental Science and Pollution Research. 2022 Aug;29(40):61065–61079.
[52] Yadav TP, Srivastava R, Awasthi K. E-Waste Management through Mechanical Milling: A Sustainable Approach. In Strategies to Achieve Sustainable Development Goals (SDGs); Srivastava, R., Ed. 2022 (pp. 1–26).
[53] Handschuh-Wang S, Wang T, Gancarz T, Liu X, Wang B, He B, Dickey MD, Wimmer GW, Stadler FJ. The liquid metal age: a transition from Hg to Ga. Advanced Materials. 2024 Nov;36(45):2408466.
[54] Wang L, Li P, Duan R, He X. Occurrence, controlling factors and health risks of Cr6+ in groundwater in the Guanzhong Basin of China. Exposure and Health. 2022 Jun;14(2):239–251.

[55] Dodd M, Amponsah LO, Grundy S, Darko G. Human health risk associated with metal exposure at Agbogbloshie e-waste site and the surrounding neighbourhood in Accra, Ghana. Environmental Geochemistry and Health. 2023 Jul;45(7):4515–4531.
[56] Wang X, Chen X, Luo J, Wang X, Chen Y, Liu X. High-temperature polymer dielectrics for printed circuit board. High Temperature Polymer Dielectrics: Fundamentals and Applications in Power Equipment. 2024 Mar; 4:181–226.
[57] Alengebawy A, Abdelkhalek ST, Qureshi SR, Wang MQ. Heavy metals and pesticides toxicity in agricultural soil and plants: ecological risks and human health implications. Toxics. 2021 Feb 25;9(3):42.
[58] Alghamdi HM, Rajeh A. Synthesis of carbon nanotubes/titanium dioxide and study of its effect on the optical, dielectric, and mechanical properties of polyvinyl alcohol/sodium alginate for energy storage devices. International Journal of Energy Research. 2022 Nov;46(14):20050–20066.
[59] Prasad KV. E-waste: A looming hazard to the environment and health-A critical review. In AIP Conference Proceedings 2025 Jul 9 (Vols. 3298, No. 1, pp. 040044). AIP Publishing LLC.
[60] Munialo S, Collins O. Impacts of electronic wastes on environment and human health. International Journal of Scientific Engineering Research. 2020;11(5):1174–1179.
[61] Parvez SM, Jahan F, Brune MN, Gorman JF, Rahman MJ, Carpenter D, Islam Z, Rahman M, Aich N, Knibbs LD, Sly PD. Health consequences of exposure to e-waste: an updated systematic review. The Lancet Planetary Health. 2021 Dec 1;5(12):e905–20.
[62] Shams S, Idris PR, Yusof I. Toxicity characterization and environmental impact of e-waste processing. Management of Electronic Waste: Resource Recovery, Technology and Regulation. 2024 Jan; 24:73–87.
[63] Gamage LE, Basak AK, Pramanik A, Prakash C, Shankar S, Debnath S, Dixit AR, Chattopadhyaya S, Ramakrishana S. Recycling of gold and silver from electronic waste – a review. Materials Circular Economy. 2023 Dec;5(1):8.
[64] Bouida L, Rafatullah M, Kerrouche A, Qutob M, Alosaimi AM, Alorfi HS, Hussein MA. A review on cadmium and lead contamination: sources, fate, mechanism, health effects and remediation methods. Water. 2022 Oct 28;14(21):3432.
[65] Pazzi G, Buiarelli F, Di Filippo P, Pomata D, Riccardi C, Lucarelli F, Giardi F, Sonego E, Galarini R, Lorenzetti S, Goracci L. Metals and organic species associated with fine and coarse aerosol particles in an electronic waste recycling plant. Air Quality, Atmosphere & Health. 2023 Apr;16(4):841–856.
[66] Tamagno WA, Freeman JL. Glutamate-mediated neural alterations in lead exposure: mechanisms, pathways, and phenotypes. Toxics. 2025 Jun 21;13(7):519.
[67] Charkiewicz AE, Omeljaniuk WJ, Nowak K, Garley M, Nikliński J. Cadmium toxicity and health effects – a brief summary. Molecules. 2023 Sep 14;28(18):6620.
[68] Kumari K, Chand GB. Effects of mercury: neurological and cellular perspective. Mercury Toxicity: Challenges and Solutions. 2023 Nov; 24:141–162.
[69] Gouesse RJ, Plante I. Environmental exposure to brominated flame retardants: unraveling endocrine and mammary gland effects that may increase disease risk. Toxicological Sciences. 2022 Apr 1;186 (2):190–207.
[70] Mathew N, Somanathan A, Tirpude A, Pillai AM, Mondal P, Arfin T. Dioxins and their impact: a review of toxicity, persistence, and novel remediation strategies. Analytical Methods. 2025.
[71] Parvez SM, Hasan SS, Knibbs LD, Jahan F, Rahman M, Raqib R, Islam N, Aich N, Moniruzzaman M, Islam Z, Fujimura M. Ecological burden of e-waste in Bangladesh – an assessment to measure the exposure to e-waste and associated health outcomes: protocol for a cross-sectional study. JMIR research protocols. 2022 Aug 16;11(8):e38201.
[72] Desye B, Tesfaye AH, Berihun G, Ademas A, Sewunet B. A systematic review of the health effects of lead exposure from electronic waste in children. Frontiers in public health. 2023 Apr 12; 11:1113561.

[73] Modak A, Bhattacharjyya P. Health and environmental effects of e-waste recycling processes: issues, challenges, and solutions. In Development in E-waste Management 2023 May 15 (pp. 107–125). CRC Press.
[74] Russo MV, Notardonato I, Rosada A, Ianiri G, Avino P. Halogenated volatile organic compounds in water samples and inorganic elements levels in ores for characterizing a high anthropogenic polluted area in the Northern Latium region (Italy). International Journal of Environmental Research and Public Health. 2021 Feb;18(4):1628.
[75] De Buyck PJ, Van Hulle SW, Dumoulin A, Rousseau DP. Roof runoff contamination: a review on pollutant nature, material leaching and deposition. Reviews in Environmental Science and Bio/ Technology. 2021 Jun;20(2):549–606.
[76] Kumar V, Umesh M, Shanmugam MK, Chakraborty P, Duhan L, Gummadi SN, Pasrija R, Jayaraj I, Dasarahally Huligowda LK. A retrospection on mercury contamination, bioaccumulation, and toxicity in diverse environments: current insights and future prospects. Sustainability. 2023 Sep 5;15 (18):13292.
[77] Xu H, Ren YA, Zhang W, Meng W, Yun X, Yu X, Li J, Zhang Y, Shen G, Ma J, Li B. Updated global black carbon emissions from 1960 to 2017: improvements, trends, and drivers. Environmental Science & Technology. 2021 Jun 7;55(12):7869–7879.
[78] Garg R, Dang S, Gauba P. Eco-disaster: The Perils of E-Waste on Plants and Human Health. In Innovative Advancements in Biotechnology: Technological Advancements in Biosciences 2025 Feb 16 (pp. 67–83). Cham: Springer Nature Switzerland.
[79] Raja V, Singh K, Qadir SU, Singh J, Kim KH. Alleviation of cadmium-induced oxidative damage through application of zinc oxide nanoparticles and strigolactones in Solanum lycopersicum L. Environmental Science: Nano. 2024;11(6):2633–2654.
[80] Banaee M, Zeidi A, Mikušková N, Faggio C. Assessing metal toxicity on crustaceans in aquatic ecosystems: a comprehensive review. Biological Trace Element Research. 2024 Dec;202 (12):5743–5761.
[81] Nnaji ND, Onyeaka H, Miri T, Ugwa C. Bioaccumulation for heavy metal removal: a review. SN Applied Sciences. 2023 May;5(5):125.
[82] Yang D, Ko E, Lim H, Lee H, Kim K, Choi M, Shin S. Persistent organic pollutants released from decomposed adipose tissue affect mitochondrial enzyme function in the brain and eyes other than the liver. Environmental Science and Pollution Research. 2024 Feb;31(7):10648–10660.
[83] Li X, Guo H, Cheng G, Song X, Ran Y, Feng M, Che T, Li X, Wang L, Duan A, Shangguan D. Polar regions are critical in achieving global sustainable development goals. Nature Communications. 2025 Apr 24;16(1):3879.
[84] Ekka P, Patra S, Upreti M, Kumar G, Kumar A, Saikia P. Land degradation and its impacts on biodiversity and ecosystem services. Land and environmental management through forestry. 2023 Jul; 4:77–101.
[85] Koli SK, Hussain A. Status of electronic waste management in India: a review. Advanced Treatment Techniques for Industrial Wastewater. 2019;238–250.
[86] Jain M, Kumar D, Chaudhary J, Kumar S, Sharma S, Verma AS. Review on e-waste management and its impact on the environment and society. Waste Management Bulletin. 2023 Dec 1;1(3):34–44.
[87] Arya S, Kumari D, Narzari R, Kumar S. A global glance on waste electrical and electronic equipments (WEEEs). In Global E-waste management strategies and future implications 2023 Jan 1 (pp. 1–11). Elsevier.
[88] Shittu OS, Williams ID, Shaw PJ. Prospecting reusable small electrical and electronic equipment (EEE) in distinct anthropogenic spaces. Resources, Conservation and Recycling. 2022 Jan 1; 176:105908.
[89] Rajamani SK, Iyer RS. Waste electrical and electronic equipment (WEEE) disposal problem in India, routes of disposal, toxicology, and impact on human health-a narrative review: waste electrical and

electronic equipment (WEEE) disposal problem in India. MMJ-A Journal by MIMER Medical College, Pune, India. 2023;7(2).

[90] Aboughaly M, Gabbar HA. Recent technologies in electronic-waste management. In E-waste Recycling and Management: Present Scenarios and Environmental Issues 2019 Jun 5 (pp. 63–80). Cham: Springer International Publishing.

[91] Jayasiri G, Herat S, Kaparaju P. Repair and reuse or recycle: what is best for small WEEE in Australia?. Sustainability. 2024 Apr 5;16(7):3035.

[92] München DD, Dias P, Veit HM. Materials used in electronic equipment and manufacturing perspectives. Electronic Waste: Recycling and Reprocessing for a Sustainable Future. 2022 Jan; 10:73–94.

[93] Kumar S, Singh H, Singh H, Soni H, Bechelany M, Singh J. Extraction of metallic nanoparticles from electronic waste for biomedical applications: pioneering sustainable technological progress. Sustainability. 2025 Feb 28;17(5):2100.

Shalini Chaudhary, Ashish GuleriaKamal Kishore, Anil Kumar*

Use of microbial biominers in bioleaching for heavy metals recovery from e-waste

Abstract: Electronic waste (e-waste) is a growing environmental issue because valuable resources, along with many toxic heavy metals, are contained in it. Both ecological and also economic concerns arise from customary metal recovery methods due to the fact that they use harsh chemicals. with a requirement of much energy. Microbial bioleaching represents an economically sustainable choice. It uses bacteria, fungi, and archaea for dissolving metals. Natural biochemical processes recover metals. This chapter examines the diverse range of microbial biominers that do include *Acidithiobacillus ferrooxidans* cyanogenic bacteria, fungi, and thermophilic archaea. It explains their operation and includes processes like acidolysis, redoxolysis, and cyanogenesis. Microbial bioleaching has benefits we stress, since it is environmentally friendly and recovers metal selectively, along with scalability. Recent improvements in bioreactor design and genetic engineering that increase efficiency are also discussed. Case studies do show bioleaching can recover precious and rare metals and this highlights its importance for sustainable e-waste management in the circular economy. Microbial bioleaching does provide some important benefits for people. It is eco-friendly. Thus, less use of toxic chemicals happens. Lowering of energy and reagent use makes it cost- effective. Selective retrieval of metals is possible in complex e-waste. This is possible through the use of this process. It is scalable from lab use to industrial use. It cuts down upon toxic emissions, and also it reduces all environmental risks via reduction of waste. Efficiency improves because of recent progress in bioreactor optimization and genetic engineering. Valuable as well as rare metals may be recovered by bioleaching, as seen in some case studies. This shows that it has an important role within sustainable e-waste management, so this supports the circular economy.

Keywords: Bioleaching, biominers, cyanogenesis, redoxolysis, risk

1 Introduction

The rapid expansion of digital technologies, along with shortened product lifespans and rising consumer demand for electronic devices, has led to a sharp increase in dis-

*Corresponding author: **Anil Kumar**, Department of Chemistry and Biochemistry, Eternal University, Baru Sahib, Sirmaur 173101, Himachal Pradesh, India, e-mail: anilthakur82@gmail.com
Shalini Chaudhary, Kamal Kishore, Department of Chemistry and Biochemistry, Eternal University, Baru Sahib, Sirmaur 173101, India
Ashish Guleria, Department of Chemistry, Chandigarh University, Mohali 140413, Punjab, India

 | https://doi.org/10.1515/9783112225509-002

carded electrical and electronic equipment (EEE, e-waste) worldwide. Studies conducted within the European Union indicate that electronic waste generation is rising at an annual rate of nearly 3–5%, which significantly exceeds the growth rate of other waste categories. Recent global assessments report that electronic waste generation reached approximately 62 million tons in 2024 and is expected to surpass 82 million tons by the year 2030 [1].

Discarded electronic materials serve as an important secondary resource, due to their substantial content of precious metals, including gold, silver, platinum, and large quantities of base metals and REE (rare earth elements), often at concentrations comparable to or exceeding those of natural ores [2, 3, 18–21]. The classification of electronic waste material is influenced by several aspects, such as material composition, production period, equipment function, manufacturer specifications, and component architecture. According to the United Nations, electronic waste is grouped into six main categories that include temperature exchange equipment (refrigerators), display devices (screens and monitors), lighting equipment (lamps), large household appliances, small consumer equipment (toasters and tools), and small information and telecommunication devices (cellular phones and routers) [4].

Conventional recovery of metals from electronic waste has primarily depended on high-temperature pyrometallurgical (smelting and high-temperature treatment) routes as well as hydrometallurgical (acid leaching, cyanidation, and solvent extraction) techniques [5]. Despite their technical efficacy, these methods are associated with high energy demand, hazardous emissions, and extensive use of hostile chemical reagents [6]. During thermal treatment of printed circuit boards (PCBs), pyrometallurgical operations can generate toxic brominated dioxins and furans, whereas hydrometallurgical processing results in the generation of strongly acidic liquid wastes and secondary residues [7]. Progressively rigorous environmental regulations across industrialized and developing nations have accelerated the quest for environmentally safer alternatives for metal extraction from secondary resources.

In recent years, biological approaches based on microbial action have gained considerable attention as environmentally benign routes for metal recovery from complex waste materials. Bioleaching organisms (biominers), a wide range of microorganisms, including acidophilic bacteria, cyanogenic bacteria, fungi, and thermophilic archaea, participate in metal solubilization through several approaches, including acidolysis, redoxolysis, cyanogenesis, complexolysis, and chelation. When compared to conventional metallurgical techniques, bioleaching systems typically operate under milder conditions, consume fewer chemical reagents, require less external energy input, produce abridged secondary pollution, and have high selectivity toward target metals [8]. The procedures for bioleaching metals from e-waste, including pretreatment, microbial inoculation, leaching, and metal recovery steps, are illustrated in Figure 1.

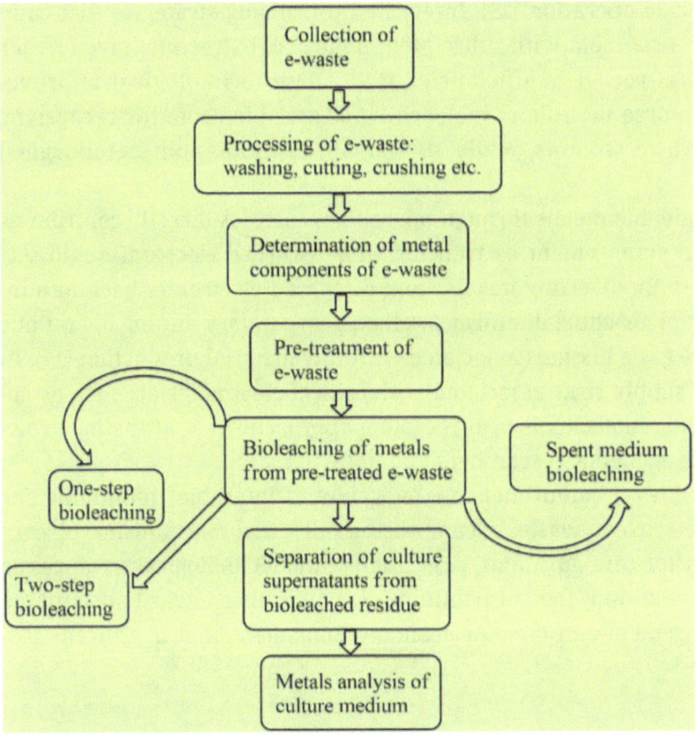

Figure 1: Schematic diagram illustrating the procedures for bioleaching metals from e-waste, including pretreatment, microbial inoculation, leaching, and metal recovery steps.

Among acidophilic microorganisms, iron and sulfur-oxidizing bacteria such as *A. ferrooxidans* and *Acidithiobacillus thiooxidans* play a major role in converting ferrous iron and elemental sulfur into ferric ions and sulfuric acid, thereby promoting dissolution of base metals like copper, zinc, and nickel [9]. Certain bacterial species such as *Pseudomonas fluorescens* and *Chromobacterium violaceum* are capable of releasing hydrogen cyanide, which forms stable complexes with gold and silver and enables their selective mobilization [10]. Fungal species such as *Aspergillus niger* and *Penicillium simplicissimum* facilitate metal dissolution by secreting organic acids such as citric, oxalic, and gluconic acid, offering an ecological alternative to mineral acids [11]. In high-temperature environments, thermophilic archaea such as *Sulfolobus metallicus* remain metabolically active and accelerate the bio-oxidation of mineral substrates that are otherwise difficult to process under mesophilic conditions [12].

Recent technological progress has significantly strengthened the industrial applicability of bioleaching systems. Key developments include the use of genetically engineered strains to enhance metal tolerance, the design of synthetic microbial consortia for cooperative leaching, and the introduction of advanced bioreactor configurations for large-scale operation [13]. Integrated processing strategies that combine mild chemical pretreatment with subsequent biological extraction have further improved overall metal recovery efficiencies [14]. These technological improvements collectively reinforce the role of bioleaching as a viable route for recovering metals from urban waste streams, while supporting low-emission metallurgical processing.

The recovery of valuable metals through microbial processes directly contributes to sustainable material management by transforming discarded electronics into secondary metal resources. By diverting metals from discarded electronics back into industrial supply chains, bioleaching diminishes reliance on primary mining operations and lowers greenhouse gas emissions associated with raw material production [15]. In the face of increasing supply risks associated with critical raw materials such as cobalt, palladium, and REE, biotechnological recycling approaches are attracting growing attention from both academic researchers and industry.

This chapter provides a comprehensive overview of microbial biomining for metal recovery from electronic waste, encompassing microbial mechanisms, operating parameters, bioreactor configurations, case studies, and technological advances as well as challenges. In addition, the contribution of bioleaching toward sustainable electronic waste management and circular economy implementation is critically discussed.

2 Composition of e-waste and target heavy metals

Electronic waste is a highly heterogeneous and diverse association of metallic, polymeric, and ceramic materials, generated from discarded EEE. According to the categorization established by the WEEE (European Union Waste Electrical and Electronic Equipment) Directive, electronic waste includes large household devices telecommunication equipment, IT, lighting devices, consumer electronics, medical instruments, monitoring systems, and automatic dispensers [14]. The composition varies significantly, depending on product type, manufacturing year, and technological design. However, most categories of e-waste share a common structural composition of nearly 40% metal, 30% oxides of different materials, and 30% plastic polymers [16]. WEEE represents a rich secondary resource to recover valuable precious metals, i.e., gold, silver, platinum, palladium, and rhodium. Among the materials recovered during recycling, steel accounts for the largest fraction, at around 50%, followed by aluminum (10–30%), and copper, along with its alloys (15–45%). The recycling process usually begins with the manual dismantling of WEEE, when steel and aluminum components are removed along with several parts containing high concentrations of precious metals. At this stage, plastic casings, kinescopes, and PCBs are separated. Plastic housings pose a significant challenge during processing because they contain complex mixtures of polymers and flame-retardant additives, making their recycling largely limited to mechanical grinding [16]. The separated scrap therefore consists mainly of aluminum, steel, and components enriched with noble metals, with PCBs being particularly rich in precious metals. These PCBs may either be reused or possibly ground, followed by the regaining of nonferrous and precious metals through conventional metallurgical routes. Such recovery technologies typically involve a combination of pyrometallurgical as well as hydrometallurgical processes [16]. Over the past few years, combined with these traditional approaches, alternative and more sustainable metal recovery strategies have been explored. Among them, bio hydrometallurgical techniques that utilize different microorganisms for the biological leaching of metals have emerged as promising future options for efficient metal recovery from electronic waste [17].

2.1 Material composition of electronic devices

The elemental composition of electronic products demonstrates substantial variation among different device categories, with respect to both base and precious metals. Television circuit boards typically contain about 30% iron (Fe), 15% aluminum (Al), 10% copper (Cu) and 28% plastic, with precious metal contents of 280 ppm silver (Ag), 20 ppm gold (Au), and 10 ppm palladium (Pd). PCBs exhibit 7% Fe, 5% Al, 18% Cu, and 23% plastic, with much higher precious metal concentrations of 900 ppm Ag, 200 ppm Au, and 80 ppm Pd.

Mobile phones contain 7% Fe, 3% Al, 13% Cu, and 43% plastic, along with extremely high precious metal contents of 3,000 ppm silver (Ag), 320 ppm gold (Au) and 120 ppm palladium (Pd). DVDs consist of approximately 62% Fe, 2% Al, 5% Cu, and 24% plastic, with 115 ppm Ag, 15 ppm Au, and 4 ppm Pd. Calculators contain 4% Fe, 5% Al, 3% Cu, and 61% plastic, with 260 ppm Ag, 50 ppm Au, and 5 ppm Pd. General e-waste has an average composition of 8.3% Fe, 0.71% Al, and 8.5% Cu, with precious metal contents of 290 ppm Ag and 124 ppm Au [18, 19].

2.2 Economic value share of selected components in electronic devices (as per year 2010 prices)

The distribution of economic value is overwhelmingly governed by precious metal content rather than bulk structural metals. In TV boards, iron contributes 4%, aluminum 14%, and copper 35%, while precious metals together account for 47% of the total value, with silver contributing 7%, gold 33%, and palladium 7%. In PCBs, iron and aluminum contribute 0% and 1%, while copper accounts for 13%, and precious metals dominate with 86% of the total value, including 5% silver, 69% gold, and 12% palladium. In mobile phones, iron and aluminum contribute 0% and copper only 6%, whereas precious metals represent 93% of the total value, with 11% silver, 71% gold, and 11% palladium. DVDs derive 15% of their value from iron, 3% from aluminum, and 30% from copper, while precious metals contribute 52%, including 5% silver, 42% gold, and 5% palladium. Calculators show 1% value contribution from iron, 4% from aluminum, 10% from copper, and 85% from precious metals, with silver, gold, and palladium contributing 6%, 76%, and 3%, respectively [20].

2.3 Mobile phone components

PCBs in mobile phones primarily consists of copper tracks deposited on epoxy resin or fiberglass substrates and protected by surface coatings and adhesive layers. These boards are commonly plated with gold to ensure high electrical conductivity and corrosion resistance, while solder joints are formed on silver-coated copper contacts. In addition to metallic elements, PCBs also incorporate various plastic, metallic, and ceramic structural components. Several valuable and potentially hazardous substances are present in PCBs, including palladium, silver, arsenic, antimony, beryllium, bromine, cadmium, lead, nickel, thallium, zinc, and chromium.

Liquid crystal displays (LCDs) are composed of liquid crystals materials sandwiched between glass layers and may contain toxic substances such as mercury; a typical mobile phone contains several milligrams of liquid crystal substances. The battery is housed in sealed polymer casings and is commonly of lithium-ion (Li-ion), nickel-metal hydride (NiMH), or nickel-cadmium (NiCd) type, with the latter largely restricted to earlier device generations. The plastic casing is typically a combination of both polycarbonate (PC) and acrylonitrile butadiene styrene. Other small-scale components such as the antenna, speaker, microphone, keypad, and accessories contain mixtures of metallic and nonmetallic elements. Chargers mainly consist of copper conductors insulated with plastic, although trace quantities of gold, cadmium, and brominated flame retardants may also be present.

2.4 Substances contained in mobile phones

Plastics account for nearly 40% of the total mass of a mobile phone and are primarily associated with the housing and PCBs, while glass and ceramics materials contribute for approximately 20%, and are concentrated in display screens and integrated circuit components. Copper-based compounds contribute around 10% of device mass and occur mainly in PCBs, interconnecting wires, connectors, and batteries. Nickel compounds make up about 2–10% and are found largely in NiCd and NiMH battery systems. Potassium hydroxide (KOH) is present at concentrations below 5% these battery types.

Cobalt contributes roughly 1–5% of the device mass and is mainly localized in lithium-ion batteries, while elemental carbon is present at levels below 5% within battery assemblies. Aluminum typically accounts for 3–20% and is distributed through the casing, frame, and battery parts. Ferrous steel components contribute close to 10% and are used in the casing, frames, charger, and battery parts. Tin (Sn) is present at approximately 1% and is mainly associated with PCB soldering. Minor constituents, in the range of 0.1–1%, include bromine, magnesium, tantalum, tungsten, and zinc, occurring predominately in PCBs. Cadmium is confined primarily to nickel-cadmium batteries, while chromium appears in the casing and frames. Lead (Pb) occurs in PCBs through tin-lead solder and in certain PVC-coated wires (2–5%) and plastic components, although its use is steadily declining and is largely absent in modern smartphone models. Lithium (Li) is present in lithium-ion batteries and silver (Ag) is found in PCBs and keypad contacts.

Trace level constituents, below 0.1%, include antimony, bismuth, calcium, barium, palladium, ruthenium, strontium, sulfur, yttrium, zirconium, arsenic (as gallium arsenide light emitting diodes), beryllium (in connectors), fluorine (lithium-ion batteries), gallium (in optoelectronic components), and gold in connectors and circuitry. Mercury is no longer used in current mobile phone designs, although some old devices may still contain mercury oxide or silver oxide-based button-cell batteries [21].

3 Overview of microbial biominers

Microbial biominers are microorganisms capable of mobilizing, oxidizing, or solubilizing metals from solid matrices such as ores, electronic waste, and industrial residues. These biominers belong to diverse taxonomic groups, including chemolithotrophic bacteria (e.g., *Leptospirillum ferriphilum*), heterotrophic bacteria, and fungi, all of which have demonstrated potential for metal recovery from e-waste. For example, *L. ferriphilum* has been shown to effectively bioleach metals from polymetallic sources [22], while two-step bioleaching strategies using acidophilic bacteria have improved recovery of copper and gold from PCBs [23]. Such microbial activities, ranging from acid production, redox reactions, to selective solubilization, enable environmentally sustainable recovery of metals from complex e-waste matrices.

3.1 Bioleaching by chemolithotrophic microorganisms

These microorganisms are commonly referred to as iron- and sulfur-oxidizing bacteria or acidophilic bacteria. During the bioleaching process, they catalyze the oxidation of ferrous ions (Fe^{2+}) into ferric ions (Fe^{3+}) and convert elemental sulfur into sulfuric acid, as described by the reactions [32–34]. The ferric iron and sulfuric acid generated through these biological reactions act as powerful oxidizing agents, or lixiviants, which promote the release of base metals from solid waste matrices through combined acidolysis and redoxolysis mechanisms:

$$2FeSO_4 + H_2SO_4 + {}^{1/2}O_2 \rightarrow Fe_2(SO_4)_3 + H_2O$$

$$S^0 + H_2O + 3/2O_2 \rightarrow H_2SO_4$$

$$H_2SO_4 + MeS \rightarrow H_2S + MeSO_4$$

$$H_2SO_4 + MeO \rightarrow H_2O + MeSO_4$$

$$Fe(SO_4)_3 + MeS + H_2O + 3/2O_2 \rightarrow Me^{2+} + SO_4{}^{2-} + 2FeSO_4 + H_2SO_4$$

In addition to their role in metal solubilization, these microorganisms utilize atmospheric carbon dioxide (CO_2) as their primary carbon source and derive metabolic energy from inorganic substrates such as ferrous iron, elemental sulfur, and a variety of reduced sulfur compounds, including polysulfides, hydrogen sulfide (H_2S), thiosulfate ($S_2O_3{}^{2-}$), and elemental sulfur (S_8). This group of microorganisms is distinct from most other microbial classes studied to date, owing to their exceptional ability to drive metal dissolution from e-waste through sequential bio-oxidation and bioleaching reactions [22, 23].

3.1.1 Chemolithotrophic bacteria (pure cultures)

Acidithiobacillus ferrooxidans has been extensively applied for metal solubilization from mobile phone PCBs, owing to its strong iron- and sulfur-oxidizing capacity. Under controlled conditions of 20–35 °C, with 5% PCB pulp density and 5% (v/v) inoculum volume, this bacterium achieved high dissolution efficiencies of 95.92% for Cu, 93.53% for Al, 92.58% for Zn, 65.27% for Ni, and 95.33% for Sn. These results confirm its high effectiveness for simultaneous multi-metal recovery from complex PCB matrices. Similarly, *A. thiooxidans* demonstrated strong leaching performance in spent mobile phone PCBs, achieving 98% Cu and 82% Ni extraction at 30 °C, with agitation at 160 rpm over 72 h. Its sulfur-oxidizing metabolism enables rapid acidification, which enhances metal solubilization.

Thiobacillus ferrooxidans has also been successfully employed for bioleaching of nickel–cadmium batteries. At pH 2.0 and 30 °C, with an incubation period of 93 days and pulp density of 0.2 g/L, complete Cd recovery (100%) was achieved along with high extraction of Ni (96.5%) and Fe (95%). This highlights its suitability for hazardous battery waste detoxification [47].

3.1.2 Chemolithotrophic bacteria (mixed cultures)

A mixed consortium of *A. ferrooxidans* and *A. thiooxidans* has shown enhanced synergistic performance in waste PCB (WPCB) bioleaching, achieving 86% Cu, 100% Zn, and 100% Ni recovery. The cooperative oxidation of Fe^{2+} and reduced sulfur species intensifies acid and ferric ion production, thereby accelerating metal dissolution. A more complex microbial consortium consisting of *A. ferrooxidans*, *Leptospirillum ferrooxidans*, and *A. thiooxidans* achieved even higher efficiencies during mobile phone PCB treatment. Under conditions of Fe^{3+} concentration of 9 g/L, pulp density of 10%, inoculum volume of 10% (v/v), and pH 1.8, recoveries of 97.3% Cu, 55.8% Al, 79.3% Ni, and 66.8% Zn were obtained. In the case of lithium-ion batteries, a thermophilic mixed culture comprising *Acidithiobacillus caldus*, *L. ferriphilum*, *Sulfobacillus* sp., and *Ferroplasma* sp. achieved extremely high extraction efficiencies of 99.9% Co, 99.7% Ni, and 84% Li at 45 °C and 130 rpm. This demonstrates the strong potential of thermophilic consortia for battery metal recovery [7, 26, 47].

3.2 Bioleaching by heterotrophic microorganisms

Heterotrophic bacteria contribute to metal mobilization from e-waste through the production of specific metabolic by-products that enhance metal solubility. A well-known example is provided by cyanogenic bacteria, which releases hydrogen cyanide (HCN) as a secondary metabolite, capable of dissolving precious metals from electronic waste – a process referred to as cyanogenesis [21]. Under alkaline conditions, these microorganisms synthesize cyanide from glycine, which serves as a metabolic precursor during the transition from the exponential growth phase to the stationary phase via oxidative decarboxylation. This reaction is catalyzed by the enzyme HCN synthase. The released cyanide solubilizes precious metals through the formation of stable metal–cyanide complexes, thereby enhancing their extraction from e-waste matrices. However, the overall recovery efficiency is often limited due to the relatively low cyanide yields produced by the bacteria and the toxic effects of e-waste components on cyanogenic microorganisms [2]. The production of cyanide is strongly influenced by operational factors such as temperature, pH, dissolved oxygen (DO) concentration, and glycine availability. The cyanide-based dissolution of precious metals proceeds through coupled anodic and cathodic electrochemical reactions represented:

$$4Au + 8CN^- \rightarrow 4Au(CN)_2^- + 4e-$$

$$O_2 + 2H_2O + 4e^- \rightarrow 4OH^-$$

$$4Au + 8CN^- + O_2 + 2H_2O_4 \rightarrow Au(CN)_2^- + 4OH^-$$

Cyanogenic microorganisms have been successfully applied for the bioleaching of valuable metals and metalloids, including gold, silver, platinum, palladium, and titanium from e-waste. In practice, base metals are typically removed prior to precious-metal bioleaching, as their high concentrations and strong complex-forming ability with microbial metabolites can otherwise hinder the efficient mobilization of noble metals [24, 25].

3.2.1 Bioleaching by heterotrophic bacteria (pure cultures)

Bacillus megaterium has been effectively applied for gold recovery from computer PCBs under alkaline conditions. At pH 10.0 and pulp density of 2 g/L, this bacterium achieved 63.8% Au recovery, indicating that heterotrophic pathways based on organic acid and cyanide-like metabolite production can support precious metal solubilization. *P. fluorescens* has also demonstrated significant gold leaching from WPCBs, achieving 54% Au extraction at pH 9.0, bacterial density of 3.33% (v/v), and pulp density of 0.33%, using glycine (0.133) with a glycine-to-methionine ratio of 1:10.*Bacillus megaterium* PTCC 1656 has shown exceptional efficiency in spent catalyst processing, achieving 15.7% Pt and 98% Re recovery using glycine at 12.8 g/L, with 4% pulp density, 30 °C temperature, 170 rpm agitation, and a bioleaching duration of 7 days [26].

3.2.2 Bioleaching by heterotrophic fungi (pure cultures)

Fungal species involved in bioleaching are known for secreting large quantities of organic acids, including carboxylic, gluconic, citric, and lactic acids, which play a major role in enhancing the mobilization of metals from e-waste. These organic acids interact with metallic constituents through multiple mechanisms such as acidolysis, redoxolysis, bioaccumulation, chelation, and complex formation, ultimately leading to increased metal solubilization. Furthermore, fungal-mediated metal dissolution can also occur through the enzymatic reduction of highly oxidized metal compounds by specific extracellular biodegradative enzymes. Due to their remarkable capacity to adapt to harsh environments and tolerate high levels of metal toxicity, fungi are considered particularly effective and resilient candidates for mineral and e-waste bioleaching applications [25].

3.3 Heterotrophic fungi (pure cultures)

Aspergillus niger has demonstrated outstanding bioleaching ability for WPCBs, achieving complete Zn recovery (100%), along with 80.39% Ni and 85.88% Cu extraction under pulp densities ranging from 0.5 to 20 g/L at 120 rpm and ambient temperature over 30 days. Organic acids such as citric and oxalic acid play a key role in this metal solubilization process.

Penicillium simplicissimum has also shown strong leaching efficiency for cell phone PCBs, achieving 90% Cu recovery at pH 7.0, with 3.3×10^7 spores, using sugar-based substrates, while 89% Ni recovery was achieved at pH 2.0 with 10^6 spores, using molasses as the carbon source.

Aspergillus niger was further applied to spent lithium-ion mobile phone batteries, yielding complete Li recovery (100%), along with 72% Mn, 45% Ni, 38% Co, 94% Cu, and 62% Al at 30 °C over 30 days. This confirms its broad-spectrum bioleaching potential for both base and critical battery metals [7, 26]. The efficiencies of various microorganisms involved in the bioleaching of metals from e-waste are summarized in Table 1

Table 1: Efficacies of some chemolithotropic bacteria (pure culture), chemolithotropic bacteria (mixed culture), heterotrophic bacteria (pure culture), and heterotrophic fungi (pure culture) in the bioleaching of metals from e-waste.

Microorganism	E-waste	Target metal	Bioleaching conditions	References
Chemolithotropic bacteria (pure culture)				
Acidithiobacillus ferrooxidans	Mobile phone printed circuit boards	Cu (95.92%), 93.53% (Al), 92.58% (Zn), 65.27% (Ni), and 95.33% (Sn)	Temperature 20–35 °C, waste PCB concentration 5%, and inoculation volume 5% (*v/v*)	[7, 26, 40, 45, 47, 51]
Acidithiobacillus thiooxidans	Spent mobile phone printed circuit boards	Cu (98%) and Ni (82%)	160 rpm; 30 °C, 72 h	[7, 26, 40, 45, 47, 51]
Thiobacillus ferrooxidans	Nickel-cadmium batteries	Cd (100%), Ni (96.5%), and Fe (95%)	pH 2.0, temperature 30 °C, incubation time 93 days, and pulp density 0.2 g/L	[7, 26, 40, 45, 47, 51]
Chemolithotropic bacteria (mixed culture)				
Acidithiobacillus ferrooxidans and *Acidithiobacillus thiooxidans*	Waste printed circuit boards	Cu (86%), Zn (100%), and Ni (100%)	Cu (86%), Zn (100%), and Ni (100%)	[7, 26, 40, 45, 47, 51]
Acidithiobacillus ferrooxidans, *Leptospirillum ferrooxidans*, and *Acidithiobacillus thiooxidans*	Mobile phone printed circuit boards	97.3% (Cu), 55.8% (Al), 79.3% (Ni), and 66.8% (Zn)	Fe^{3+} concentration 9 g/L, pulp density 10%, and inoculum volume 10% (*v/v*); pH 1.8	[7, 26, 40, 45, 47, 51]

Table 1 (continued)

Microorganism	E-waste	Target metal	Bioleaching conditions	References
Acidithiobacillus caldus, *Leptospirillum ferriphilum*, *Sulfobacillus* sp., and *Ferroplasma* sp.	Lithium-ion batteries	Co (99.9%), Ni (99.7%), and Li (84%)	45 °C, 130 rpm	[7, 26, 40, 45, 47, 51]
Heterotrophic bacteria (pure culture)				
Bacillus megaterium	Computer printed circuit boards	63.8% (Au)	Pulp density 2 g/L, pH 10.0	[7, 26, 40, 45, 47, 51]
Pseudomonas fluorescens	Waste printed circuit boards	54% (Au)	pH 9.0, bacterial density 3.33% (*v/v*); pulp density 0.33%, glycine 0.133, glycine, methionine ratio 1:10	[7, 26, 40, 45, 47, 51]
Bacillus megaterium PTCC 1656	Spent catalyst	15.7% (Pt) and 98% (Re)	Glycine 12.8 g/L, pulp density 4%, 30 °C, 170 rpm; 7 d	[7, 26, 40, 45, 47, 51]
Heterotrophic fungi (pure culture)				
Aspergillus niger	Waste printed circuit boards	100% (Zn), 80.39% (Ni), and 85.88% (Cu)	Pulp density (0.5–20 g/L), 120 rpm; ambient temperature, 30 d	[7, 26, 40, 45, 47, 51]
Penicillium simplicissimum	Cell phone printed circuit boards	90% (Cu) and 89% (Ni)	Cu: pH 7.0; 3.3×10^7 spores, sugarNi: pH 2.0; 10^6 spores, molasses	[7, 26, 40, 45, 47, 51]
Aspergillus niger	Spent lithium-ion mobile phone batteries	100% (Li), 72% (Mn), 45% (Ni), 38% (Co), 94% (Cu), and 62% (Al)	30 °C; 30 d	[7, 26, 40, 45, 47, 51]

4 Mechanisms of microbial bioleaching

Microbial bioleaching refers to the solubilization of metals through biochemical reactions carried out by bacteria, fungi, and archaea. These microorganisms facilitate metal mobilization through several mechanisms, including acidolysis, redoxolysis, cyanogenesis, complexolysis, chelation, and biosorption, which collectively enhance the dissolution of metals from electronic waste matrices such as PCBs, lithium-ion batteries, and high-tech alloys. Because e-waste contains metals in diverse chemical forms (metallic, oxide, alloy, or sulfide), multiple microbial mechanisms may act simultaneously or sequentially.

Microbial bioleaching is governed by a series of biochemical mechanisms that enable microorganisms to solubilize metals from complex electronic waste matrices. The most dominant mechanism involved in base metal dissolution is redoxolysis, which is primarily driven by acidophilic iron- and sulfur-oxidizing bacteria such as *A. ferrooxidans*. These bacteria oxidize ferrous ions (Fe^{2+}) to ferric ions (Fe^{3+}), which act as strong oxidizing agents for metallic copper and other base metals present in PCBs, thereby enhancing metal solubilization efficiency [8, 19, 43, 48]. Simultaneously, sulfur oxidation leads to the generation of sulfuric acid, which maintains the acidic conditions required for sustained metal dissolution.

Cyanogenesis represents another important mechanism, particularly for precious metal recovery. Certain heterotrophic bacteria such as *C. violaceum*, and *Pseudomonas* species produce hydrogen cyanide as a metabolic by-product, which complexes with gold and silver to form soluble metal-cyanide complexes [9, 24, 33]. This biological cyanidation pathway provides a lower-toxicity alternative to conventional chemical cyanidation used in gold extraction.

Fungal bioleaching operates primarily through acidolysis and complexolysis. Filamentous fungi such as *A. niger* and *Phanerochaete chrysosporium* secrete organic acids, including citric, oxalic, and gluconic acids that chelate metal ions and enhance their solubility [25, 50, 53]. These organic acids also reduce solution pH and promote structural weakening of electronic substrates, facilitating subsequent leaching stages.

In addition to active leaching mechanisms, biosorption plays a significant role in metal uptake. Functional groups present on microbial cell walls – such as carboxyl, hydroxyl, and phosphate groups – bind metal ions through surface complexation and ion-exchange reactions [28, 58]. This passive accumulation process is particularly important for the removal of low-concentration heavy metals during post-leaching treatment.

Collectively, these microbial mechanisms form the fundamental basis for the sustainable recovery of copper, gold, silver, nickel, cobalt, and REEs from electronic waste. Their successful application has been widely demonstrated in both batch and continuous reactor systems, highlighting the technical feasibility of bioleaching-based urban mining strategies [26, 27, 39, 46].

4.1 E-waste preparation

1. In most bioleaching studies, electronic waste is subjected to systematic preparation and, in many cases, pretreatment prior to metal recovery. Typically, elec-

tronic devices are first manually dismantled, and the target components are carefully separated.
2. For battery waste, the anode and cathode materials are commonly isolated and utilized as feedstock for bioleaching experiments. The selected fractions are then crushed and pulverized using mechanical grinders such as blenders or ball mills to find a fine and homogeneous powder.
3. This material is subsequently sieved to achieve a uniform particle size distribution. In certain studies, drying of the powdered sample is carried out before bioleaching, to remove residual moisture. Prior to microbial processing, the prepared samples are characterized using several analytical techniques.
4. The qualitative and quantitative metal composition is commonly determined using inductively coupled plasma-optical emission spectrometry, while X-ray diffraction analysis is active to identify the crystalline phases present in the powdered material [17].

4.2 Redoxolysis

Mobilization through redoxolysis occurs through the transfer of electrons from mineral phases to microorganisms, primarily through the biological oxidation of ferrous iron (Fe^{2+}) to ferric iron (Fe^{3+}). Key microorganisms actively involved in this redox-driven process include *A. thiooxidans*, *A. ferrooxidans*, and *L. ferrooxidans*.

Microbial interactions with mineral substrates take place through either contact or noncontact mechanisms. In the contact mode, microorganisms secrete extracellular polymeric substances (EPSs), which act as reactive interfaces where oxidation reactions occur. Two distinct contact-based operational strategies are commonly applied, namely the one-step and two-step leaching processes, reliant on the timing between microbial growth and exposure to the mineral substrate. In the one-step approach, both the REE-bearing material and the microbial inoculum are introduced simultaneously into the same growth medium. In contrast, the two-step method involves introducing the REE-containing material into a culture in which the microorganisms have already been pre-grown. Typically, the one-step process is favored for systems with low mineral concentrations, while the two-step method is more suitable for substrates containing higher REE levels.

The presence of insoluble REE-bearing matrices, such as red mud, within the microbial growth environment can impose significant toxicity, thereby suppressing microbial metabolism and reducing the production of organic acids. In the noncontact mode bioleaching, the solid mineral matrix does not directly interact with the microorganisms. Instead, a leaching medium enriched with biologically generated lixiviants from a pre-cultured microbial system is applied to the mineral substrate to facilitate metal solubilization.

In the redoxolysis mechanism, microorganisms generate reactive catalytic species that control and adjust the oxidation–reduction potential of the leaching system. The overall efficiency and kinetics of this process are strongly influenced by the mineralogical structure, the specific metal involved, and its oxidation state. Among all redox agents, ferric iron (Fe^{3+}) plays a central role in bioleaching systems. It is continuously produced by iron-oxidizing microorganisms and functions as a powerful oxidant that converts metallic species into their soluble ionic forms. During the leaching reaction, Fe^{3+} is reduced to ferrous iron (Fe^{2+}), subsequently re-oxidized back to Fe^{3+} by iron-oxidizing bacteria, thereby maintaining a self-sustaining cyclic redox process.

The thermodynamic feasibility of metal dissolution through ferric iron oxidation has been clearly demonstrated for copper, nickel, and zinc based on standard enthalpy and Gibbs free energy values under ambient conditions, as represented by the following reactions:

$$Cu^0 + 2Fe^{3+} \rightarrow Cu^{2+} + 2Fe^{2+} \quad \Delta G^\circ = -347.1 \ kJ$$

$$Ni^0 + 2Fe^{3+} \rightarrow Ni^{2+} + 2Fe^{2+} \quad \Delta G^\circ = -822.6 \ kJ$$

$$Zn^0 + 2Fe^{2+} \rightarrow Zn^{2+} + 2Fe^{2+} \quad \Delta G^\circ = -1{,}235.9 \ kJ$$

These highly negative Gibbs free energy values confirm that Fe^{3+} acts as a strong oxidizing agent, readily promoting the dissolution of metals. In addition to aerobic reoxidation, several microbial strains are also capable of reducing Fe^{3+} to Fe^{2+} under anaerobic conditions. The Fe^{3+}/Fe^{2+} redox couple exhibits a standard reduction potential of 0.77 V relative to the standard hydrogen electrode (SHE), making it a highly effective electron transfer system. Redox potential thus represents a critical parameter for assessing iron speciation and redox behavior in bioleaching environments. Due to the continuous activity of iron-oxidizing microorganisms, elevated redox potentials are typically observed in bioleaching systems. Furthermore, for mesophilic acidophilic iron oxidizers, the complete redox pathway from ferrous iron (Fe^{2+}) to the terminal electron acceptor oxygen has been well established [8, 12, 23].

4.3 Acidolysis

Acidolysis is primarily mediated by sulfur-oxidizing and phosphate-oxidizing bacteria during the bioleaching of REEs. In sulfur-driven acidolysis, sulfur-oxidizing microorganisms oxidize sulfide minerals to yield sulfuric acid, which subsequently dissolves the REE-bearing phases. Major sulfur-oxidizing bacteria reported to participate in this mechanism include *Alicyclosbacillus disulfidooxidans*, *A. ferrooxidans*, and *A. thiooxidans*.

In phosphate-mediated acidolysis, phosphate-solubilizing bacteria facilitate REE dissolution by releasing phosphate from mineral lattices, leading to metal mobilization through acidification and complexation. A wide range of bacterial genera have been identified as effective phosphate oxidizers in this process, including *Bacillus, Klebsiella, Agrobacterium, Acetobacter, Burkholderia, Serratia, Pseudomonas, Clavibacter, Thiobacillus, Flavobacterium, Enterobacter, Erwinia, Micrococcus,* and *Streptomyces.* The collective activity of these microorganisms significantly enhances REE liberation through biologically induced acid dissolution pathways [7, 26, 27].

4.4 Complexolysis

In the complexolysis mechanism, metal solubilization occurs through the action of biologically produced ligands that enhance metal mobility by forming soluble complexes through chelation and coordination reactions. These biogenic complexing agents bind strongly with metal ions, replacing stable bonds within the solid matrix and thereby promoting metal dissolution. Among such ligands, siderophores secreted by several bacterial and fungal species are recognized as some of the most powerful natural iron-binding compounds, particularly for ferric iron (Fe^{3+}), which is otherwise poorly soluble under natural aqueous conditions. In addition to iron, siderophores are also capable of binding and mobilizing a wide range of other metals [27].

Complexolysis is predominantly driven by heterotrophic cyanide-producing microorganisms. Cyanide refers to a group of compounds containing the cyano functional group (CN^-), and trace amounts of its detoxified form, thiocyanate (SCN^-), naturally occur in human biological fluids such as saliva, urine, and gastric secretions. During bioleaching, cyanide acts as a highly effective complexing agent that reacts with metals present in waste materials to form stable, water-soluble metal-cyanide complexes. This mechanism is particularly important for the recovery of noble metals from secondary resources, as valuable metals such as gold, platinum, and palladium exhibit exceptional chemical stability and react with only a limited number of lixiviants. Importantly, safety concerns associated with cyanide are significantly reduced in biological systems because cyanogenic microorganisms naturally metabolize cyanide into nontoxic β-cyanoalanine. Consequently, biologically mediated cyanide production allows the development of safer recovery systems, without the need for additional chemical detoxification steps. Once formed, cyanide–metal complexes can be efficiently recovered using conventional downstream techniques, for instance – activated carbon adsorption, cementation, electrowinning, and electrorefining [26, 27, 38].

5 Biorecovery of metals

A wide range of well-established conventional separation techniques – such as solvent extraction, cementation, ion exchange, chemical precipitation, adsorption, and electrowinning – are routinely employed for the selective recovery of metals from leachate solutions. These methods have been successfully applied for the targeted removal and recovery of toxic and valuable metals, including As, Cd, Ni, Zn, Cr, Se, Cu, Fe, and Pb from domestic wastewater, industrial effluents, and solid waste residues. In parallel, growing scientific attention has been directed toward the use of biosorbents for the recovery of precious metals from aqueous systems. Several biologically driven processes – such as biosorption, bioreduction, biomineralization, and bioprecipitation – have emerged as promising alternative strategies for metal recovery. The choice of an appropriate recovery technique is primarily governed by key process parameters, particularly the metal concentration in the leachate as well as selectivity characteristics of the chosen method. At the industrial scale, metal recovery operations commonly integrate multiple conventional techniques into hybrid process flowsheets to enhance overall efficiency [27].

Biosorption is a process in which metal ions are passively bound to biological materials, such as bacteria, fungi, algae, or their extracellular components, without the need for metabolic energy. This process occurs mainly through mechanisms like ion exchange, complexation, or chelation with functional groups (carboxyl, hydroxyl, sulfhydryl, phosphoryl, or amino), physical adsorption, and in some cases, local precipitation on the biomass surface. Biosorption is particularly effective for low-to-moderate concentrations of metals, allowing rapid and reversible metal removal, and can utilize inexpensive or waste biomass. Its applications extend to the recovery of metals from aqueous solutions, including electronic waste leachates, where metals concentrated on biomass can be subsequently desorbed and recovered. However, biosorption efficiency is influenced by competing ions, pH, ionic strength, and the presence of multiple metals, which can limit capacity and selectivity [28].

5.1 Biomineralization

Biomineralization, in contrast, involves the microbial-induced development of stable mineral phases, such as carbonates, phosphates, oxides, or hydroxides, which immobilize or precipitate metals from solution. Microbial-induced carbonate precipitation is mediated by ureolytic bacteria that hydrolyze urea to produce carbonate ions, which then react with divalent cations, including certain heavy metals, to form insoluble carbonates. Similarly, microbial-induced phosphate precipitation (MIPP) is driven by phosphate-solubilizing bacteria, which release inorganic phosphate to precipitate metals as metal-phosphate minerals, such as lead or cadmium phosphates.

Biomineralization can occur via other microbial pathways as well, including iron oxidation, which produces iron oxyhydroxides, capable of adsorbing or co-precipitating metals. The major advantage of biomineralization is the production of highly stable mineral forms that immobilize metals for long-term stability, making it particularly suitable for environmental remediation and recovery applications. Nonetheless, efficiency depends on specific environmental conditions, availability of precursors (carbonate or phosphate), metal chemistry, and selectivity in mixed-metal systems. Studies on *Cupriavidus metallidurans* have shown that this bacterium protects itself from the toxic effects of soluble gold by internalizing and accumulating inert gold nanoparticles within its cytoplasm. Similarly, the gold-tolerant bacterium, *Delftia acidovorans*, secretes a unique metabolite known as delftibactin, which enhances the resistance to gold toxicity. It is proposed that the secretion of such protective metabolites not only mitigates metal toxicity but also promotes the biomineralization of elemental gold (Au^0).

The discovery of a previously unknown biomineralization pathway is accountable for the reduction of Pd(II) in aerobically grown *Escherichia coli* cultures, primarily mediated by molybdenum-dependent enzyme systems. Interestingly, even mutant strains lacking all molybdoenzymes were still capable of palladium reduction, although the process required approximately 7 h when compared to less than 30 min in wild-type strains. Real-time monitoring confirmed that the formation of biogenic Pd(0) occurred extracellularly.

The biomineralization of platinum group metals from electronic waste has recently emerged as an important area of biotechnological research. This approach exhibits an auspicious and selective alternative for recovering precious metals from complex electronic waste leachates. Additionally, the inherent metal-specific selectivity of microbial cells provides a major advantage over conventional physicochemical separation technologies [26, 27, 32, 34, 58].

5.2 Bioprecipitation

Biogenic sulfide precipitation, or bioprecipitation, is a microbial process commonly carried out by sulfate-reducing bacteria (SRB) below anaerobic conditions. These bacteria decrease sulfate to generate sulfide ions (H_2S or HS^-), which react with dissolved metal ions to form highly insoluble metal sulfides, such as CuS, ZnS, PbS, or CdS. This process is effective in removing and immobilizing metals from wastewater, sediments, and industrial effluents, yielding stable sulfide precipitates. Biogenic sulfide precipitation is particularly advantageous for anaerobic environments and can achieve high removal efficiencies; however, it is sensitive to metal toxicity and pH, and requires careful control of sulfide production and precipitation processes [30, 31].

Collectively, these three biological mechanisms offer complementary approaches for metal recovery. In the context of e-waste, where complex metal mixtures are present, hybrid strategies combining bioleaching to solubilize metals with biosorption, biomineralization, or sulfide precipitation, to recover or immobilize them, are increasingly explored. Such integrated approaches provide environmentally friendly alternatives to conventional chemical methods, enabling efficient recovery of both base as well as precious metals from electronic waste leachates [29, 30–32].

In addition to SRB, other microbial pathways such as biomineralization and biosorption have been efficiently employed to recover metals from electronic waste. Biomineralization includes microbial-mediated precipitation of metals as carbonates, phosphates, and oxalates, which stabilizes metals in solid forms, appropriate for downstream recovery. Biosorption, alternatively exploits the affinity of microbial cell walls, EPSs, or dead biomass for metal ions, enabling efficient capture of metals even at low concentrations. Integrating these mechanisms with bioleaching can improve overall metal recovery, decrease environmental toxicity, and enhance process sustainability, especially when dealing with complex e-waste streams containing both base and precious metals [7, 26, 27, 32, 58].

6 Genetic engineering for the recovery of metals

Genetically engineered microorganisms have developed as a promising strategy to improve the recovery of metals, particularly precious metals, from electronic waste. A notable example involves *C. violaceum*, which was genetically modified by inserting a duplicated copy of the hcnABC operon under strong exogenous promoters, thereby boosting cyanide production, the primary lixiviant for gold solubilization. This metabolic engineering led to approximately a 70% increase in cyanide output, compared to wild-type strains, resulting in gold recoveries exceeding 30% from electronic scrap, in contrast to only ~11% recovery with unmodified bacteria under identical conditions. Proteomic analyses confirmed that the engineered strain maintained high cyanide availability without significant upregulation of detoxifying pathways, highlighting the efficiency of this approach [33, 34].

Beyond genetic modification, mutation and adaptive evolution strategies have also been employed to enhance metal tolerance and process efficiency. For instance, alkali-tolerant strains of *C. violaceum*, evolved to survive at pH 9–10, achieved gold recoveries of 18–22.5% under alkaline conditions, compared with ~11% for wild-type strains at neutral pH, demonstrating the potential of adaptive evolution in improving bioleaching performance in harsh, real-world e-waste environments. In addition to genetic modification and adaptive evolution, selecting or engineering microbial consortia or extremophilic strains has shown promise in improving metal recovery from e-waste. Such strategies enhance tolerance to toxic metals, pH fluctuations, and variable compositions, typical of heterogeneous electronic waste, thereby improving leaching efficiency and stability. These approaches complement single-strain engineering by broadening the functional capabilities of the microbial system and enabling more robust bioleaching processes under challenging conditions [56, 58].

These approaches collectively demonstrate that combining metabolic engineering or adaptive evolution with microbial bioleaching can pointedly improve the efficacy, kinetics, and robustness of metal recovery from complex e-waste matrices. Reviews in the field further emphasize that genetically engineered microbes offer a sustainable and scalable route for biometallurgical applications, providing a viable alternative to conventional chemical or hydrometallurgical recovery methods, while also addressing challenges associated with heterogeneous waste compositions and toxic metal concentrations. However, practical deployment requires careful consideration of environmental biosafety, downstream processing for metal extraction, and the performance of engineered strains in mixed-metal e-waste streams [33, 34].

7 Bioleaching processes

7.1 One-step bioleaching

In the one-step bioleaching approach (Figure 2), microbial inoculum (cells) and the e-waste (or other metal-bearing waste) are presented together into a fresh growth/leaching medium. Thus, microbial growth, production of biolixiviants (e.g., acids, ferric ions, and protons), and dissolution of metals from the waste proceed simultaneously. Because the waste (e-waste) typically contains toxic components – heavy metals, metals in mixed phases, resins, plastics, solder materials – these can inhibit microbial growth or metabolism before sufficient lixiviant is produced.

As a result, one-step bioleaching often leads to poor metal recovery, especially when the pulp density (i.e., fraction/weight of waste in medium) is high; typically, this limits efficient operation to relatively low pulp densities (e.g., 1–10% w/v) for many microbes. Therefore, while this method is conceptually simple and easy to implement (one mixing step), its practical use for complex waste like e-waste is often constrained by microbial toxicity and sub-optimal metal solubilization [6, 10, 43, 47, 48].

To overcome the limitations of one-step bioleaching, researchers have discovered approaches such as the gradual adaptation of microbes to higher pulp densities, use of tolerant strains, or preconditioning the waste to reduce toxicity. For instance, pretreatment of e-waste to eliminate or neutralize inhibitory components, or adjusting growth conditions (pH, nutrient supplementation, or aeration), can improve microbial survival and lixiviant production. These strategies help maintain microbial activity long enough for effective metal dissolution, improving recovery yields even in complex and high-metal-content wastes [6, 10, 47].

7.2 Two-step bioleaching

The two-step bioleaching (Figure 2) approach seeks to overcome the limitations of one-step by decoupling microbial growth/lixiviant production from metal dissolution. In this method, microorganisms are first cultured in a growth medium, without the metal-bearing waste, allowing them to proliferate under optimal conditions and produce leaching agents (e.g., ferric ions, acids, or protons) without stress. Once the culture reaches maximum growth (or optimal lixiviant concentration), the e-waste is then added as a second step; metal solubilization proceeds from that point. This separation reduces the inhibitory effects of toxic waste on microbial viability and lixiviant production, improving the likelihood of effective metal dissolution.

Indeed, studies on PCBs have demonstrated that two-step bioleaching using iron- and sulfur-oxidizing acidophiles can achieve substantial extraction efficiencies for multiple metals (e.g., Cu, Al, Fe, Mn, Ni, Pb, Zn, and Sn) – depending on variables such as particle size, pulp density, and leaching duration [10]. Moreover, recent work using biogenic ferric-ion leaching from discarded mobile phone PCBs at 10% pulp density reported copper extraction exceeding 99% within a few days, when optimized – showing that two-step bioleaching can be both efficient and relatively rapid under favorable conditions. The separation of growth and leaching phases also allows better control of conditions (pH, redox, and lixiviant concentration) and is often more scalable than one-step, when handling more concentrated wastes [12].

7.3 Spent-medium (cell-free) bioleaching

In the spent-medium bioleaching (Figure 2) (also called "cell-free" bioleaching) method, microorganisms are the first to be grown in a suitable medium (without e-waste) up to the point when they reach stationary phase and have secreted a maximum amount of lixiviants (acids, ferric ions, cyanide, organic acids, etc.). Then, the microbial cells are removed (e.g., by centrifugation or filtration), leaving a cell-free "spent medium" rich in biogenic lixiviants; the e-waste is then introduced into this medium for leaching. Because no live cells are present during leaching, the toxic effects of the waste do not affect microbial survival, allowing leaching under more aggressive conditions (higher pulp densities, lower pH, and stronger lixiviant concentration), which often enhances metal extraction yields [36].

For example, in a work with *C. violaceum* (a cyanogenic bacterium) on electronic scrap material (ESM), spent-medium leaching achieved ~18% gold recovery at 0.5% w/v pulp density; and when the pH of the spent medium was adjusted to favor cyanide ion formation, gold recovery increased to ~30%, substantially higher than the ~11% obtained by two-step bioleaching under similar conditions [37]. This shows that spent-medium bioleaching can outperform contact methods under certain conditions, especially for precious-metal recovery (e.g., gold) via ligand-mediated leaching (e.g., cyanide complexation). Because there are no living cells in contact with e-waste, spent-medium bioleaching also makes downstream separation and purification simpler (less biological waste and fewer biosafety concerns), and can be more readily scaled or optimized independently for lixiviant production and leaching chemistry [10].

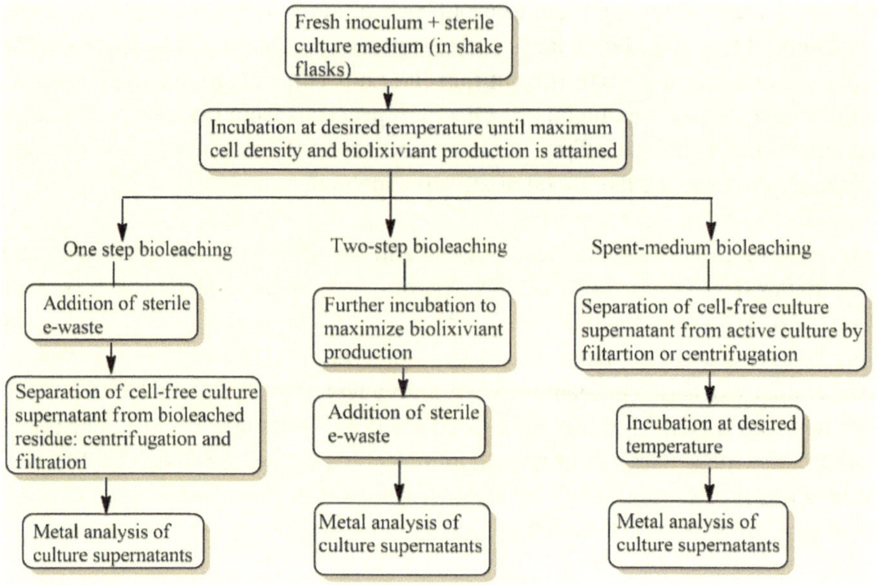

Figure 2: Schematic representation of the three bioleaching process configurations used for metal recovery from electronic waste.

8 Bioleaching process parameters

Bioleaching efficiency in electronic waste (e-waste) recycling depends strongly on several operational and environmental parameters that influence microbial growth, metabolic activity, and metal solubilization kinetics. Optimizing these parameters is essential for achieving high metal recovery yields, maintaining microbial stability, and scaling up the process from laboratory to industrial levels. Critical process variables include pH, temperature, pulp density, particle size, DO levels, aeration/agitation, inoculum size, nutrient availability, and pretreatment of e-waste. This section discusses each parameter in detail.

8.1 pH

- The pH of the leaching medium is possibly the most critical factor: it controls microbial viability/metabolism, acid-mediated metal dissolution, and prevents formation of passivating compounds [38].
- For acidophilic bioleaching bacteria (e.g., *A. ferrooxidans* and similar iron- or sulfur-oxidizers), optimal pH values lie typically around 2.0–2.5 for iron/sulfide oxidation.

- In studies on WPCBs, metal dissolution (Cu, Ni) was best when the initial pH was set at ~1.8–2.0. Increases in pH, above ~2.0–2.25, often triggered formation of secondary precipitates (e.g., jarosite), which hinder leaching [39].
- For non-acidophilic or heterotrophic/cyanogenic microbial processes (e.g., for precious metal leaching), the optimum pH may differ, and pH control must be adapted accordingly [40].

8.2 Temperature

- Temperature governs microbial growth rates, metabolic activity, redox chemistry, and overall kinetics of leaching [38].
- For mesophilic acidophiles, commonly used in e-waste bioleaching, the optimum temperature is typically in the range 25–30 °C.
- Some studies also explore moderate thermophiles/thermophiles – demonstrating that bioleaching at higher temperatures (e.g., 40–45 °C or more) can work in certain contexts.
- The choice of temperature must balance microbial tolerance and kinetics: too low → slow leaching; too high → microbial stress or denaturation [7].

8.3 Pulp density (solid loading)/solid:liquid ratio

- Pulp density: the ratio of solid waste (e-waste powder or particles) to liquid leaching medium strongly influences leaching yields, microbial viability, mass transfer, and process scalability [7].
- For many PCB bioleaching studies: pulp densities in the range of 5–15% (w/v) (or equivalent g/L solids loading) are common.
- Lower pulp densities facilitate better oxygen, nutrient and acid transfer – improving microbial activity and metal dissolution; but they reduce metal yield per batch, affecting throughput. High pulp densities may yield more metal per batch but risk microbial inhibition, poor mixing, or reduced oxygen transfer [41].
- Some studies, optimizing bioleaching (e.g., via statistical methods), report optimum pulp densities, often in the lower-moderate range for maximum recovery [42].

8.4 Particle size/pretreatment of e-waste material

- The particle size of shredded/crushed e-waste (e.g., PCB fragments, metal concentrate powders) affects surface area available for microbial attack and acid/oxidant access – finer particles generally lead to higher leaching efficiency [6].

- Example: In one study on PCB metal concentrates, optimum leaching occurred at 60–80 mesh (i.e., moderate particle size), with 96.8% Cu extraction within 45 h [43].
- However, excessively fine particles may cause issues, e.g., slurry thickening, decreased aeration/mixing, microbial stress or inhibition, reduced mass transfer – which can degrade performance [6].

8.5 Aeration/agitation/redox conditions/oxygen supply

- Many bioleaching microorganisms (acidophiles, sulfur or iron oxidizers) are **aerobic** and depend on adequate oxygen supply and redox cycling (e.g., $Fe^{2+} \rightarrow Fe^{3+}$) to drive metal oxidation and solubilization [41].
- Agitation (stirring) helps maintain homogeneity, prevents sedimentation, ensures gas transfer (O_2/CO_2), and disperses microbes and wastes – typical lab-scale parameters: 120–180 rpm stirring rate in flasks/reactors.
- Redox potential (ORP) is also a key physicochemical parameter that reflects the oxidative environment; shifts in ORP can influence oxidation of metal sulfides/metal-bearing phases and enhance dissolution [6].

8.6 Microbial strain/consortium/type of microorganism and inoculum conditions

- The choice of microbe (species, metabolic type – chemolithotrophic iron/sulfur oxidizers, heterotrophic, cyanogenic, fungal, etc.) strongly determines which metals can be leached and under what conditions [39].
- Mixed microbial consortia (e.g., combinations of iron- and sulfur-oxidizers, or bacteria + fungi) often show improved stability, wider metabolic capacities, and sometimes better metal recovery, compared to pure cultures [44].
- Inoculum density (initial microbial load), nutrient supplementation (energy sources such as ferrous iron, sulfur, organic substrates for heterotrophs/fungi), can affect leaching kinetics and yields [43].

8.7 Leaching duration/retention time/kinetics

- The duration of bioleaching must be optimized: insufficient time → incomplete metal dissolution; excessively long → microbial stress, toxicity accumulation, passivation, or diminishing returns.

- Many published PCB bioleaching studies report durations of **several days to a few weeks** (e.g., 5–25 days), depending on waste type, microbial system, and process parameters [39].

8.8 Other physicochemical factors: redox potential (ORP), dissolved oxygen (DO), nutrients, metal toxicity, and waste matrix composition

- Redox potential (ORP), DO, presence of electron donors/acceptors, and composition of waste matrix (metals + non-metals: plastics, resins, and fiberglass) influence the bioavailability of metals and microbial health [44].
- Nonmetal matrix components (plastics, resins, fiberglass, or organic binders) in e-waste often interfere with microbial colonization, acid diffusion, and mass transfer – making pretreatment (shredding or decoupling metals) important [39].

9 Bioreactor systems for e-waste bioleaching

The rapid global increase in electronic waste (e-waste), particularly PCBs, has intensified the need for sustainable and economically viable metal recovery technologies. While shake-flask bioleaching experiments have demonstrated feasibility at the laboratory scale, industrial application requires greater control and efficiency. Bioreactor-based bioleaching systems therefore play a critical role in scaling up the process. Bioreactors provide precise regulation of temperature, pH, aeration, agitation, and pulp density, which significantly enhance microbial activity and metal solubilization efficiency. Studies have shown that controlled bioreactor systems consistently outperform conventional flask operations in terms of metal recovery rates and process stability.

Several types of bioreactors have been successfully applied for e-waste bioleaching, including stirred-tank reactors, bubble-column reactors, and fluidized-bed reactors. For instance, bubble-column bioreactors, operated with fungal cultures, have achieved copper and nickel extraction efficiencies as high as 96.9% and 71.5%, respectively, from mobile phone PCBs using low cost carbon sources such as molasses [45]. Fluidized-bed reactors have also demonstrated excellent performance, with sequential batch processing enabling cumulative copper recoveries exceeding 99%, along with considerable dissolution of silver and gold from PCB waste [46]. Similarly, stirred-tank reactors, inoculated with iron-oxidizing bacteria, have achieved zinc and copper recoveries of over 90% and 70%, respectively, under optimized aeration and redox control.

Compared with shake-flask systems, bioreactors offer several operational advantages. Enhanced oxygen transfer and mixing improve microbial metabolism and ferric iron regeneration, which are essential for sulfide mineral oxidation and indirect metal dissolution [7].

Bioreactors also allow operation at higher pulp densities, enabling higher waste throughput and making the process more economically attractive at scale. Furthermore, precise control of temperature, pH, redox potential, and nutrient feeding ensure stable microbial performance and reproducible metal recovery, which is difficult to maintain in conventional batch flasks.

Despite these advantages, bioreactor-based bioleaching of e-waste still faces several technical challenges. High concentrations of dissolved heavy metals, brominated flame retardants, and other toxic additives present in e-waste can inhibit microbial growth and activity. Elevated pulp densities often result in prolonged microbial adaptation phases and reduced leaching kinetics due to toxicity and mass-transfer limitations [47]. Moreover, the heterogeneous nature of e-waste – consisting of metals, plastics, ceramics, and glass – complicates reactor hydrodynamics and may cause clogging, sedimentation, or uneven biofilm formation during long-term operation.

To address these constraints, several optimization strategies have been developed. Adaptive evolution of acidophilic microbial consortia through successive subculturing in metal-rich environments has proven effective in enhancing tolerance to high metal loads and improving leaching efficiency [47]. Two-step bioreactor bioleaching systems, in which microbes first generate lixiviants such as ferric iron and sulfuric acid, followed by metal dissolution in a separate stage, have demonstrated accelerated copper recovery and reduced microbial inhibition. Additionally, heterotrophic bacteria and fungi have gained attention for e-waste bioleaching due to their ability to tolerate broader pH ranges and complex organic matrices [7].

With rising demand for critical metals and growing pressure for sustainable recycling technologies, bioreactor-assisted bioleaching represents a promising pathway for large-scale e-waste valorization. Laboratory- and pilot-scale studies reporting copper recoveries above 95–99% demonstrate strong potential for industrial implementation, when integrated with downstream processes such as solvent extraction and electrowinning. Because bioreactor bioleaching significantly reduces chemical consumption and energy input compared to conventional pyrometallurgical routes, it aligns well with circular economy principles and environmental sustainability goals. Future research should therefore focus on reactor design optimization, microbial strain engineering, process intensification, and techno-economic assessment to enable commercial-scale deployment of this green technology [45, 52].

10 Case studies of metal recovery from e-waste

Bioleaching has emerged as an auspicious route for recovering metals from various categories of electronic waste, especially PCBs, but also other discarded components such as display panels and battery materials. Among these, WPCBs are the most studied, owing to their high metal content (often 20–40 wt% metals embedded in fiberglass/polymer matrices).

In classical bacterial bioleaching, acidophilic iron-oxidizing prokaryotes, for instance *A. ferrooxidans* (and in some studies, together with *A. thiooxidans* or other acidophiles), have demonstrated efficient recovery of base metals from shredded PCBs. For instance, in a study on WPCBs at 15 g/L pulp density, *A. ferrooxidans* achieved ~96.8% Cu, ~83.8% Zn, and ~75.4% Al recovery, within 72 h, under optimized conditions [48]. In another work using mixed *A. ferrooxidans* + *A. thiooxidans* on shredded PCBs (pulp density 8%, 308 K, and pH ~2), ~ 93% copper recovery was reported over 10 days – significantly outperforming a fungal strain tested under similar conditions (~ 66% Cu) [49].

Moreover, column-bioleaching of WPCBs by *A. ferrooxidans* yielded approximately 94.8% copper recovery after 28 days, demonstrating that bacterial bioleaching can be effective not only in shake-flasks but also in reactor-like configurations.

Beyond bacteria, fungal bioleaching – especially using *A. niger* and other acid-producing fungi – has been explored as an alternative. A recent review compiling multiple studies reports wide ranges of metal recovery efficiencies from WPCBs: Cu (~34–100%), Ni (~8–95%), Zn (~54–99%), Mn (~70–95%), Pb (~27–95%), and Co (~45–90%) [50]. In one experimental study, *A. niger* was able to leach both copper and gold from PCBs: under a "one-step" process, copper leaching reached 100% and gold nearly 41%, while under a "two-step" method (using spent medium), copper remained fully leached and gold nearly 42.5% [51]. This demonstrates that fungal leaching via organic acid production (citric, oxalic, or gluconic acids) can mobilize both base and precious metals, although recovery yields (particularly for precious metals) are generally lower and more variable than bacterial systems [40].

However, shifting from base metals (Cu, Zn, Ni, or Al) to valuable metals (Au or Ag) or strategic/critical metals (e.g., rare metals and battery metals), reveals limitations. In a stirred-tank reactor experiment using an acidophilic iron-oxidizing bacterial consortium (comprising mainly of iron-oxidizers such as *Leptospirillum* and other acidophiles) on real PCB waste (10 g/L) over 18 days, the authors recorded 69% Cu and 91% Zn recovery, but only approximately 28% for gold, nearly 16% for tin, and with negligible (<0.25%) recovery of silver or lead [52]. This indicates that while bacterial consortia can excel in mobilizing base metals, they are far less effective at leaching precious or noble metals under the same conditions, likely due to metal speciation, mineral matrix, and bioavailability constraints.

Moreover, microbial bioleaching of non-PCB e-waste, such as display panels (LCD/LED), lithium-ion battery black mass, or rare-earth-bearing components remains far less documented in peer-reviewed literature, compared to PCBs. A comprehensive review on fungal bioleaching of WEEE emphasizes that although fungi are versatile and can produce a variety of organic acids and complexing agents, most studies focus on base and common metals; systematic data on lithium, cobalt, indium, and rare earth metals recovery are sparse and inconsistent [53].

Similarly, the general review of bioleaching methods for e-waste points out that while bioleaching is fascinating due to lower energy consumption and minimal use of toxic reagents (compared to pyrometallurgy or harsh chemical leaching), the large fraction of nonmetallic matrix (plastics, resins, and fiberglass) remains a challenge: after metal extraction, substantial nonmetal residue persists and may need additional processing [39].

Consequently, while laboratory- and bench-scale studies influentially demonstrate that microbial bioleaching can efficiently recover base metals (especially Cu, Zn, Ni, and Al) from shredded PCBs – and in some cases, mobilize precious metals (Au) – the generalizability of this approach to all types of e-waste (batteries, displays, rare-metal components) remains limited. The variability of recovery yields, depending on microbial strain, leaching conditions (pH, pulp density, and particle size), waste composition, and scale (shake-flask vs reactor vs column), underscores the need for cautious interpretation and further optimization before industrial-scale deployment.

In summary, microbial bioleaching offers a sustainable, low-energy, potentially cost-effective alternative to conventional smelting or hydrometallurgical methods – particularly for base-metal recovery from high-metal-content wastes like PCBs – but its application for precious metals, strategic metals, or mixed e-waste streams remains a research challenge. Future work should focus on (i) optimizing consortia/fungal strains, tailored to specific waste types; (ii) scaling up from lab to pilot and industrial scale; (iii) improving downstream purification and separation of recovered metals; and (iv) integrating bioleaching with complementary processes (mechanical pretreatment, hydrometallurgy, and biosorption) to handle nonmetallic residues and refine metal recovery [35, 43, 47–49].

11 Advantages, challenges, and future prospects of microbial bioleaching

Microbial bioleaching is progressively observed as a sustainable and ecological route for recovering metals from electronic waste, as it avoids many of the shortcomings of conventional methods like smelting or strong-acid leaching. The microbes (bacteria, fungi, or even mixed consortia) secrete lixiviants (e.g., acids, cyanide, and chelating agents) or oxidize components in the waste matrix, thereby converting insoluble metals into soluble forms amenable to recovery. This biological route significantly reduces energy demand and avoids high-temperature processes, lowering greenhouse gas emissions and minimizing generation of secondary toxic wastes. Such features make bioleaching well-suited for complex, heterogeneous waste like e-waste – where metals are embedded in polymers, glass, plastics, and mixed substrates – which are often uneconomical or environmentally hazardous to process via conventional metallurgy.

However, microbial bioleaching also has important limitations and practical challenges. Chief among these is slow kinetics: the rate of metal solubilization is typically much slower than chemical leaching or smelting, which constrains throughput and raises concerns about viability for large-scale recycling [54]. The heterogeneity of e-waste presence of plastics, flame retardants, and mixed materials can impede microbial access to metal surfaces or inhibit microbial metabolism, reducing recovery efficiency [55]. Additionally, downstream recovery of metals from the leachate (after solubilization) often requires further treatment steps (precipitation, biosorption, or chemical separation), which can complicate the process and diminish overall yield [54]. Finally, different metals may respond variably, and obtaining high recovery yields for some critical or precious metals remains a challenge under current bioleaching protocols [56].

Further, research is pointing toward promising directions that could overcome these challenges and improve the viability of microbial bioleaching at scale. One exciting avenue is the use of engineered microbial consortia or genetically optimized strains, microbes modified or selected for higher tolerance to toxic e-waste components, enhanced lixiviant production, or improved contact with metal-bearing substrates. Another promising trend is the integration of bioleaching with complementary recovery methods (e.g., bioelectrochemical methods, biosorption, and chemical precipitation), forming hybrid flowsheets that maximize metal recovery while minimizing environmental impact [57]. Further, optimizing process parameters (pH, pulp density, particle size, and incubation conditions) and developing robust reactor designs, tailored for heterogeneous e-waste, can help scale up bioleaching toward commercial applicability [58].

Bioleaching is an emerging and rapidly evolving green technology that is gradually being adopted worldwide for the retrieval of metals from abandoned mine wastes and for the elimination of toxic pollutants from polluted mining sites. It can be described as the acceleration of a natural ecological process, in which microorganisms solubilize or mobilize metals from sulfide minerals. Although the process is relatively slow, it has proven to be economically viable, and at present, several commercial operations successfully employ bioleaching for copper extraction from low-grade ores. From the perspective of valuable metal recovery, bioleaching is no longer viewed solely as a metal dissolution technique, but also as a lucrative and environmentally sustainable alternative to conventional metallurgical processes. Continuous technological advancement is essential in both engineering and biological aspects, particularly in enhancing metal leaching rates and improving microbial tolerance to high metal concentrations. Genetic modification of bioleaching microorganisms, either through mutation, as well as selection, or via genetic engineering, offers faster and more targeted improvements than conventional strain-screening approaches. Notably, significant progress has been achieved in developing genetic systems for *Thiobacillus (Acidithiobacillus) ferrooxidans*. Subsequently, bioleaching is now known as an eco-friendly technology for commercial metal recovery from ores, with growing interest in extending its application to the elimination of heavy metals from solid wastes [59]. Collectively, these research directions suggest that microbial bioleaching with continued innovation could become a key pillar of sustainable, circular economy approaches for metal recovery from e-waste.

12 Conclusion

This chapter comprehensively examined the potential of microbial biominers as sustainable tools for metal recovery from electronic waste. The complex and heterogeneous nature of e-waste, enriched with high-value base metals, precious metals, and critical raw materials, presents both a major environmental challenge and a significant secondary resource opportunity. Conventional pyrometallurgical and hydrometallurgical approaches, although industrially established, are associated with high energy consumption, toxic emissions, and secondary waste generation. In contrast, bioleaching offers a low-energy, environmentally benign, and selective alternative that aligns strongly with circular economy principles and sustainable resource management.

The fundamental mechanisms driving microbial metal solubilization – acidolysis, redoxolysis, complexolysis, cyanogenesis, biosorption, and biomineralization – demonstrate the remarkable biochemical versatility of bacteria, fungi, and archaea in mobilizing metals from solid electronic matrices. Acidophilic iron- and sulfur-oxidizing bacteria enable efficient dissolution of base metals, while cyanogenic bacteria and organic-acid-producing fungi facilitate selective recovery of precious metals such as gold and silver. The diversity of microbial metabolisms allows bioleaching to be tailored to a wide range of e-waste streams, including PCBs, mobile phones, lithium-ion batteries, and mixed electronic assemblies.

Process configurations such as one-step bioleaching, two-step bioleaching, and spent-medium bioleaching have shown distinct advantages, depending on the nature of the waste material and microbial system employed. While one-step systems offer operational simplicity, two-step and spent-medium approaches allow greater control over microbial growth and lixiviant production, leading to improved metal recovery and reduced microbial toxicity. The use of stirred-tank, airlift, column, and heap-type bioreactors further supports the scalability of bioleaching technologies, demonstrating their potential transition from laboratory to pilot and industrial scales.

Recent advancements, including genetic engineering of metal-tolerant strains, synthetic microbial consortia, hybrid chemical-biological pretreatments, and high-performance bioreactor designs, have significantly enhanced the efficiency and robustness of e-waste bioleaching. Case studies discussed in this chapter clearly demonstrate that high recovery yields of copper, gold, silver, cobalt, nickel, and REEs are technically feasible under optimized biological conditions. However, challenges related to slow kinetics, toxic metal inhibition, heterogeneous feedstock composition, and downstream metal purification still limit large-scale industrial implementation.

Overall, microbial biomining represents a transformative approach for sustainable urban mining of electronic waste. By integrating biological innovation with process engineering and circular economy strategies, bioleaching can significantly reduce dependence on primary mining, lower carbon emissions, and promote closed-loop material cycles. Continued research focusing on microbial adaptation, reactor optimization, techno-economic assessment, and life cycle sustainability will be critical in advancing bioleaching from an emerging green technology to a fully industrialized solution for global e-waste management.

References

[1] Baldé CP, Kuehr R, Yamamoto T, McDonald R, D'Angelo E, Althaf S, Bel G, Deubzer O, Fernandez-Cubillo E, Forti V, Gray V. Global E-waste Monitor 2024. In United Nations University, International Telecommunication Union, International Solid Waste Association 2024.

[2] Davis C, Heart S. Electronic waste: The local government perspective in Queensland. Resources, Conservation and Recycling. 2008;52:1031–1039.

[3] Van Yken J, Boxall NJ, Cheng KY, Nikoloski AN, Moheimani NR, Kaksonen AH. E-waste recycling and resource recovery: A review on technologies, barriers and enablers with a focus on Oceania. Metals. 2021;11:1313.

[4] Debnath B, Chowdhury R, Ghosh SK. Sustainability of metal recovery from e-waste. Frontiers of Environmental Science & Engineering. 2018;12(6):2.

[5] Wan J, Sun J, Zhao XL, Le AS, Ren PB, Zhan MX, Chen T. Emission of brominated pollutants from waste printed circuit boards during thermal treatment: A review. Aerosol and Air Quality Research. 2023;23(12):230135.

[6] Magoda K, Mekuto L. Biohydrometallurgical recovery of metals from waste electronic equipment: Current status and proposed process. Recycling. 2022;7(5):67.

[7] Adetunji AI, Oberholster PJ, Erasmus M. Bioleaching of metals from e-waste using microorganisms: A review. Minerals. 2023;13(6).828.

[8] Valdés J, Pedroso I, Quatrini R, Dodson RJ, Tettelin H, Blake R, Holmes DS. Acidithiobacillus ferrooxidans metabolism: From genome sequence to industrial applications. BMC Genomics. 2008;9:597.

[9] Pradhan JK, Kumar S. Metals bioleaching from electronic waste by Chromobacterium violaceum and Pseudomonads sp. Waste Management & Research. 2012;30(11):1151–1159.

[10] Tezyapar Kara I, Kremser K, Wagland ST, Coulon F. Bioleaching metal-bearing wastes and by-products for resource recovery: A review. Environmental Chemistry Letters. 2023;21(6):3329–3350.

[11] Tiwari S, Aachhera S, Verma P, Gahan CS. Bioremediation of e-waste: A sustainable approach. In Biotechnology for Environmental Sustainability 2025 (pp. 709–724). Springer Nature Singapore.

[12] Aachhera S, Tiwari S, Verma P, Gahan CS. E-waste recycling: A comprehensive review on metal recovery and the factors affecting bioleaching. Biotechnology for Environmental Sustainability. 2025;209–243.

[13] European Commission. A new circular economy action plan: For a cleaner and more competitive Europe. European Commission; 2020.

[14] Villa Gomez D, Whitworth AJ, Vaughan J, Sultana U, Ledezma P, Parbhakar-Fox A. Review on developments in technologies for critical metal recovery from mining and processing wastes. Mineral Processing and Extractive Metallurgy Review. 2025;46(7):751–770.

[15] Sahajwalla V, Hossain R. The science of microrecycling: A review of selective synthesis of materials from electronic waste. Materials Today Sustainability. 2020;9:100040.

[16] Willner J, Fornalczyk A. Electronic scraps as a source of precious metals. Przemysł Chemiczny. 2012;91(4):517–522.

[17] Cui J, Zhang L. Metallurgical recovery of metals from electronic waste: A review. Journal of Hazardous Materials. 2008;158(2–3):228–256.

[18] Hagelüken C. Improving metal returns and eco-efficiency in electronics recycling: A holistic approach for interface optimisation between pre-processing and integrated metals smelting and refining. In Proc IEEE Int Symp Electron Environ 2006 (pp. 218–223).

[19] Ilyas S, Anwar MA, Niazi SB, Ghauri MA. Bioleaching of metals from electronic scrap by moderately thermophilic acidophilic bacteria. Hydrometallurgy. 2007;88(1–4):180–188.

[20] Hagelüken C, Corti CW. Recycling of gold from electronics: Cost-effective use through "design for recycling". Gold Bulletin. 2010;43(3):209–220.

[21] Ylä-Mella J, Keiski R, Fornalczyk A, Willner J, Francuz K, Cebulski J. E-waste as a source of valuable metals. Archives of Materials Science and Engineering. 2013;63(2):87–92.

[22] Vardanyan A, Khachatryan A, Castro L, Willscher S, Gaydardzhiev S, Zhang R, Vardanyan N. Bioleaching of sulfide minerals by Leptospirillum ferriphilum CC from polymetallic mine (Armenia). Minerals. 2023;13:243.

[23] Işildar A, Van de vossenberg J, Rene ER, van Hullebusch ED, Lens PNL. Two-step bioleaching of copper and gold from discarded printed circuit boards (PCB). Waste Management. 2016;57:149–157.

[24] Campbell SC, Olson GJ, Clark TR, McFeters G. Biogenic production of cyanide and its application in gold recovery. Journal of Industrial Microbiology and Biotechnology. 2001;26:134–139.

[25] Liu Q, Bai JF, Gu WH, Peng SJ, Wang LC, Wang JW, Li HX. Leaching of copper from waste printed circuit boards using Phanerochaete chrysosporium fungi. Hydrometallurgy. 2020;196:105427.

[26] Baniasadi M, Vakilchap F, Bahaloo-Horeh N, Mousavi SM, Farnaud S. Advances in bioleaching as a sustainable method for metal recovery from e-waste: A review. Journal of Industrial and Engineering Chemistry. 2019;76:75–90.

[27] Işıldar A, van de Vossenberg J, Rene ER, van Hullebusch ED, Lens PN. Biorecovery of metals from electronic waste. In Sustainable Heavy Metal Remediation 2017 (Vol. 2, pp. 241–278). Cham: Springer Int Publ.

[28] Volesky B, Holan ZR. Biosorption of heavy metals. Biotechnology Progress. 1995;11(3):235–250. Wang J, Chen C. Biosorbents for heavy metals removal and their future. Biotechnol Adv 2009, 27(2), 195–226.

[29] Mobley HL, Hausinger RP. Microbial ureases: Significance, regulation, and molecular characterization. Microbiological Reviews. 1989;53(1):85–108.

[30] Muyzer G, Stams AJ. The ecology and biotechnology of sulphate-reducing bacteria. Nature Reviews Microbiology. 2008;6(6):441–454.

[31] Lens PN, Meulepas RJ, Sampaio R, Vallero M, Esposito G. Bioprocess engineering of sulfate reduction for environmental technology. In Microbial Sulfur Metabolism 2008 (pp. 285–295). Berlin: Springer.
[32] Shreya S, Nigam VK, Sampath MK. Biorecovery of metals from e-waste: An elaborate study. In Extraction of Precious Metals from Industrial Waste Using Microbial Technology 2025 (pp. 69–86). Springer Nature Switzerland.
[33] Tay SB, Natarajan G, Rahim MNBA, Tan HT, Chung MCM, Ting YP, Yew WS. Enhancing gold recovery from electronic waste via lixiviant metabolic engineering in Chromobacterium violaceum. Scientific Reports. 2013;3:2236.
[34] Banerjee A, Jhariya MK, Yadav DK, Raj A. Micro-remediation of metals: A new frontier in bioremediation. In Handbook of Environmental Materials Management 2018 (pp. 1–36). Cham: Springer.
[35] Aachhera S, Tiwari S, Verma P, Gahan CS. Two-step leaching of metal values from discarded mobile phone printed circuit board using biogenic ferric ion with copper leaching kinetics. Journal of Microbiology, Biotechnology and Food Sciences. 2025;15(2):e12378.
[36] Naseri T, Pourhossein F, Mousavi SM, Kaksonen AH, Kuchta K. Manganese bioleaching: An emerging approach for manganese recovery from spent batteries. Reviews in Environmental Science and Biotechnology. 2022;21(2):447–468.
[37] Natarajan G, Ting YP. Gold biorecovery from e-waste: An improved strategy through spent medium leaching with pH modification. Chemosphere. 2015;136:232–238.
[38] Bosecker K. Bioleaching: Metal solubilization by microorganisms. FEMS Microbiology Reviews. 1997;20(3–4):591–604.
[39] Ji X, Yang M, Wan A, Yu S, Yao Z. Bioleaching of typical electronic waste – printed circuit boards (WPCBs): A short review. International Journal of Environmental Research and Public Health. 2022;19(12):7508.
[40] Mishra S, Panda S, Akcil A, Dembele S, Agcasulu I. Chemical versus microbial leaching of electronic wastes with emphasis on base metals dissolution. Minerals. 2021;11(11):1255.
[41] Ilkhani Z, Aiouache F. Bioleaching of gold from printed circuit boards: Potential sustainability of thiosulphate. Recycling. 2025;10(3):87.
[42] Naseri T, Beiki V, Mousavi SM, Farnaud S. Bioleaching optimization by statistical approaches: Recycling mechanisms, affecting factors, challenges, and sustainability. RSC Advances. 2023;13(34):23570–23589.
[43] Zhu N, Xiang Y, Zhang T, Wu P, Dang Z, Li P, Wu J. Bioleaching of metal concentrates of waste printed circuit boards by mixed acidophilic bacteria. Journal of Hazardous Materials. 2011;192(2):614–619.
[44] Bayat B, Sari B. Comparative evaluation of microbial and chemical leaching for heavy metal removal from metal plating sludge. Journal of Hazardous Materials. 2010;174(1–3):763–769.
[45] Saidutta MB Bioleaching of metals from electronic waste by heterotrophic bacteria in fluidized-bed bioreactor. PhD Thesis, NITK Surathkal, India, 2023.
[46] Jaiswal M, Srivastava S. Sustainable approach of bioleaching of precious metals from electronic wastes: A review. Journal of Hazardous Materials Advances. 2024;14:100435.
[47] Xiang Y, Wu P, Zhu N, Zhang T, Liu W, Wu J, Li P. Bioleaching of copper from waste printed circuit boards by bacterial consortium from acid mine drainage. Journal of Hazardous Materials. 2010;184(1–3):812–818.
[48] Yang Y, Chen S, Li S, Chen M, Chen H, Liu B. Bioleaching of waste printed circuit boards by Acidithiobacillus ferrooxidans and its kinetics. Journal of Biotechnology. 2014;173:24–30.
[49] Abhilash TS, Ghosh A, Meshram P, Van hullebusch ED. Microbial processing of shredded PCBs for copper extraction: Bacterial vs fungal leaching. Metals. 2021;11(2):317.

[50] Trivedi A, Vishwakarma A, Saawarn B, Mahanty B, Hait S. Fungal biotechnology for urban mining of metals from waste printed circuit boards. Journal of Environmental Management. 2022;323:116133.

[51] Rodrigues ÉF, Cesa Rovaris B, Valerio A, De oliveira D, Hotza D. Bioleaching of printed circuit board waste to obtain metallic nanoparticles. Sustainability. 2024;16(22):9837.

[52] Tapia J, Dueñas A, Cheje N, Soclle G, Patiño N, Ancalla W, Lazarte A. Bioleaching of heavy metals from printed circuit boards using iron-oxidizing consortium in STRs. Bioengineering. 2022;9(2):79.

[53] Li J, Xu T, Liu J, Wen J, Gong S. Bioleaching metals from WEEE by Aspergillus niger: A review. Environmental Science and Pollution Research. 2021;28(33):44622–44637.

[54] Gulliani S, Volpe M, Messineo A, Volpe R. Recovery of metals and valuable chemicals from WEEE: A critical review. RSC Sustainability. 2023;1(5):1085–1108.

[55] Sneha JA, Kalaichelvan G. Microbiological approach for leaching metallic elements from e-waste. Journal of Resources and Ecology. 2023;14(3):667–674.

[56] Kim JY, Hwang JH, Lee T, Kim Y, Bae S, Kim D, Kwon EE. Sustainable metal recovery from e-waste via bioleaching: Current status and strategies. Chemical Engineering Journal. 2025;164149.

[57] Pines Pozo MT, Lopez Fernandez E, Villasenor J, Leon-Fernandez LF, Fernandez-Morales FJ. Metal recovery from wastes using bioelectrochemical systems. Applied Sciences. 2025;15(3):1456.

[58] Kaur P, Sharma S, Albarakaty FM, Kalia A, Hassan MM, Abd-Elsalam KA. Biosorption and bioleaching of heavy metals from electronic waste. Sustainability. 2022;14(2):935.

[59] Gao X, Jiang L, Mao Y, Yao B, Jiang P. Bioleaching of heavy metals from mine tailings: Progress and challenges. Adsorption Science & Technology. 2021;2021:9941979.

Chahat Sharma, Jyoti R.*

Bioleaching and biosorption of metals from e-wastes using microorganisms: a sustainable approach for resource recovery

Abstract: Electronic waste has been growing rapidly alongside technological advancement, leading to significant challenges in environmental and resource management. E-waste contains substantial quantities of valuable metals, including precious, base, and rare earth elements, as well as hazardous components that can cause severe ecological damage if not handled properly. Conventional metal recovery methods are often efficient but tend to be energy-intensive and involve the use of toxic chemicals, making them unsuitable for sustainable, long-term use. Bioleaching is a biological method that employs the action of microorganisms and has emerged as a green alternative for recovering metals from e-waste. In this sustainable approach, microorganisms facilitate the solubilization of metals through biochemical reactions. Biosorption, in contrast, relies on microbial biomass to absorb and bind metals via surface functional groups. These biological methods offer several advantages, including lower energy requirements and minimal chemical usage. Their application and efficiency have been enhanced by recent advances in microbial biotechnology, process optimization, and the development of hybrid biological chemical systems. This chapter presents a comprehensive review of the principles, mechanisms, and microorganisms involved in bioleaching and biosorption. It also discusses recent technological developments and potential collaborative opportunities with other recovery processes. Challenges such as scalability, microbial inhibition, and policy considerations are addressed, alongside future perspectives involving synthetic biology and computational modeling.

Keywords: Microbial metal recovery, e-waste metal extraction, bioleaching, biosorption, green circular technologies

***Corresponding author: Jyoti R**, University Centre for Research and Development, Chandigarh University, Mohali 140413, Punjab, India, e-mail: jyotiyogi1222@gmail.com
Chahat Sharma, Department of Environment Studies, Panjab University, Chandigarh, India

 | https://doi.org/10.1515/9783112225509-003

1 Introduction

1.1 Overview of e-waste generation and composition

The global generation of electric and electronic waste (e-waste) has increased significantly in recent decades. This sudden rise in e-waste is due to accelerated device turnover, shorter life span of products, and consumption of electrical gadgets worldwide. A study conducted in 2022 outlined the average per capita generation of e-waste to be 7.6 kg per person and cites a global total of 57.4 million tons (MT) [1]. E-waste is a heterogeneous composition of electronic and electrical equipment (EEE) that contains diverse components, including plastics, metals, glass, etc. The metallic assemblies often comprise precious metals such as gold, silver, platinum, and some rare earth elements (REEs), which classify e-waste as an "urban mine" for secondary material recovery [2]. Despite the material and economic potential, the literature highlights a significant gap in the systematic collection and eco-friendly recycling of e-waste. A large portion of e-waste goes undocumented or ends up being exported because proper treatment facilities are lacking. These practices result in resource loss and environmental pollution, as open burning, acid extraction, and informal dismantling release toxic metals and persistent organic pollutants into the environment [3].

E-waste contains a complex mix of materials that offers valuable opportunities for recovering critical resources, yet it also presents serious environmental risks when mishandled or when toxic components are released. This calls for a dual approach of research that covers efficiently recovering precious materials from heterogeneous matrices, along with recycling the material left behind in the process [4]. The following sections examine biological methods of recovery, promising an environmentally sustainable circular economy (CE).

1.2 Importance of metal recovery for sustainability and the circular economy

Metal recovery from end-of-life e-waste is an integral part of decoupling material supply from primary extraction. This step reduces the reliance on primary ore extraction approach and the substantial environmental burdens associated with mining, including habitat destruction, ore beneficiation, and excessive energy use. All these cons together contribute to lower life cycle greenhouse gas (GHG) emissions and a decline in land-use impacts when implemented properly [5]. Practicing urban mining in the context of resource security is particularly important for critical raw materials (CRMs) that have concentrated demand in clean-energy and digital technologies. Other than this perspective, it is also equally crucial to consider the primary geopolitical constraint. Recent reviews emphasize that the e-waste contains economically notable quantities of CRMs. It is also to be noted that the recovery of REEs and platinum-group elements holds the power to bring down supply-chain vulnerability for technologies like EV motors and wind-power generators [6].

Process-level trade-offs moderate the claims for secondary recovery. Consequently, life cycle assessment (LCA) and technoeconomic analysis (TEA) are required to determine sustainability outcomes. Comparative LCA/TEA projects demonstrate how integrated recovery chains that combine biological steps like bioleaching with downstream hydrometallurgical purification can achieve environmental footprints relative to pyrometallurgy. This can especially be done when energy demand and consumption are minimized via optimization [7]. Concrete techno-economic studies show that bio-based routes can be viable under particular feedstock and scale conditions. For instance, TEA analysis indicates that process economics improve when bioleaching is applied to metal-rich fractions and when downstream metal separation is built to maximize product value. Research demonstrates how targeted recovery strategies can significantly influence net impacts. Life-cycle-informed process design for precious metal recovery shows that combining low-temperature selective dissolution with efficient downstream separation can markedly reduce energy intensity compared to high-temperature routes. This approach can yield metal of comparable quality and holds strategic significance for both policy development and industrial planning [8].

1.3 Challenges in conventional recovery technologies and the shift toward biorecovery

Conventional metallurgical processes for recovering metals from e-waste are based on pyrometallurgy and hydrometallurgy. These techniques are still prevalent in the industry; however, they exhibit significant economic and environmental constraints when utilized for heterogeneous streams. Prior studies indicate that the substantial energy and chemical demands of conventional technologies significantly restrict their scalability and long-term viability. Pyrometallurgical processing is effective for bulk separation along with treatment of mixed and contaminated feedstocks. However, this demands high temperatures and often leads to the formation of slag phases that interfere with the recovery selectivity for some valuable fractions. A study provides significant insights into the material losses and secondary waste generation caused by furnace-based processing systems [9].

Studies on process optimization show that the high thermal intensity of pyrometallurgy contributes to life cycle GHGs. Some organics may be lost or transformed into toxic by-products during thermal treatment. Hydrometallurgical techniques improve selectivity through the use of aqueous leachates but often depend on strong acids or oxidants, resulting in substantial secondary wastewater generation. This wastewater is then required to be neutralized and treated. Processing of waste of electrical and electronic equipment (WEEE) burdens the associated cost and environmental trade-offs. LCA studies comparing different routes for recovering copper and other metals show that lower energy consumption and the use of cleaner energy sources can dramatically reduce the overall carbon footprint. The downside remains that the conventional high-temperature pathways may produce significant CO_2-e if run on fossil fuel. A recent review analyzing copper recovery cases shows that carbon intensity drops significantly when greener energy sources or cleaner process alternatives are adopted. Heterogeneity of e-waste, for example, mixed polymers, PCBs, alloys and other components may further complicate the process of chemical separation. In such cases, pretreatment is done, which increases CAPEX and OPEX, thereby driving up the overall cost of conventional treatment [10].

Biorecovery is a sustainable and greener alternative, using a lower temperature and is a less chemical-intensive process, which has emerged as a promising approach. Microorganisms and their metabolites can selectively mobilize or bind metals under ambient conditions, reducing the need for high energy inputs and harsh chemicals. A focused review on bioleaching of metals from e-waste shows solid evidence of how bioroutes are a viable choice compared to conventional processes. Biological routes can be tuned by adjusting the physical conditions (pH, temperature, redox potential, and nutrient supply) to favor solubilization of base metals (e.g., Cu and Ni) before targeting the recoverable elements [11]. This enables selective downstream purification similar to what is achieved through aggressive chemical leaching. From the perspective of scalability, it is indicated that bioleaching can be the financially right route to opt for when it comes to metal-rich concentrates or where process intensification (e.g., aerated bioreactors and higher pulp densities) is applied.

Nonetheless, biorecovery faces certain challenges that must be overcome to realize industrial deployment. These include: (i) inhibition of microbial activity due to high concentrations of toxic metals or contaminants in the leachate of e-waste, (ii) slow kinetics of refractory materials, (iii) need for a robust design for bioreactors for optimal control, and (iv) combination of downstream separation to obtain quality products [12, 13].

2 Principles of bioleaching

2.1 Definition and mechanism of bioleaching

Bio-hydrometallurgy, better known as bioleaching, is a process by which microorganisms mobilize metals from solid materials such as ores, residues, or end-of-life e-waste, either by their own or through their metabolic products. To better understand, Figure 1 shows a step wise approach of the the mechanism of Bioleaching. This process is done in an aqueous medium, thereby enabling subsequent recovery of those metals [14]. According to previous research, the process involves the microbial

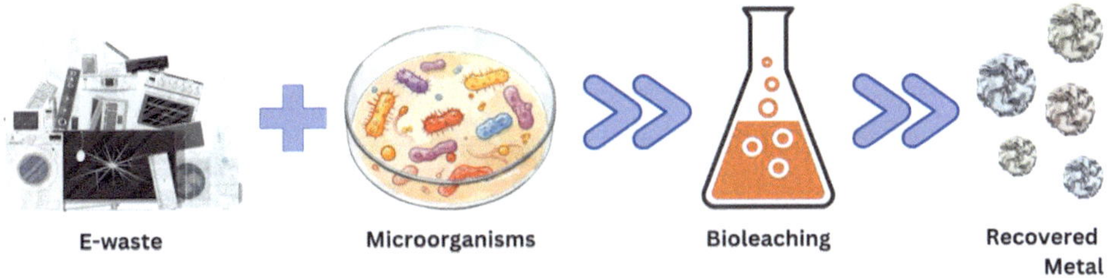

Figure 1: Schematic diagram of bioleaching.

conversion of insoluble metal compounds within a solid matrix into their soluble forms. This category of biorecovery has expanded in the mining sector to include complex secondary streams, such as e-waste, driven by the need for more sustainable, energy-efficient recovery routes. Bioleaching is the opposite of conventional methods, which involve a high amount of reagents, as it proceeds at moderate temperatures and relies on microbial oxidation rather than smelting [15].

Bioleaching primarily constitutes three action pathways: (i) acidolysis, (ii) redoxolysis, and (iii) complexolysis. For instance, a review draws out a comparison between chemical and microbial leaching of e-waste. It indicated that acidolysis involves protonation of oxygen atoms on metal oxides and dissolution of metals, redoxolysis involves oxidation of $Fe^{2+} \rightarrow Fe^{3+}$ or $S^0 \rightarrow SO_4^{2-}$, which generate oxidants that dissolve metals, and complexolysis involves production of microbial chelators which solubilize metals by complex formation [16].

2.1.1 Acidolysis

During acidolysis, microorganisms secrete inorganic acids like H_2SO_4 or organic acids like gluconic and oxalic acids. These acids lower the pH in the environment and promote dissolution of metal-oxygen bonds in metal oxides or metal sulfides. For instance, a study shows the protonation of oxygen atoms and how it reduces the strength of metal-oxygen bonds to release ions. Another study reports that fungi and heterotrophic bacteria produce organic acids that help maintain a low pH, thereby enhancing the solubilization of metals from complex e-waste matrices.

2.1.2 Redoxolysis

The mechanism here is relevant for systems that involve autotroph microbes that oxidize iron or sulfur. The series of reactions goes as follows:

$$S^0 + 1.5O_2 + H_2O \rightarrow 2H^+ + SO_4{}^{2-}$$

$$4Fe^{2+} + O_2 + 2H^+ \rightarrow 4Fe^{3+} + 2OH^-$$

$$Cu^0 + 2Fe^{3+} \rightarrow Cu^{2+} + 2Fe^{2+}$$

The microbe does not directly attack the metal substrate but oxidizes a secondary substrate. This generates a strong chemical oxidant (e.g., Fe^{3+}), which then acts on the metallic substrate. This indirect reaction for leaching is commonly observed in mineral bioleaching and adapted in e-waste streams.

2.1.3 Complexolysis

Complexolysis includes microbial production of ligands like cyanide that bind to metal ions or complexes to facilitate their solubilization. For instance, cyanogenic bacteria produce HCN, which forms metal-cyanide complexes, thus dissolving precious metals from scrap. Redoxolysis and complexolysis often occur simultaneously in real systems [17, 18].

2.2 Direct versus indirect bioleaching

Literature distinguishes between direct (contact) and indirect (non-contact) bioleaching mechanisms. Direct bioleaching requires microorganisms to attach to the solid surface of the substrate and form a biofilm. This biofilm acts as an interface to mediate electron transfer directly to the substrate. This contact mode of mechanism is typically seen in ores, but is widely studied in regard to e-waste. On the other hand, unlike the direct mechanism, the indirect mechanism is predominant in the e-waste industry. Within this process, oxidation of the substrate via microbial action produces chemical oxidants. These oxidants then act on solid matrix, decoupling the microbial cell from directly being in contact with the unwanted material. Indirect bioleaching is highly effective when the composition of the substrate is heterogeneous. This mechanism is particularly useful when the mixture contains materials such as glass, ceramics, or metals that can inhibit direct contact [19].

2.3 Microbe-substrate interactions

In a typical bioleaching process, the first step is substrate comminution and conditioning, which involves size reduction and elimination of plastics or unwanted coatings. The second step is preparation of the leaching medium, wherein, nutrient solution is formed and adjustments in pH are made. As per the requirement, ferrous sulfate or elemental sulfur is added to the prepared solution before it is inoculated with microbial culture and incubated under controlled physical conditions. Lastly, downstream separation is opted to recover metal ions after the dissolution of metals into the leachate takes place. The exopolysaccharide (Exo-PS) layer, which is secreted by the microbes, aids in adhesion and provides an environment for redox reactions to occur.

2.4 Factors influencing mechanism and efficiency

There are certain biological factors that may influence the mechanism of bioleaching. External factors include microbial strain used, the metabolic type of the microbe under action, pH, redox potential of the system, pulp density, rate of aeration, composition, and size of the substrate particle [20]. Other factors include secondary mineral phases or coatings, and the presence of inhibitory metals or organics. Processes discussed before (acidolysis, redoxolysis, and complexolysis) determine the mobility efficiency of target metals. Previous studies indicate that e-waste lacks a natural substrate that provides energy to microbes, so adding them to the system is necessary to maintain a high redox potential. Moreover, biofilm formation enhances the dissolution of metal that comes in contact with systems by maintaining high concentration of oxidants and ligands [21].

2.5 Role of acidophilic microorganisms

Acidophilic chemolithoautotrophic microorganisms are central to bioleaching due to their unique metabolism to oxidize ferrous iron (Fe^{2+}) and elemental sulfur (S^0) under highly acidic conditions ($pH \leq 2$). Among these, *Acidithiobacillus ferrooxidans* is widely studied as it is highly adaptive in acidic and metallic environments. It holds the capacity to catalyze the generation of F^{3+} and H^+ to drive the mobilization of target metals from solid matrices [22].

2.6 Application to e-waste matrices

Acidophiles are capable of tolerating heterogeneous substrates, coatings, mixed alloys, and ceramic or glass-based supports. In a study using strains of *A. ferrooxidans* and *Acidiphilium* on pulverized PCNs over 18 days, metal recoveries of 96% Cu, 94.5% Zn, 75% Ni, and 74.5% Pb were achieved. This demonstrated the effectiveness of acidophiles in complex EEE matrices. Another study investigates the existence of a delay induced by toxicity on the onset of Fe^{2+} oxidation when PCBs were at high loadings [23].

3 Principles of biosorption

Biosorption is defined as the passive, physicochemical binding of metal ions or other ionic species onto the surface or inside the biological materials. These materials could be alive or dead biomass [24]. The bonding between the two is generally via multiple surface and bulk interaction mechanisms such as adsorption, ion exchange, complexation, and microprecipitation [25]. In this section of the chapter, all the mentioned mechanisms related to biosorption are briefly discussed. Figure 2 shows an overview of the process of biosorption.

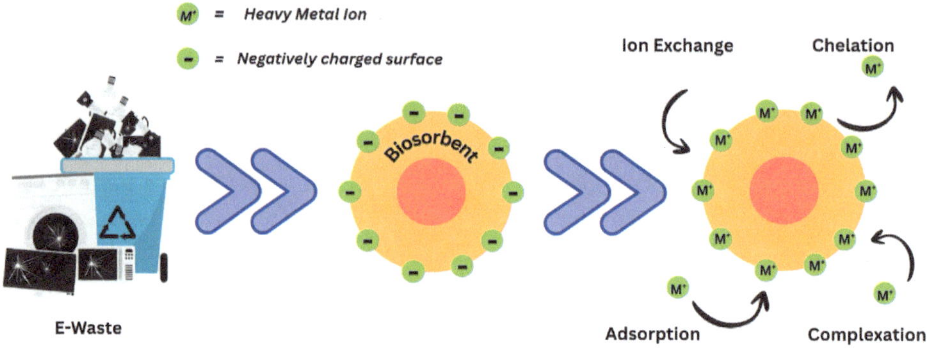

Figure 2: Schematic diagram of biosorption.

3.1 Mechanisms of biosorption

3.1.1 Adsorption

Physical adsorption, also known as physisorption, refers to the accumulation of metal ions on the surface of the biosorbent through weak forces such as van der Waals, electrostatic attraction, or others. The effectiveness of this mechanism depends on the surface area of the biosorbent, the charge of the biomass surface, and external conditions such as pH and ionic strength [26]. In biomass, which is derived from microbes or algae, the negatively charged functional groups (e.g., deprotonated carboxyl, hydroxyl, or phosphate moieties) attract positively charged metal cations, thus enabling rapid initial uptake via electrostatic forces. Adsorption lacks selectivity but still ensures the initial collection of metal ions over the sorbent surface [27].

3.1.2 Ion exchange

Ion exchange, as the name suggests, involves the replacement of ions originally present on the surface (e.g., Na^+, K^+, Ca^{2+}, and Mg^{2+} bound to carboxylate or phosphate groups) by metal cations in the solution (e.g., Pb^{2+}, Cd^{2+}, and Cu^{2+}) [28]. This mechanism is particularly effective when the biosorbent possesses exchangeable cations associated with functional groups and the target metal has a higher affinity for binding sites. This mechanism is strongly influenced by: (i) the pH, since the functional groups must be deprotonated, (ii) ionic strength, and (iii) temperature, which influences the ion mobility. In many biosorption studies, the rapid uptake phase is attributed to ion exchange, followed by slower processes such as complexation or precipitation [29].

3.1.3 Complexation

Complexation, aka surface chelation, involves the formation of coordinate bonds between metal ions in solution and functional groups on the biosorbent surface (such as carboxyl, amine, hydroxyl, and phosphate). The interactions may include inner-sphere or outer-sphere complexes, which are generally stronger and more selective than adsorption. The functional groups present in microbial or algal cell walls offer lone-pair electrons to coordinate metal ions. This mechanism often dominates at moderate or near-neutral pH values. This is when the surface of the biosorbent is negatively charged, and the metal ion can form stable chelates. It is especially important when targeting metals of high affinity [30].

3.1.4 Micro-precipitation

The in situ formation of insoluble metal compounds (hydroxides, sulfides, carbonates, phosphates) within the biosorbent structure or on its surface, effectively immobilizing the metal from the solution phase, is referred to as micro-precipitation [31]. Biosorbents may facilitate micro-precipitation by creating micro-environments. This can be via shifts in the pH, release of local ligand, changes in redox reactions, and other factors that may lead to the precipitation of target metal species. For instance, microbial biomass may release biogenic sulfide or carbonate, or increase the pH near the surface of the cell. This may cause the metal ion species to precipitate as sulfide or hydroxide directly within the Exo-PS layer or the cell wall. This process is considered as the secondary mechanism in biosorption. After initial binding or accumulation, conditions change so that the metal ions are converted to an insoluble form, becoming sequestered within the biosorbent matrix [32].

3.1.5 Interplay of mechanisms and practical implications

The discussed mechanisms of biosorption rarely ever operate in isolation. Instead, there is often an overlapping sequence wherein there is an initial rapid uptake via electrostatic adsorption or ion exchange, followed by slower complexation and finally micro-precipitation. A microalgal biosorbent study found that binding capacities vary in the order Pb(II) > Zn(II) > Cu(II) > Ni(II) > Cd(II) for *Chlorella vulgaris*. The study uses isothermal and kinetic modeling to show that the dominant mechanism was monolayer adsorption with contributions from ion exchange and complexation [33]. The relationship between the biosorbent and the operating conditions heavily influences the mechanism. For example, in low pH (≤3) systems ion exchange and adsorption might dominate, whereas at a higher pH value (≥6), precipitation mechanisms may occur. The role of biosorption in metal recovery from e-waste streams is significant, as the process operates with low energy demand and limited chemical usage. Furthermore, the selection of biomass with an appropriate functional group enables the selective binding of target metals.

3.2 Role of microbial biomass

Biomass, be it living or nonliving cells, offers cost-effective and large surface area materials embedded with functional groups, which help in metal binding. Biomass includes various kinds of biosorbent surfaces, like those of bacteria, fungi, or algae.

3.2.1 Bacterial biomass

Bacteria are among the most widely used biomass materials due to their huge range of properties, including rapid growth rates, easy cultivation, and a well-formed cell wall. Gram-positive bacterial species (like *Bacillus* sp.) and gram-negative species (like *Pseudomonas* sp.) have demonstrated a significant rate of biosorption for metals such as Pb^{2+} and Cu^{2+} [34]. Ion exchange is the key mechanism associated with bacterial biomass. Surface-bound cations exchange with metal cations, and complexation with cell wall ligands takes place. Eventually, the physical adsorption of metals onto the Exo-PS layer (rich in polysaccharides and proteins) occurs [35].

3.2.2 Fungal biomass

Fungal biomass, like filamentous fungi (e.g., *Aspergillus*, *Penicillium*, and *Trichoderma*) and yeasts, provide advantages toward biosorption. Due to their high-surface-area hyphal networks, abundant cell wall functional groups (such as chitin and glucans), and strong tolerance to heavy metal stress, fungal biomasses have emerged as highly effective candidates for metal biosorption [36]. Dead fungal biomass has the added benefit of being non-metabolically active, which simplifies handling and reduces nutrient requirements. Fungal biomass binds metals through a combination of adsorption, ion exchange, chelation, and micro-precipitation within cell wall matrices.

3.2.3 Algal biomass

Microalgae and macroalgae emerge as highly effective biosorbents as they possess high growth rates, large biomass yields, and rich cell wall chemistry. According to a 2023 review, brown algae demonstrate outstanding biosorption capabilities, frequently outperforming bacteria and fungi in specific metal uptake experiments. Algal biomass can be deployed as dead dried material or as live cultures. Additionally, algae can act as biosorbents integrated in photobioreactor systems, enabling continuous biosorption along with biomass production [37].

3.3 Importance of cell wall functional groups in metal binding

The biosorption capacity of microbes is largely influenced by the nature, quantity, and chemical state of cell wall and Exo-PS-associated functional groups, which act as binding sites for metal cations.

3.3.1 Chemical nature of binding moieties

Microbial surfaces consist of anionic functional groups that deprotonate under particular biosorption conditions. This leads to the generation of negatively charged sites that attract metal cations. Such functional groups originate from the core cell wall biopolymers. For bacteria, including gram-positive and gram-negative ones, there are groups like carboxyl, phosphate, and hydroxyl groups, whereas, for fungi, chitin, mannans, etc., amine and hydroxyl ligands contribute. For algae, polysaccharides and sulfonated groups provide binding capacity [38, 39].

3.3.2 Role of extracellular polymeric substances (EPS) and biofilm matrices

EPS is secreted by microorganisms, which provides a rich matrix of polysaccharides, proteins, and lipids. These matrices have binding sites in them. Bacterial studies show that EPS functional groups enhance metal biosorption by increasing the density at the site [40]. Then it enables micro-precipitation within biofilm textures. In e-waste systems, EPS networks trap solubilized metals from leachates and localize them near biomass and eventually immobilize them [41].

4 Influence of physical conditions and metal speciation

4.1 pH

pH is one of the most critical variables in both bioleaching and biosorption. In bioleaching systems, acidophilic microorganisms typically operate at a very low pH (~1.0–3.0), which supports the oxidation of ferrous iron and elemental sulfur and thus maintains metal solubilization [42]. A study found the optimum bioleaching pH values of near 2.0 for many iron and sulfur-oxidizing systems. In biosorption, an increase in pH often leads to more deprotonation of biosorbent functional groups, thereby increasing available negative binding sites for metal cations. A review of marine microalgae shows that metal uptake improved as pH increases from very acidic toward neutral conditions. However, pH may also lead to unwanted precipitation of metal hydroxides rather than biosorption onto biomass [43].

4.2 Temperature

In specific experimental studies, thermoacidophilic archaea were applied to metal-rich steel waste products. Slight shifts in temperature didn't dramatically alter the solubilization of certain elements. Suboptimal temperature conditions reduce reaction kinetics and contribute to the accumulation of undesirable by-products [44].

4.3 Metal speciation

Chemical forms of metal ions in solution critically affect both leaching and biosorption kinetics and yield. The choice of pH and redox conditions influences speciation. At low pH, metal ions may remain as free cations, promoting biosorption via electrostatic or ion exchange mechanisms. Conversely, at a higher pH, metals may form stable complexes that reduce their bioavailability. This may also be the case when complexing agents are present in the system. A study concludes that under suboptimal leaching conditions, metal ions can be immobilized, thus reducing recovery efficiency [45].

5 Technological and computational insights

Designing and engineering of bioprocesses for metal recovery from e-waste by microbial routes involves merging microbial ecology, mass transfer, hydrodynamics, and configuration of the machine. Several types of reactors have been employed in the biorecovery of metals. Some of them are batch, continuous tanks, packed-bed/column reactors, air-lift reactors, and heap or dump-type systems. For metal recovery from e-waste, column reactors loaded with crushed PCB feed and inoculated with bacteria give encouraging results. For example, it has been reported that a column-based system generated 14–22 g Cu/L in the bioleachate, with more than 99% of the copper electrowin within 3.3 h. This outcome underscores the importance of reactor hydrodynamics and precise control of residence time [46].

A review of bioreactor systems highlights that maintaining effective control over mixing time, oxygen saturation, temperature, and sensor integration becomes increasingly challenging as the operational scale expands. In bioleaching of e-waste, the feed heterogeneity makes scaling up difficult, as zones for penetration are limited, and the contact is poor, which leads to reduced recovery yield. Bioprocess engineering increasingly uses digital sensor networks and machine learning to predict performance and obtain optimization.

Reactor design for metal biorecovery should be configured to integrate downstream operations such as solvent extraction, electrowinning, or biosorption columns, which follow the microbial solubilization step [47]. One example of an integrated system involves a bioleaching reactor that generates a high-concentration Cu leachate, subsequently processed through electrowinning to obtain a copper deposit with >99% purity. Hybrid reactor configurations have also been widely explored to enhance overall recovery efficiency, minimize waste residue, and improve resource circularity.

6 Environmental and economic perspectives

The transition toward bio-based metal recovery systems is closely tied to both environmental sustainability and economic feasibility. Biorecovery technologies, such as bioleaching and biosorption, significantly reduce the carbon footprint compared to pyrometallurgical and hydrometallurgical processes [47]. Both of these are energy-intensive and generate hazardous effluents. These biological routes operate under mild temperature and pressure, use renewable feedstocks, and minimize secondary pollution. Microbial biorecovery aligns with the United Nations Sustainable Development Goals (SDGs), particularly SDG 12 (Responsible Consumption and Production) and SDG 13 (Climate Action) [48]. Carpen et al. 2022 showed that bioleaching of nickel laterite with *Burkholderia* reached a maximum Ni solubilization of ~39.7% under the tested conditions at low temperature (mesophilic) and far lower thermal energy input than pyrometallurgical processing. Although explicit percentage results regarding TEA or LCA are not reported in this study, the authors emphasize that such low-temperature bioleaching has clear potential to reduce operational energy demand

and associated GHG emissions compared to conventional high-temperature routes. However, scaling up remains a constraint due to challenges like slower kinetics, metal selectivity, and bioreactor scalability [49].

6.1 Life cycle and technoeconomic evaluation of biorecovery

A LCA provides a quantitative means to evaluate the environmental performance of biorecovery systems across their production, operation, and disposal phases. Figure 3 shows a pictorial flow diagram of LCA of biorecovery of metals. Studies comparing bioleaching and chemical leaching have revealed that biological routes exhibit up to 70% lower global warming potential and 80% lower human toxicity impacts. This is mostly due to reduced chemical reagent use and lower demand for energy. Experimental studies report metal recoveries >90% in some *A. ferrooxidans* bioleaching trials, and TEA work on hybrid bio/hydro-flowsheets shows such routes can be cost-competitive under favorable feed and energy scenarios. The combined use of LCA and TEA provides a holistic decision-making framework, guiding policymakers and industries toward practicing sustainable technologies [50].

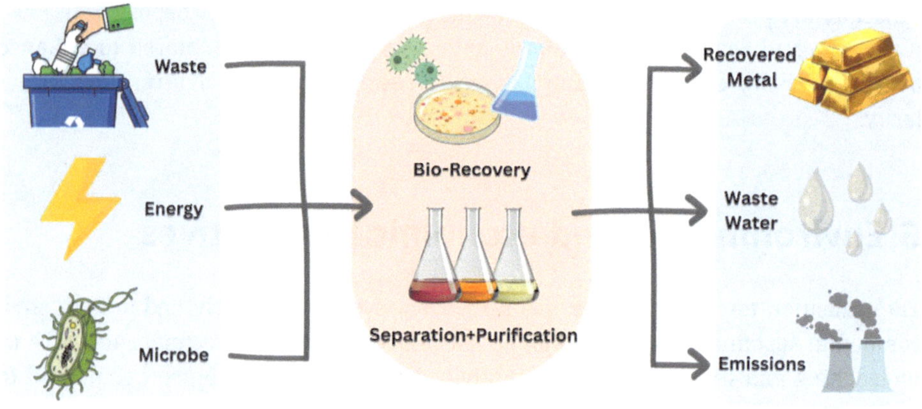

Figure 3: LCA of microbial metal recovery.

6.2 Integration into circular economy frameworks and policy outlook

Integrating microbial metal-recovery technologies like bioleaching and biosorption into CE frameworks is a transformative step toward achieving sustainable industrial systems. CE aims to focus on eliminating the linear model by promoting resource efficiency, material reuse, and waste valorization. CE thus decouples economic growth from raw material consumption. Laboratory-scale studies employing bacterial-free cultural supernatants from *Acidithiobacillus* iron-sulfur oxidizers have achieved near-complete dissolution of copper from pulverized waste printed circuit boards, as proof of the strong potential of biological lixiviants in high-efficiency metal recovery [51]. The recovered metals can be reintroduced in battery manufacturing, electronic components, and catalysis sectors. These applications thus close the material loop while minimizing the environmental footprint. The widespread adoption of microbial metal-recovery technologies depends strongly on policy instruments and frameworks that talk about CE principles. Extended producer responsibility (EPR) systems were implemented in the European Union and the E-waste (Management) Rules, 2022 [52]. These mandates require producers to manage end-of-life products in a sustainable manner. At a global level, the European Commission's Circular Economy Action Plan (2023) promotes bio-innovation for sustainable materials management, recognizing microbial recovery as a pathway to achieve strategic raw material resilience [53]. The standardization and certification protocols for biorecovered metals ensure quality consistency and enhance their market acceptance. Even after all the advantages that the current biorecovery options offer, challenges persist. These challenges are made up by slow kinetics, limited scalability, and regulatory ambiguity regarding the classification of bio-derived metals.

7 Conclusion and future prospects

Biological pathways for metal recovery provide a more sustainable and resource-efficient option, offering an environmentally responsible alternative to traditional metallurgical methods. By leveraging the metabolic activity and surface chemistry of microorganisms, these methods enable selective extraction of metals from mining residues, industrial effluents, and e-waste. This can be achieved under well-controlled conditions and typically generates significantly lower emissions. The alignment with sustainability goals and CE principles underscores the increasing relevance in modern resource management. Life cycle and technoeconomic studies consistently show that microbial routes can reduce environmental burdens and offer competitive operational costs when compared to conventional methods. Overall, the next decade is expected to see substantial advances in scaling microbial metal recovery, driven by interdisciplinary collaboration, supportive regulatory frameworks, and continued technological innovation. When this technology is combined with CE systems, it holds the capacity to strengthen critical metal supply chains, reduce economic impacts and contribute to a more resilient and sustainable industrial mechanism.

References

[1] Murthy V, Ramakrishna S. A review on global e-waste management: Urban mining toward a sustainable future and circular economy. Sustainability. 2022;14(2):647.

[2] Seif R, Salem FZ, Allam NK. E-waste recycled materials as efficient catalysts for renewable energy technologies and better environmental sustainability. Environment, Development and Sustainability. 2023;1.

[3] Shahabuddin M, Uddin MN, Chowdhury JI, Ahmed SF, Uddin MN, Mofijur M, Uddin MA. A review of the recent development, challenges, and opportunities of electronic waste (e-waste). International Journal of Environmental Science and Technology. 2023;20(4):4513–4520.

[4] Xavier LH, Giese EC, Ribeiro-Duthie AC, Lins FAF. Sustainability and the circular economy: A theoretical approach focused on e-waste urban mining. Resources Policy. 2021;74:101467.

[5] Dutta D, Rautela R, Gujjala LKS, Kundu D, Sharma P, Tembhare M, Kumar S. A review on recovery processes of metals from e-waste: A green perspective. Sci. Total Environ. 2023;859:160391.

[6] Adetunji AI, Oberholster PJ, Erasmus M. Bioleaching of metals from e-waste using microorganisms: A review. Minerals. 2023;13(6):828.

[7] Cottes M, Mainardis M, Simeoni P. Assessing the techno-economic feasibility of waste electric and electronic equipment treatment plant: A multi-decisional modeling approach. Sustainability. 2023;15 (23):16248.

[8] Van Yken J, Boxall NJ, Cheng KY, Nikoloski AN, Moheimani N, Kaksonen AH. Techno-economic analysis of an integrated bio- and hydrometallurgical process for base and precious metal recovery from waste printed circuit boards. Hydrometallurgy. 2023;222:106193.

[9] Dutta D, Rautela R, Gujjala LKS, Kundu D, Sharma P, Tembhare M, Kumar S. A review on recovery processes of metals from e-waste: A green perspective. Sci. Total Environ. 2023;859:160391.

[10] Torrubia J, Parvez AM, Sajjad M, Paz FAG, Van den boogaart KG. Recovery of copper from electronic waste: An energy transition approach to decarbonise the industry. Journal of Cleaner Production. 2024;485:144349.

[11] Chaturvedi K, Singhwane A, Dhangar M, Raghuwanshi S, Tak D, Srivastava AK, Verma S. State-of-the-art review on the potentiality of microorganisms for extracting metals from e-waste i.e., PCBs of mobile phones and computers. Environmental Technology Reviews. 2024;13(1):186–213.

[12] Tezyapar Kara I, Kremser K, Wagland ST, Coulon F. Bioleaching metal-bearing wastes and by-products for resource recovery: A review. Environmental Chemistry Letters. 2023;21(6):3329–3350.

[13] Cozma P, Bețianu C, Hlihor RM, Simion IM, Gavrilescu M. Bio-recovery of metals through biomining within circularity-based solutions. Processes. 2024;12(9):1793.

[14] Jaiswal M, Srivastava S. A review on sustainable approach of bioleaching of precious metals from electronic wastes. Journal of Hazardous Materials Advances. 2024;14:100435.

[15] Olson GJ, Brierley JA, Brierley CL. Bioleaching review part B: Progress in bioleaching: Applications of microbial processes by the minerals industries. Applied Microbiology and Biotechnology. 2003;63 (3):249–257.

[16] Van Rompaey K, Van Ranst E, Verdoodt A, De Coninck F. Use of the test-mineral technique to distinguish simple acidolysis from acido-complexolysis in a podzol profile. Geoderma. 2007;137(3–4):293–299.

[17] Bhandari G, Gupta S, Chaudhary P, Chaudhary S, Gangola S. Bioleaching: A Sustainable Resource Recovery Strategy for Urban Mining of E-Waste. In Microbial Technology for Sustainable E-Waste Management 2023 (pp. 157–175). Cham: Springer International Publishing.

[18] Mishra S, Panda S, Akcil A, Dembele S, Agcasulu I. A review on chemical versus microbial leaching of electronic wastes with emphasis on base metals dissolution. Minerals. 2021;11(11):1255.

[19] Kumar PS, Yaashikaa PR. Recent Trends and Challenges in Bioleaching Technologies. In Biovalorisation of Wastes to Renewable Chemicals and Biofuels 2020 (pp. 373–388).

[20] Sarkodie EK, Jiang L, Li K, Yang J, Guo Z, Shi J, Liu X. A review on the bioleaching of toxic metal(loid)s from contaminated soil: Insight into the mechanism of action and the role of influencing factors. Frontiers in Microbiology. 2022;13:1049277.
[21] Jasu A, Ray RR. Biofilm-mediated strategies to mitigate heavy metal pollution: A critical review in metal bioremediation. Biocatalysis and Agricultural Biotechnology. 2021;37:102183.
[22] Tonietti L, Esposito M, Cascone M, Barosa B, Fiscale S, Muscari Tomajoli MT, Giovannelli D. Unveiling the bioleaching versatility of Acidithiobacillus ferrooxidans. Microorganisms. 2024;12(12):2407.
[23] Priya A, Hait S. Biometallurgical recovery of metals from waste printed circuit boards using pure and mixed strains of Acidithiobacillus ferrooxidans and Acidiphilium acidophilum. Process Safety and Environmental Protection. 2020;143:262–272.
[24] Ali Redha A. Removal of heavy metals from aqueous media by biosorption. Arab Journal of Basic and Applied Sciences. 2020;27(1):183–193.
[25] Priya AK, Gnanasekaran L, Dutta K, Rajendran S, Balakrishnan D, Soto-Moscoso M. Biosorption of heavy metals by microorganisms: Evaluation of different underlying mechanisms. Chemosphere. 2022;307:135957.
[26] Raji Z, Karim A, Karam A, Khalloufi S. Adsorption of heavy metals: Mechanisms, kinetics, and applications of various adsorbents in wastewater remediation – a review. Waste. 2023;1(3):775–805.
[27] Chakraborty R, Asthana A, Singh AK, Jain B, Susan ABH. Adsorption of heavy metal ions by various low-cost adsorbents: A review. International Journal of Environmental Analytical Chemistry. 2022;102 (2):342–379.
[28] Vijayan PP, Chithra PG, Krishna SV, Ansar A, B. E, Parameswaranpillai J. Development and current trends on ion exchange materials. Sep. Purification Reviews. 2024;53(1):40–60.
[29] Parmar KS, Patel KM. Biosorption and bioremediation of heavy metal ions from wastewater using algae: A comprehensive review. World Journal of Microbiology and Biotechnology. 2025;41(7):262.
[30] Salehi M. Surface complexation at charged organic surfaces. Reviews in Mineralogy and Geochemistry. 2025;91(1):149–173.
[31] Staven V, Wang S, Grønlie I, Tho I. Physical stability of an all-in-one parenteral nutrition admixture for preterm infants upon mixing with micronutrients and drugs. European Journal of Hospital Pharmacy. 2020;27(1):36–42.
[32] Torres E. Biosorption: A review of the latest advances. Processes. 2020;8(12):1584.
[33] Gu S, Lan CQ. Mechanism of heavy metal ion biosorption by microalgal cells: A mathematic approach. Journal of Hazardous Materials. 2024;463:132875.
[34] El-Gendy MMAA, Abdel-Moniem SM, Ammar NS, El-bondk ly AMA. Bioremoval of heavy metals from aqueous solution using dead biomass of indigenous fungi derived from fertilizer industry effluents: Isotherm models evaluation and batch optimization. Biometals. 2023;36(6):1307–1329.
[35] Younas H, Nazir A, Latif Z, Thies JE, Shafiq M, Bareen FE. Biosorption potential and molecular characterization of metal-resistant autochthonous microbes from tannery solid waste. Archives of Microbiology. 2022;204(10):651.
[36] Isaza-Pérez F, Ramírez-Carmona M, Rendón-Castrillón L, Ocampo-López C. Potential of residual fungal biomass: A review. Environmental Science and Pollution Research. 2020;27(12):13019–13031.
[37] Ordóñez JI, Cortés S, Maluenda P, Soto I. Biosorption of heavy metals with algae: Critical review of its application in real effluents. Sustainability. 2023;15(6):5521.
[38] Wood PL. Metabolic and lipid biomarkers for pathogenic algae, fungi, cyanobacteria, mycobacteria, gram-positive bacteria, and gram-negative bacteria. Metabolites. 2024;14(7):378.
[39] Niu C, Liu Y, Yang Y, Wang R, Li T. Advances in sulfonated modification and bioactivity of polysaccharides. International Journal of Biological Macromolecules. 2023;253:126400.
[40] Cui D, Tan C, Deng H, Gu X, Pi S, Chen T, Li A. Biosorption mechanism of aqueous Pb^{2+}, Cd^{2+}, and Ni^{2+} ions on extracellular polymeric substances (EPS). Archaea. 2020;2020(1):8891543.

[41] Bharathi SD, Dilshani A, Rishivanthi S, Khaitan P, Vamsidhar A. Jacob, S. Resource recycling, recovery, and xenobiotic remediation from e-wastes through biofilm technology: A review. Applied Biochemistry and Biotechnology. 2023;195(9):5669–5692.
[42] López-Martínez A, Martínez-Prado MA, Núñez-Ramírez DM, Medina-Torres L, Rojas-Contreras JA, Anguiano-Vega GA, Soto-Cruz NO. Acidophilic bacteria for metal extraction: Biotechnological characteristics and applications. Brazilian Journal of Chemical Engineering. 2025;42(1):31–52.
[43] Jones S, Santini JM. Mechanisms of bioleaching: Iron and sulfur oxidation by acidophilic microorganisms. Essays in Biochemistry. 2023;67(4):685–699.
[44] Panyushkina A, Muravyov M. Biomining nickel for a greener low-carbon future: Progress in bioleaching technologies and acidophilic microbiology. Reviews in Environmental Science and Biotechnology. 2025;1–51.
[45] Xu YN, Chen Y. Advances in heavy metal removal by sulfate-reducing bacteria. Water Science and Technology. 2020;81(9):1797–1827.
[46] Pakostova E, Herath A. A bioleaching process for sustainable recycling of complex structures with multi-metal layers. Sustainability. 2023;15(19):14068.
[47] Rendón-Castrillón L, Ramírez-Carmona M, Ocampo-López C, Gómez-Arroyave L. Bioleaching techniques for sustainable recovery of metals from solid matrices. Sustainability. 2023;15(13):10222.
[48] Crowther TW, Rappuoli R, Corinaldesi C, Danovaro R, Donohue TJ, Huisman J, Van galen LG et al. Scientists' call to action: Microbes, planetary health, and the sustainable development goals. Cell. 2024;187(19):5195–5216.
[49] Carpen HL, Gerson AR, Bælum J, Keegan H, Summerton M. Analysis of nickel laterite ore bioleaching using Burkholderia sp. Applied Water Science. 2022;12:177.
[50] Pakostova E, Herath A. A bioleaching process for sustainable recycling of complex structures with multi-metal layers. Sustainability. 2023;15(19):14068.
[51] Wu W, Liu X, Zhang X, Zhu M, Tan W. Bioleaching of copper from waste printed circuit boards by bacteria-free cultural supernatant of iron–sulfur-oxidizing bacteria. Bioresources and Bioprocessing. 2018;5:1–13.
[52] Thapa K, Vermeulen WJ, Deutz P, Olayide O. Ultimate producer responsibility for e-waste management: A proposal for just transition in the circular economy based on the case of used European electronic equipment exported to Nigeria. Business Strategy and Development. 2023;6 (1):33–52.
[53] Dietz T, Bogdanski A, Boldt C, Börner J, Von braun J, O'Concubhair C, Werheim P et al. Bioeconomy Globalization: Recent Trends and Drivers of National Programs and Policies. 2024.

S. Sanjupriya, R. Pavithra, M. Poonkothai*

Green pathways in e-waste management: exploring bioleaching and biosorption for efficient metal recovery

Abstract: Microbial methods such as bioleaching and biosorption are emerging as promising, eco-friendly, and sustainable alternatives for recovering valuable metals from electronic waste. E-waste contains precious metals such as gold and silver, base metals such as copper and aluminum, and hazardous metals including lead and cadmium, all of which pose environmental and health risks if improperly managed. Conventional recovery techniques like pyrometallurgy and hydrometallurgy often face limitations related to cost, environmental impact, and inefficiency. Bioleaching employs acidophilic bacteria and fungi to solubilize metals through processes such as acidolysis, redoxolysis, and complexolysis, while biosorption utilizes microbial biomass to passively bind metals through ion exchange and complexation mechanisms. Factors such as pH, temperature, metal concentration, and microbial growth influence recovery efficiency. Combining bioleaching and biosorption can enhance recovery rates for multi-metal systems. Advantages of microbial approaches include low energy demand, cost-effectiveness, minimal environmental damage, and potential scalability. However, challenges remain, such as long processing times, toxicity of metals to microbes, and difficulties in scaling up. Recent advances include the use of genetically engineered microbes, nanobiotechnology, and immobilized microbial systems. Future prospects emphasize hybrid chemical-biological processes, AI-guided optimization, and industrial-scale applications. Overall, microbial bioleaching and biosorption offer a green pathway toward efficient and sustainable metal recovery from e-waste, reducing environmental hazards while promoting resource circularity.

Keywords: Bioleaching, biosorption, e-waste, microorganisms, solubilization, complexation, scalability, sustainability

***Corresponding author: M. Poonkothai,** Department of Zoology, Avinashilingam Institute for Home Science and Higher Education for Women, Coimbatore 641 043, Tamil Nadu, India, e-mail: poonkothaii_zoo@avinuty.ac.in

S. Sanjupriya, R. Pavithra, Department of Zoology, Avinashilingam Institute for Home Science and Higher Education for Women, Coimbatore 641 043, Tamil Nadu, India, poonkothaii_zoo@avinuty.ac.in

 | https://doi.org/10.1515/9783112225509-004

1 Introduction

The rapid growth of technology and consumer electronics has led to an alarming rise in electronic waste (e-waste) generation worldwide. Short product life cycles, frequent upgrades, and improper disposal have resulted in millions of discarded devices each year, creating serious environmental and health concerns. E-waste contains both valuable metals such as gold, silver, and copper, and toxic substances such as lead, mercury, and cadmium. Improper handling leads to soil and water contamination, air pollution, and the release of hazardous compounds, underscoring the urgent need for sustainable management approaches [1]. Recovering valuable metals from e-waste not only conserves natural resources but also minimizes environmental degradation associated with conventional mining. Traditional recovery methods, such as pyrometallurgy and hydrometallurgy, though effective, involve high energy consumption, strong chemicals, and toxic emissions, making them environmentally and economically unsustainable. In response, microbial technologies such as bioleaching and biosorption have emerged as green and efficient alternatives [2]. Bioleaching employs bacteria and fungi to solubilize metals through biochemical oxidation or acid production, enabling low-cost, low-energy extraction. Biosorption uses microbial biomass or biopolymers to absorb and concentrate metal ions from e-waste or contaminated solutions. These methods are eco-friendly, operate under mild conditions, and produce minimal secondary waste. Integrating such biological techniques into e-waste management supports circular economy principles by enabling resource recovery, waste reduction, and sustainable production. Overall, microbial approaches represent promising green pathways for efficient and environmentally responsible metal recovery from e-waste [3].

2 Overview of valuable metals in e-waste

E-waste is a rich secondary source of valuable metals that can be recovered and reused, reducing dependence on primary mining and minimizing environmental degradation. The composition of e-waste varies with device type and age, but it typically contains a complex mix of precious, base, and hazardous metals embedded within plastics, ceramics, and glass matrices. Precious metals such as gold (Au), silver (Ag), and platinum (Pt) are present in small quantities but hold significant economic value due to their excellent electrical conductivity, corrosion resistance, and catalytic properties. Gold and silver are commonly found in printed circuit boards, connectors, and microprocessors, while platinum is used in sensors and hard drives. Recovering these metals is essential for resource conservation and high economic returns [4].

Base metals, including copper (Cu), aluminum (Al), and nickel (Ni), are present in much larger quantities and form the structural and conductive components of electronic devices. Copper is abundant in wiring and circuit boards, aluminum is used in casings and heat sinks, and nickel appears in batteries and magnetic components. Their recovery helps reduce the environmental burden of mining and supports metal recycling industries [5]. Hazardous metals like lead (Pb), cadmium (Cd), and mercury (Hg) are toxic and pose severe health risks when released into the environment. Managing these metals safely during recovery is crucial to prevent contamination. Understanding the distribution and recovery potential of these metals is vital for designing efficient, eco-friendly recovery methods such as bioleaching and biosorption, which offer selective, low-cost, and sustainable alternatives to conventional extraction processes [6].

3 Microbial processes for metal recovery

Microbial processes offer environmentally friendly pathways for recovering metals from e-waste, providing sustainable alternatives to conventional chemical and thermal methods. Among these, bioleaching and biosorption utilize the natural capabilities of microorganisms to solubilize or capture metal ions, enabling efficient recovery under mild conditions. These biologically driven techniques reduce pollution, lower energy consumption, and support circular resource recovery from complex e-wastes [7].

3.1 Principles of bioleaching and biosorption

Bioleaching is a biologically mediated process in which microorganisms, mainly bacteria and fungi, solubilize metals from solid waste through biochemical reactions. Acidophilic bacteria such as *Acidithiobacillus ferrooxidans* and fungi like *Aspergillus niger* produce organic acids or oxidizing agents that break down metal-bearing components of e-waste. Through mechanisms such as redox reactions, acidolysis, and complexation, metals are converted into soluble forms that can be recovered from solution [8]. Biosorption, by contrast, relies on the ability of microbial biomass, living or dead, to bind and concentrate metal ions from aqueous solutions. Cell wall components, such as polysaccharides, proteins, and functional groups (carboxyl, hydroxyl, and phosphate), act as natural sorbents. Metals are captured through mechanisms that include ion exchange, adsorption, chelation, and microprecipitation. Together, these processes provide low-energy, selective, and eco-friendly routes for metal extraction from complex e-waste matrices [9].

3.2 Comparison between bioleaching and biosorption approaches

Bioleaching and biosorption differ in mechanisms, applications, and operational conditions, though both rely on microbial activity. Bioleaching is ideal for solubilizing metals directly from solid e-waste materials such as printed circuit boards, where microorganisms break down complex structures to release metals into solution. It is slower but highly effective for large-scale extraction of metals such as copper, nickel, and gold. Conversely, biosorption is typically used after metals have been leached into solution, functioning as a polishing or concentration step [10]. It is rapid, does not require microbial growth, and works efficiently even with low metal concentrations. While bioleaching requires controlled conditions such as pH, aeration, and nutrients, biosorption can operate under a wide range of environmental parameters. Bioleaching is more suited for bulk metal recovery, whereas biosorption excels in selective metal uptake and wastewater treatment. Together, they can form an integrated, sustainable recovery system [11].

3.3 Advantages over chemical and physical methods

Microbial metal recovery offers several advantages over traditional chemical and physical methods, such as hydrometallurgy, pyrometallurgy, and mechanical processing. Bioleaching and biosorption operate under mild temperatures and pressures, drastically reducing energy consumption. They eliminate or minimize the use of strong acids, cyanides, and other hazardous reagents, thereby lowering the risk of secondary pollution and toxic emissions. These biological approaches are also cost-effective, relying on naturally occurring microorganisms or inexpensive biomass instead of costly chemicals [12]. Additionally, microbial processes exhibit high selectivity for specific metals, improving recovery efficiency and reducing the need for complex separation steps. Their scalability and compatibility with existing waste management systems make them suitable for both small and large operations. Importantly, these methods support circular economy goals by enabling sustainable resource recovery while mitigating environmental impacts. Through reduced ecological footprints and enhanced metal recovery potential, microbial techniques represent a green and viable alternative to conventional extraction practices [13].

4 Bioleaching: mechanisms and microorganisms

Bioleaching is a microbe-mediated process in which metals are solubilized from solid e-waste using biochemical reactions. This approach relies on the metabolic activities of bacteria and fungi that generate acids, oxidizing agents, or chelating compounds capable of breaking down metal-bearing components of e-waste (Figure 1). Bioleaching offers a low-cost, energy-efficient, and environmentally friendly alternative to conventional extraction methods, making it a promising tool for sustainable metal recovery. Its success depends on selecting suitable microorganisms, understanding underlying mechanisms, and optimizing operating conditions [14].

4.1 Role of acidophilic bacteria

Acidophilic bacteria are the most widely used microorganisms in bioleaching due to their ability to thrive in extremely low pH and high-metal environments. *Acidithiobacillus ferrooxidans* and *Leptospirillum ferrooxidans* play key roles in oxidizing ferrous iron and reduced sulfur compounds, generating ferric ions and sulfuric acid – powerful leaching agents responsible for metal solubilization. These bacteria attach to e-waste surfaces and create localized acidic microenvironments that enhance metal dissolution. Their metabolic activity accelerates the breakdown of metal sulfides and complex electronic components, facilitating the release of copper, zinc, nickel, and trace precious metals. Acidophilic bacteria operate efficiently in continuous or batch bioreactors and are particularly effective in indirect bioleaching systems where their oxidizing products act on metal-bearing substrates [15].

4.2 Role of fungi

Fungi play an important role in bioleaching due to their ability to produce large amounts of organic acids, chelating compounds, and extracellular enzymes. Species such as *Aspergillus niger* and *Penicillium* spp. generate citric, oxalic, and gluconic acids, which dissolve metals through complexation and acidolysis. Fungal hyphae can penetrate porous e-waste materials, increasing contact between leaching agents and embedded metals. Their tolerance to high metal concentrations and variable pH conditions makes them suitable for solubilizing metals like aluminum, copper, cadmium, and rare earth elements. Fungal bioleaching is especially effective in non-sulfide materials commonly found in printed circuit boards. Additionally, fungal biomass can contribute to biosorption, enabling a combined approach where fungi both solubilize metals and bind dissolved ions from leachates [16].

Discarded electronic components serve as rich sources of metals. Acidophilic bacteria facilitate metal solubilization, thereby lowering pH and enhancing the dissolution of metal compounds. Simultaneously, fungi such as *Aspergillus Niger* and *Penicillium* species contribute through the production of organic acids that chelate and mobilize metals, as well as through various biosorption by their hyphal structures that bind metal ions. These combined biological actions release metals into solution, from which they can be recovered as usable metal concentrates.

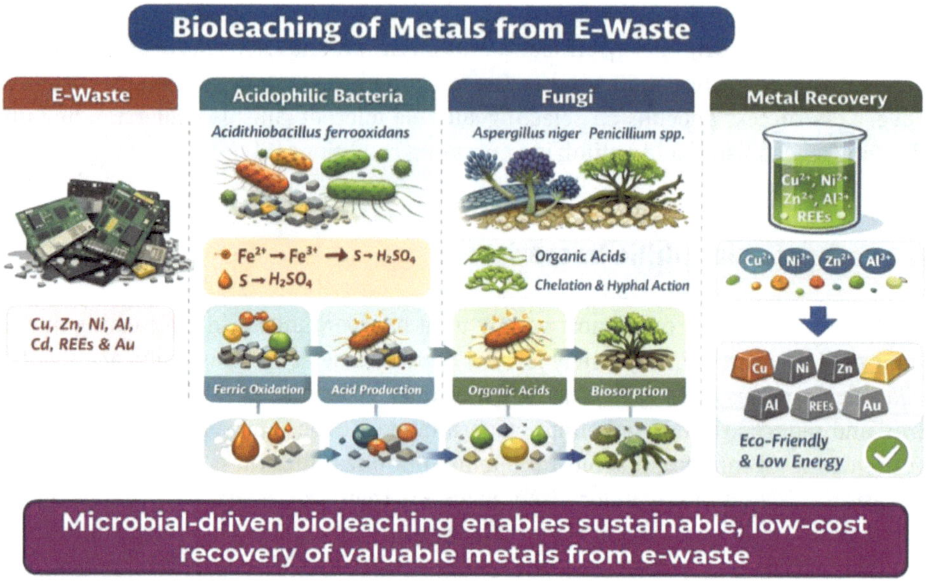

Figure 1: Microbial bioleaching of metals from e-wastes.

4.3 Mechanisms

4.3.1 Direct and indirect bioleaching

Direct bioleaching occurs when microorganisms attach directly to the metal-bearing surface and catalyze oxidation reactions through enzymatic activity. In this mechanism, microbial cells generate local acidic and oxidative conditions that break down metal matrices. This is common in systems involving sulfur-oxidizing or iron-oxidizing bacteria.Indirect bioleaching does not require physical contact. Instead, microorganisms produce leaching agents such as ferric ions, sulfuric acid, or organic acids that chemically dissolve metals from e-waste substrates [17]. Iron-oxidizing bacteria regenerate ferric ions in a cyclic process, while fungi supply organic acids that complex with metals. Indirect mechanisms are dominant in fungal bioleaching and many bacterial systems, offering greater flexibility and easier control in bioreactors. Bioleaching operates through three major chemical pathways. Acidolysis involves proton attack by mineral or organic acids produced by microbes, which break down metal compounds through acid dissolution [18]. This mechanism is common in fungal systems that produce oxalic or citric acids. Redoxolysis relies on microbial oxidation-reduction reactions, where metal sulfides or metallic components are oxidized by ferric ions or microbial enzymes, enabling their solubilization. Iron-oxidizing bacteria play a central role in this process. Complexolysis occurs when microbial metabolites, such as organic acids or siderophores, form stable complexes with metal ions, increasing their solubility and mobilizing them from solid matrices. These three mechanisms often act simultaneously, making bioleaching efficient for diverse metal types [19].

4.4 Factors affecting bioleaching efficiency

Bioleaching efficiency is strongly influenced by environmental and operational parameters. pH is critical, as most acidophilic bacteria function optimally between pH 1.5 and 3.0, while fungi prefer slightly higher acidic ranges; deviations reduce microbial activity and metal solubilization. Temperature affects metabolic rates, with bacteria thriving between 25 and 35 °C and some thermophiles operating at higher temperatures to accelerate oxidation reactions [20]. Pulp density, or the proportion of solid e-waste in the medium, must be optimized because high densities inhibit microbial growth through metal toxicity and poor mass transfer. Oxygen supply is essential for aerobic microorganisms, particularly iron- and sulfur-oxidizing bacteria, which require sufficient aeration to sustain redox reactions. Balancing these parameters ensures stable microbial activity and maximizes metal recovery [21].

5 Biosorption: mechanisms and microorganisms

Biosorption is a biological process in which microbial biomass binds and concentrates metal ions from liquid solutions without requiring metabolic activity. This technique relies on natural components of cell walls such as polysaccharides, proteins, and lipids that contain functional groups capable of attracting and immobilizing metals. Biosorption is rapid, cost-effective, and efficient even at low metal concentrations, making it suitable for recovering dissolved metals generated during e-waste processing. Because biomass can be regenerated and reused, biosorption supports sustainable and circular resource recovery [22].

5.1 Biosorption principles: passive uptake of metals

Biosorption operates through the passive uptake of metal ions by microbial biomass, relying on physicochemical interactions rather than active metabolism. Metals bind to the surface of dead or living cells through mechanisms such as adsorption, ion exchange, and complexation. Because the process does not depend on microbial growth, biosorption remains effective under harsh conditions, including high metal concentrations or toxic environments often associated with e-waste leachates [23]. Functional groups on cell walls carboxyl, hydroxyl, amino, and phosphate groups serve as natural binding sites that attract positively charged metal ions. This passive accumulation enables the efficient capture of metals such as copper, zinc, nickel, cadmium, and rare earth elements. Biosorption's simplicity and low operational cost make it a promising step for polishing and concentrating metals following bioleaching [24].

5.2 Microbial biomass used

Bacterial, fungal, yeast, and algal biomasses serve as highly effective biosorbents due to their diverse cell wall compositions and strong metal-binding capabilities. Bacterial species such as *Bacillus* and *Pseudomonas* possess peptidoglycan-rich walls and extracellular polymeric substances (EPSs) that offer abundant carboxyl, phosphate, and amine groups for metal uptake. Their resilience to extreme pH and temperature makes them suitable for treating acidic e-waste leachates, while their rapid adsorption kinetics enhances recovery efficiency.Fungi and yeasts, including *Saccharomyces cerevisiae*, have complex cell walls composed of chitin, glucans, and mannans, enabling efficient metal binding through ion exchange and complexation [25]. Their low cost, availability, and tolerance to high metal concentrations make them suitable for large-scale applications, with filamentous fungi like *Aspergillus* and *Penicillium* offering additional structural advantages. Algae and cyanobacteria feature polysaccharide-rich surfaces with sulfate, hydroxyl, and carboxyl groups that strongly interact with metals such as copper, nickel, cobalt, and rare earth elements. Their easy cultivation and ability to operate under extreme conditions make them highly sustainable biosorbents. When immobilized on supportive matrices, algal and cyanobacterial biomasses gain enhanced stability and reusability, making them valuable tools for environmentally friendly metal recovery from e-waste [26].

5.3 Mechanisms of metal binding

Metal binding during biosorption involves several interconnected mechanisms. Ion exchange occurs when metal ions replace naturally occurring ions such as calcium, magnesium, or hydrogen on the biomass surface. Complexation involves the formation of stable coordination bonds between metal ions and functional groups like amino, carboxyl, and phosphate groups [27]. Precipitation results when metals react with biomolecules to form insoluble compounds that deposit on the cell surface. In micro-precipitation, microbial metabolites create localized conditions that induce the formation of fine metal precipitates. These mechanisms may act simultaneously, contributing to the high efficiency and selectivity of biosorption. Their combined action enables microorganisms to capture diverse metals from e-waste leachates, supporting effective downstream recovery [28].

5.4 Factors influencing biosorption

Biosorption efficiency depends strongly on the nature of the biomass, solution pH, and duration of contact with metal ions. Different microorganisms vary in cell wall composition, influencing their affinity for specific metals. pH affects surface charge and metal speciation; low pH favors protonation of binding sites, while moderate pH conditions enhance adsorption of positively charged metals. Optimal pH ranges differ among bacteria, fungi, and algae. Contact time determines the extent of adsorption, with rapid initial binding followed by slower saturation phases [29]. Many biosorbents reach equilibrium within minutes to hours. Maintaining sufficient interaction time ensures maximum uptake, while avoiding excessive exposure that could cause desorption or biomass degradation. Properly optimizing these parameters enhances metal recovery efficiency in integrated e-waste treatment systems [30].

6 Factors affecting metal recovery efficiency

The efficiency of metal recovery through bioleaching and biosorption depends on several interconnected factors that influence metal solubilization, binding, and microbial performance. Proper pretreatment of e-waste determines the accessibility of metals, while suitable microbial growth conditions ensure optimal metabolic activity or biosorption capacity. Additionally, the chemical form and concentration of metals affect their mobility, reactivity, and ease of extraction. Understanding these factors allows the development of more efficient, scalable, and environmentally sustainable metal recovery systems for e-waste management [31].

6.1 E-waste pretreatment

Pretreatment of e-waste is a critical step that significantly enhances metal recovery efficiency. Mechanical crushing, shredding, and size reduction increase the surface area of electronic components, improving the contact between microorganisms or biosorbents and metal-bearing substrates. Finer particle sizes expose embedded metals in printed circuit boards, capacitors, and wiring, enabling more effective leaching by acidophilic bacteria or organic acids produced by fungi [32]. Pretreatment also helps in separating plastics, glass, and ferrous components, creating a more homogeneous material suitable for bioprocessing. Additionally, reducing particle size lowers mass-transfer resistance, allowing leaching agents to penetrate more easily into complex matrices. However, excessively fine particles may inhibit microbial activity by releasing toxic compounds or causing poor aeration. Therefore, optimizing crushing and size reduction is essential to maximize metal solubilization and ensure smooth downstream biosorption or recovery operations [33].

6.2 Microbial adaptation and growth conditions

Microbial adaptation plays a vital role in successful bioleaching and biosorption, as microorganisms must tolerate high metal concentrations, acidic environments, and potentially toxic compounds present in e-waste leachates. Acidophilic bacteria such as *Acidithiobacillus ferrooxidans* require controlled conditions of pH, temperature, aeration, and nutrient supply to maintain efficient oxidation reactions. Similarly, fungi and biosorbent biomass require appropriate carbon sources, moisture levels, and incubation conditions to produce organic acids or maintain active binding sites [34]. Some microbes undergo adaptive evolution, improving their tolerance to specific metals, thereby enhancing leaching performance. Maintaining microbial growth in optimal ranges ensures high metabolic activity, faster metal solubilization, and consistent biosorption capacity. Any deviation, such as pH shifts, oxygen limitation, or nutrient depletion, can significantly reduce efficiency. Thus, careful monitoring and adjustment of microbial growth parameters are essential for achieving stable and high-yield metal recovery processes [35].

6.3 Metal speciation and concentration

The chemical form and concentration of metals in e-waste leachates strongly influence their recovery efficiency. Metal speciation determines solubility, mobility, and reactivity, affecting both bioleaching and biosorption outcomes. Metals present as oxides, sulfides, or complex alloys require different biochemical reactions for solubilization, while ionic species are more readily adsorbed by microbial biomass. High metal concentrations may enhance driving forces for adsorption but can also impose toxicity on microbial cells, inhibiting growth and metabolic activity [36]. Conversely, extremely low concentrations limit biosorption capacity because binding sites remain underutilized. pH further alters speciation by affecting ionization states and precipitation tendencies. Understanding speciation helps select appropriate microorganisms, tailor leaching conditions, and design selective biosorbents. By managing concentration and chemical form, recovery processes can be optimized to achieve maximum extraction efficiency in environmentally friendly metal recovery systems [37].

7 Integration of bioleaching and biosorption

Integration of bioleaching and biosorption offers a sustainable approach for recovering valuable metals from e-waste. Bioleaching employs specific microorganisms to solubilize metals from e-waste, while biosorption uses biological materials to bind and separate metals from solution. Combining these methods can improve selectivity, recover efficiency, and process economics, supporting greener pathways in e-waste management and reducing environmental impacts [38].

7.1 Sequential and combined approaches for enhanced recovery

Sequential and combined approaches integrating bioleaching and biosorption have shown promise for enhanced metal recovery from e-waste. In sequential processes, bioleaching is initially used to solubilize metals from substrates such as printed circuit boards, followed by biosorption with microbial or plant-based biosorbents to selectively recover dissolved metals from leachates [39]. Recent studies demonstrate that combined strategies can significantly improve overall recovery rates for copper, gold, and other valuable metals due to increased process selectivity and efficiency. These hybrid methods also reduce toxic reagent consumption and environmental risks compared to conventional-only routes, while facilitating more sustainable resource recovery. Advances include the optimization of operating conditions, cell immobilization, and the application of engineered microbes and statistical methods for process design, all contributing to higher recovery yields and lower ecological impact [40].

7.2 Synergistic effect on recovery of multi-metal systems

Recent studies demonstrate that combining bioleaching and biosorption produces a synergistic effect in the recovery of multiple metals from e-waste streams, especially printed circuit boards. Sequential integration leverages microbial consortia to dissolve metals like copper, gold, and nickel efficiently, followed by biosorption using engineered biosorbents for selective metal uptake from leachates. Hybrid systems enhance process selectivity and overall yield, and reduce energy as well as reagent consumption compared to traditional single-step recovery [33]. Optimization of microbial strains and biosorbent properties, alongside process parameter-tuning enabled by data-driven models, facilitates high and simultaneous recovery rates for precious and base metals, supporting greener urban mining routes. Advancements include the use of two-stage processes, cell immobilization, and machine learning for operational improvement, contributing to sustainable multi-metal recovery and significant reduction in environmental impact [41].

8 Advantages of microbial approaches

Microbial approaches are gaining attention as sustainable solutions for e-waste management due to their unique ability to recover valuable metals using natural biological processes. Techniques such as bioleaching and biosorption utilize the metabolic activities of microorganisms to extract and bind metals from e-waste, offering an eco-friendly alternative to conventional methods. These strategies are recognized for their cost-effectiveness, lower energy requirements, and potential to minimize environmental pollution, making them integral to green pathways in e-waste management [42].

8.1 Low cost and energy requirement

Bioleaching and biosorption as microbial technologies have demonstrated significant advantages in terms of low cost and minimal energy demands for e-waste management. [43] reported that bioleaching relies on natural microbial processes to extract metals from waste printed circuit boards without the need for high temperatures or hazardous chemicals, reducing both energy consumption and secondary pollution. The bioleaching method utilizes the metabolic activity of microbes for metal solubilization, avoiding the need for high temperatures or hazardous reagents. Similarly, [38] reviewed that biosorption utilizes microbial biomass under ambient conditions to capture metals, proving more economically viable than traditional methods. These findings confirm microbial approaches not only provide efficient metal recovery but also contribute to sustainable e-waste solutions.

8.2 Eco-friendly and sustainable process

Microbial bioleaching and biosorption are recognized as eco-friendly and sustainable processes for metal recovery from e-waste. These methods leverage the natural metabolic capabilities of microorganisms to extract and absorb metals without producing toxic byproducts or consuming excessive energy, unlike traditional physical and chemical methods [44]. According to a recent review, bioleaching involves acidolysis, redoxolysis, and complexolysis mechanisms where bacteria and fungi mobilize metals efficiently while maintaining environmental safety. Biosorption utilizes microbial biomass as a natural adsorbent under ambient conditions, promoting sustainability with low environmental impact. These biological techniques offer significant promise, reducing pollution and supporting circular economy goals [21].

8.3 Scalability and resource circularity

Microbial approaches such as bioleaching and biosorption offer promising scalability and promote resource circularity in e-waste management. A study on *Methylobacterium extorquens* AM1 demonstrated scalable bioleaching up to 10 liters with consistent metal recovery from e-waste without harsh chemicals or extreme conditions [45]. This natural production of metal-binding ligands and phosphate storage granules by microbes facilitates selective metal accumulation and recovery, supporting circular economy principles by recycling critical elements. Furthermore, hybrid and genetically engineered microbial systems are being developed to enhance efficiency and adaptability for large-scale applications. These processes reduce reliance on conventional mining and toxic reagents, offering sustainable metal recovery pathways [46].

9 Limitations and challenges

Green methods such as bioleaching and biosorption use microorganisms to recover metals from e-waste. These techniques offer environmentally friendly alternatives to traditional methods but face multiple challenges, including lengthy processing times, metal toxicity to microbes, scale-up difficulties, and the need for optimized conditions to enhance efficiency. Recent research underscores that bioleaching and biosorption have great promise for eco-efficient recovery of metals from waste printed circuit boards [47]. However, the slow kinetics of bioleaching remains a significant bottleneck, as the toxicity of heavy metals present in e-waste hinders the metabolic activity of bacteria. Optimization of parameters like pH, temperature, aeration, and nutrient supply is crucial for enhancing microbial growth and metal solubilization. Scale-up barriers arise from inconsistent performance in variable waste compositions and the sensitivity of microorganisms to extreme conditions [48]. Hybrid strategies combining bioleaching with biosorption or other biotechnologies are gaining attention for improved selectivity and recovery rates. Despite advances, industrial implementation is limited due to the complexity of biological systems and operational costs [49].

10 Recent advances and emerging technologies

Recent advances in green pathways for e-waste management focus on utilizing biological methods such as bioleaching and biosorption to recover valuable metals efficiently. These eco-friendly approaches harness microorganisms to extract metals from e-waste, offering sustainable alternatives to conventional recycling techniques. Such biotechnological innovations not only reduce environmental pollution but also support the economic recovery of critical and precious metals from complex electronic materials [50].

10.1 Genetically engineered microbes

Genetically engineered microbes have emerged as promising tools in e-waste management to enhance metal recovery via bioleaching and biosorption. These microbes are modified using synthetic biology tools to improve enzyme activity, substrate specificity and toxicity tolerance, enabling efficient breakdown of e-waste components. For example, *Chromobacterium violaceum* has been engineered to increase cyanide production, enhancing gold bioleaching rates. Despite promising lab-scale results, challenges include the complex and toxic nature of e-waste and limited molecular toolkits for some key microbes. Commercialization efforts are underway, but ethical and environmental safety issues related to the release of genetically modified organisms must be addressed [51].

10.2 Nanobiotechnology-assisted recovery

Nanobiotechnology-assisted recovery in e-waste management combines nanotechnology and biotechnology to enhance metal recovery efficiency. This approach uses nanoparticles synthesized via green methods, including microbial bioleaching, to improve metal ion concentration and transformation from waste printed circuit boards. Nanoparticles such as CuO, Au, and SnO_2 are produced through eco-friendly synthesis and applied for selective metal extraction and purification [52]. These nanomaterials exhibit high surface area and reactivity, facilitating effective leaching and adsorption processes. Nanobiotechnology thus offers a sustainable pathway for converting e-waste into value-added materials with applications in electronics, catalysis, and environmental remediation [53].

10.3 Immobilized microbial systems

Immobilized microbial systems have gained attention for continuous biosorption in e-waste metal recovery due to their improved stability, reusability, and efficiency compared to free biomass [54]. Immobilizing microbial biomass on carriers like alginate, silica gel, or magnetic nanoparticles enhances mechanical strength, facilitates separation, and allows their use in fixed-bed reactors for continuous processing. This technique addresses challenges such as biomass loss and metal ion desorption difficulties in free systems. Despite advantages, immobilization may increase costs and reduce eco-friendliness [55].

11 Policy, economics, and circular economy linkage

The regulatory framework for e-waste bioprocessing has evolved significantly, focusing on sustainable resource recovery and circular economy principles aligned with Sustainable Development Goals (SDGs). India's 2025 E-Waste Management Rules emphasize extended producer responsibility (EPR), mandating producers, importers, and brand owners to register with the Central Pollution Control Board (CPCB) and meet annual recycling targets, currently set at 60% of the product weight sold [33]. Compliance involves digital tracking, purchase of EPR certificates, and partnerships with authorized recyclers to ensure environmentally sound recycling. Economic feasibility studies highlight bioprocessing as cost-effective due to lower energy consumption and reduced secondary pollution compared to conventional methods. This approach supports multiple SDGs, including responsible consumption, environmental sustainability, and economic growth by promoting a circular economy where e-waste is transformed into valuable resources [56].

12 Future prospects

Hybrid processes combining chemical and biological methods have shown significant promise in e-waste metal recovery due to their complementary advantages. Chemical leaching offers rapid and efficient metal extraction but often involves harsh reagents causing environmental concerns. Biological leaching is more ecofriendly and cost-effective but slower and less complete on its own [4]. Integrating both methods optimizes metal recovery by using mild chemicals with tailored microbial consortia to enhance bioleaching efficiency and selectivity. Additionally, advances in artificial intelligence and bioprocess modeling enable optimization of parameters such as pH, temperature, and microbial metabolism to improve recovery rates and reduce costs. These hybrid methods hold strong global potential for industrial-scale adoption, fostering sustainable, circular economy models aligned with environmental goals [57].

13 Conclusion

In conclusion, bioleaching and biosorption stand out as promising green technologies for metal recovery from e-waste due to their eco-friendly, cost-effective, and scalable nature. These microbial processes employ bacteria, fungi, and other biomass to efficiently solubilize and bind valuable metals, such as gold, copper, and nickel, while minimizing the use of hazardous chemicals and excessive energy. When integrated in sequential or hybrid systems, bioleaching and biosorption can offer enhanced selectivity and yield, supporting the circular economy by reducing reliance on primary mining and mitigating e-waste's environmental impacts. Recent innovations, including genetically engineered microbes, nanobiotechnology, and immobilized microbial systems, hold the potential to overcome major hurdles such as slow processing rates, microbial toxicity, and scale-up challenges. The future of sustainable e-waste management lies in the optimization and industrial implementation of these green technologies guided by advances in artificial intelligence and combined chemical-biological protocols. Altogether, adopting microbial approaches for e-waste recycling aligns with global environmental goals, promotes safer resource recovery, and advances a circular economy model that transforms waste into value.

References

[1] Serpe A, Purchase D, Bisschop L, Chatterjee D, De Gioannis G, Garelick H, Verbeek S. 2002–2022: 20 years of e-waste regulation in the European Union and the worldwide trends in legislation and innovation technologies for a circular economy. RSC Sustain. 2025;3:1039–1083.

[2] Malik AH, Rehan M, Hussain A, Ur-Rehman A, Kashif M. Balancing innovation and sustainability: The e-waste challenge in achieving SDG 13 across EEA countries. Technology in Society. 2025;103119.

[3] Rastogi S. e-waste collection, sorting, and handling practices. In E-Waste: From Generation to Management–Practices, Challenges, and Solutions 2025 (pp. 81–103). Springer Nature Singapore.

[4] Bhattacharjee J, De A, Kamila B, Mandal A. Recent scenario of e-waste recycling: Chemical engineering. International Journal of Chemical Reactor Engineering. 2025;23:675–700.

[5] Sarkhoshkalat MM, Afkham A, Bonyadi Manesh M, Sarkhosh M. Circular economy and the recycling of e-waste. In New Technologies for Energy Transition Based on Sustainable Development Goals: Factors Contributing to Global Warming 2024 (pp. 319–354). Springer Nature Singapore.

[6] Hossain R, Sahajwalla V. Current recycling innovations to utilize e-waste in sustainable green metal manufacturing. Philosophical Transactions of the Royal Society A. 2024;382:20230239.

[7] Bai Y, Su J, Wang F, Cui H, Zhang H, Liu K. Harnessing synthetic biology for sustainable recovery of critical metal materials from electronic waste. Advanced Functional Materials. 2025;e09900.

[8] Jaiswal M, Srivastava S. A review on sustainable approach of bioleaching of precious metals from electronic wastes. Journal of Hazardous Materials Advances. 2024;14:100435.

[9] Aachhera S, Tiwari S, Verma P, Gahan CS. e-waste recycling: A comprehensive review on metal recovery and the factors affecting bioleaching. In Biotechnology for Environmental Sustainability 2025 (pp. 209–243). Springer Nature Singapore.

[10] Kompalitch AH, Van hullebusch ED. Integration of bioleaching and biorefinery technologies for the recovery of base and critical elements from electronic waste. In Biotechnological Innovations in the Mineral-Metal Industry 2024 (pp. 179–206). Springer International Publishing.

[11] Shreya S, Nigam VK, Sampath MK. Biorecovery of metals from e-waste: an elaborate study. In Extraction of Precious Metals from Industrial Waste Using Microbial Technology 2025 (pp. 69–86). Springer Nature Switzerland.

[12] Kiran NS, Yashaswini C, Chatterjee A, Shah MP. Biotechnological approaches for metal recovery from electronic wastes. Current Microbiology. 2024;81:419.

[13] Bajpai M, Nigam M, Singh LK, Awasthi G. Electronic waste: an emerging pollutant of human health concern and biological techniques for their treatment. In Environmental Toxicants and Lifestyle Diseases 2025 (pp. 141–156). Springer Nature Singapore.

[14] Selvakumar P, Dahiya R, Manjunath TC, Nathsharma SK. Chemical and biological processes for e-waste recycling. In Integrated Approaches for Sustainable E-Waste Management 2025 (pp. 151–176). IGI Global Scientific Publishing.

[15] Rehman H, Debik E, Ulucan-Altuntas K, Manav-Demir N, Canci B, Iqbal M, Khan AHA. Bioleaching of waste-derived rare earth elements: An integrated approach with meta-analysis and predictive analytics for scale-up. Results in Engineering. 2025;107720.

[16] Ghosh S, Chakraborty S, Bloch K. Microbe mediated metal recovery: a sustainable e-waste management approach. In Extraction of Precious Metals from Industrial Waste Using Microbial Technology 2025 (pp. 1–19). Springer Nature Switzerland.

[17] Baez AG, Muñoz LP, Timmermans MJ, Garelick H, Purchase D. Molding the future: Optimization of bioleaching of rare earth elements from electronic waste by Penicillium expansum and insights into its mechanism. Bioresource Technology. 2024;402:130750.

[18] Mathivanan K, Zhang R, Chandirika JU, Mathimani T, Wang C, Duan J. Bacterial biofilm-based bioleaching: Sustainable mitigation and potential management of e-waste pollution. Waste Management. 2025;193:221–236.

[19] Wu X, Roy V, Shakelly N, Sutherland JW, Zhao F. Bioleaching gallium from e-waste: comparative techno-economic analysis. In International Manufacturing Science and Engineering Conference 2025 (Vol. 89022, p. V002T10A004). ASME.

[20] Jaiswal M, Srivastava S. A review on sustainable approach of bioleaching of precious metals from electronic wastes. Journal of Hazardous Materials Advances. 2024;14:100435.

[21] Ferreira-Filipe DA, Duarte AC, Hursthouse AS, Rocha-Santos T, Silva ALP. Biobased strategies for e-waste metal recovery: A critical overview of recent advances. Environments. 2025;12:26.

[22] Dev RK, Yadav SN, Magar N, Ghimire S, Koirala M, Giri R, Bhattarai A. Recovery of rare earth elements (REEs) from different sources of e-waste and their potential applications: A focused review. Geological Journal. 2025.

[23] Aslam A, Kanwal F, Javied S, Nisar N, Torriero AAJ. Microbial biosorption: A sustainable approach for metal removal and environmental remediation. International Journal of Environmental Science and Technology. 2025;1–32.

[24] Khan HA, Sher S, Bukhari DA, Rehman A. Bioremediation of heavy metals from electronic waste dumping sites with bacteria. Current Research in Biotechnology. 2025;100309.

[25] Tiwari S, Aachhera S, Verma P, Gahan CS. Bioremediation of e-waste: a sustainable approach. In Biotechnology for Environmental Sustainability 2025 (pp. 709–724). Springer Nature Singapore.

[26] Ahuja V, Bhatt AK, Sharma V, Rathour RK, Rana N, Bhatia RK, Bhatia SK. Advances in glucosamine production from waste biomass and microbial fermentation technology and its applications. Biomass Conversion and Biorefinery. 2025;15:1643–1665.

[27] Eze C, Vinken M. E-waste: Mechanisms of toxicity and safety testing. FEBS Open Bio. 2024;14:1420–1440.

[28] Qin J, Liu H, Fang Z, Pei J, Yin K, Fu K, Luo J. Selective gold extraction from e-waste leachate via sulfur-redox mechanisms using sulfhydryl-functionalized MOFs. Water Research. 2025;275:123170.

[29] Al-Gailani A, Bongosia JG, Hornsby K, Taylor MJ. Maximising e-waste leachate adsorption: Multi-component isotherm models and mechanisms with scalable mesoporous sugarcane bagasse-derived biochar. Journal of Water Process Engineering. 2025;75:107979.

[30] Chakraborty A, Bhakat K, Islam E, Murmu R. Arthrobacter sp. mediated chromium remediation and copper accumulation from leached liquor for e-waste management. Microbe. 2025;6:100277.

[31] Javed A, Singh J. Process intensification for sustainable extraction of metals from e-waste: Challenges and opportunities. Environmental Science and Pollution Research. 2024;31:9886–9919.

[32] Niu B, Shanshan E, Song Q, Xu Z, Han B, Qin Y. Physicochemical reactions in e-waste recycling. Nature Reviews Chemistry. 2024;8:569–586.

[33] Phogat P, Kumar S, Wan M. A scientometrics study of advancing sustainable metal recovery from e-waste: Processes, challenges, and future directions. RSC Sustain. 2025.
[34] Maluleke MD, Kotsiopoulos A, Govender-Opitz E, Harrison ST. Exploring microbial adaptation of immobilised acidophilic cultures to improve microbial oxidation rates and copper tolerance in e-waste bioleaching. Minerals Engineering. 2024;207:108560.
[35] Priyadarsini S, Das AP. Isolation and identification of native lithium-resistant bacteria from electronic waste contaminated sites: Tolerance assessment and potential lithium bio-recovery application. Geomicrobiology Journal. 2025;1–15.
[36] Yao W, Ge B, Chen Y, Jiao W, Han H, Song X, Liang Z. Gold nanoparticles recovered from e-waste by one-dimensional covalent organic framework for efficient reduction of nitrophenols. Chemical Engineering Journal. 2025;170535.
[37] Fang J, Xu M, Tan Z, Guo J, Tao H, Ruan J. Bioaccessibility and health risk assessment of heavy metals in dust from electronic waste recycling in China. Process Safety and Environmental Protection. 2025;107550.
[38] Hu S, Wang H, Li X, He W, Ma J, Xu Y, Ming W. Recent advances in bioleaching and biosorption of metals from waste printed circuit boards: A review. Journal of Environmental Management. 2024;371:123008.
[39] Anaya-Garzon J, Hubau A, Joulian C, Guezennec AG. Bioleaching of e-waste: Influence of printed circuit boards on the activity of acidophilic iron-oxidizing bacteria. Frontiers in Microbiology. 2021;12:669738.
[40] Kim JY, Hwang JH, Lee T, Kim Y, Bae S, Kim D, Kwon EE. A review of sustainable metal recovery from e-waste via bioleaching: Current status and strategies for improvement. Chemical Engineering Journal. 2025;164149.
[41] Mourdikoudis S, Dominguez-Benetton X. Physicochemical vs electrochemical technologies for metal recovery: Insights, comparison, complementarity and challenges. Chemical Methods. 2025;5: e202400046.
[42] Adetunji AI, Oberholster PJ, Erasmus M. Bioleaching of metals from e-waste using microorganisms: A review. Minerals. 2023;13:828.
[43] Ji X, Yang M, Wan A, Yu S, Yao Z. Bioleaching of typical electronic waste-printed circuit boards: A short review. International Journal of Environmental Research and Public Health. 2022;19:7508.
[44] Kaur P, Sharma S, Albarakaty FM, Kalia A, Hassan MM, Abd-Elsalam KA. Biosorption and bioleaching of heavy metals from electronic waste varied with microbial genera. Sustainability. 2022;14:935.
[45] Good NM, Kang-Yun CS, Su MZ, Zytnick AM, Barber CC, Vu HN, Martinez-Gomez NC. Scalable microbial platform for rare earth element leaching and recovery from waste sources. Environmental Science & Technology. 2023;58:570–579.
[46] Arya S, Kumar S. Bioleaching: Urban mining option to curb the menace of e-waste challenge. Bioengineered. 2020;11:640–660.
[47] Waghmode MS, Gunjal AB, Patil NN. Bioleaching of electronic waste. Pollution. 2021;7:141–152.
[48] Zhang X, Shi H, Tan N, Zhu M, Tan W, Daramola D, Gu T. Advances in bioleaching of waste lithium batteries under metal ion stress. Bioresources and Bioprocessing. 2023;10:19.
[49] Rendón-Castrillón L, Ramírez-Carmona M, Ocampo-López C, Gómez-Arroyave L. Bioleaching techniques for sustainable recovery of metals from solid matrices. Sustainability. 2023;15:10222.
[50] Nagarajan N, Panchatcharam P. A comprehensive review on sustainable metal recovery from e-waste based on physiochemical and biotechnological methods. Engineering Science. 2023;22:844.
[51] Han Y, Shevchenko T, Qu D, Li G. Smart e-waste management in China: a review. In Congress on Intelligent Systems: CIS 2021 Proceedings, Volume 2 2022 (pp. 515–533). Springer Nature Singapore.
[52] Pineda-Vásquez T, Rendón-Castrillón L, Ramírez-Carmona M, Ocampo-López C. From e-waste to high-value materials: Sustainable synthesis of metal, metal oxide and MOF nanoparticles from waste PCB. Nanomaterials. 2023;14:69.

[53] Liu YH, Wiratama BS, Chao CJ, Wang MH, Chen RS, Saleh W, Pai CW. Unhelmeted riding, drunk riding and unlicensed riding among motorcyclists in Taiwan. International Journal of Environmental Research and Public Health. 2023;20:1412.
[54] Maluleke MD, Kotsiopoulos A, Govender-Opitz E, Harrison ST. Exploring microbial adaptation of immobilised acidophilic cultures to improve microbial oxidation rates and copper tolerance in e-waste bioleaching. Minerals Engineering. 2024;207:108560.
[55] Najim AA, Radeef AY, Al-Doori I, Jabbar ZH. Immobilization: Technique to protect and increase efficiency of microorganisms in contaminant removal. Journal of Chemical Technology and Biotechnology. 2024;99:1707–1733.
[56] Hobé TR, Vermeulen WJ. Analysing quality assurance levels in EPR-regulated recycling sectors. Circular Economy and Sustainability. 2024;4:3025–3057.
[57] Sahoo S, Bhol NK, Kanungo N, Majhi S, Pradhan J, Dandapat J. Artificial intelligence in bio-recovery of noble metals: A sustainable strategy for circular economy. International Journal of Bioinformatics and Intelligent Computing. 2025;4:101–128.

Tijo Cherian*, G. Balasubramani, Teena Merlin, R. Mohanraju, Jayalakshmi Sanal, Shehin Sadaka M.C., Shibin Eranhottu, Fahmeeda Parveen P.S.

BioGold rush: microbial extraction of metals from e-waste

Abstract: In the prevailing waste management environment, the mounting volumes of e-waste provide a substantial focus, mainly because of the complicated contents and components of electronic devices as well as their enormous manufacture and turnovers. Conventional e-waste treatment methods prioritize the recovery of metals through expensive, energy-intensive, and ecologically dangerous procedures like hydrometallurgical and pyrometallurgical methods, frequently ignoring other components of e-waste. Bio-based alternatives to these conventional methods are being studied more and more, as efforts are focused on developing a more circular and sustainable economic model. This critical assessment only takes into account research that employ e-waste sources and concentrates on recent developments in sustainable e-waste treatment. It critically examines methods utilizing live biomass, inactive biomass, and biogenic chemicals, highlighting the variety of ways and going over reaction variables, benefits and drawbacks, obstacles, and the possibility of valuing the byproducts produced. Bioprocessing techniques still have a lot of promise for recovering metal from e-waste, even though current research is concentrated on improving operational timeframes and metal recovery efficiencies. A more circular economy in the electronics sector is made possible by these strategies, which also lessen the environmental impact by using less energy and treating wastewater and metal recovery from complicated e-waste streams.

Keywords: Biogenic lixiviants, bioleaching, bioremediation, biosorption, e-waste

***Corresponding author: Tijo Cherian**, Post Graduate Department of Food and Industrial Microbiology, Bishop Kurialacherry College for Women, Amalagiri 686561, Kottayam, Kerala, India, e-mail: tvarghese891@gmail.com

Jayalakshmi Sanal, Post Graduate Department of Food and Industrial Microbiology, Bishop Kurialacherry College for Women, Amalagiri 686561, Kottayam, Kerala, India

G. Balasubramani, Department of Biotechnology, Saveetha Institute of Medical and Technical Sciences (SIMATS), SIMATS engineering, Thandalam, Chennai 602105, Tamil Nadu, India

Teena Merlin, School of Biosciences, Mar Athanasios College for Advanced Studies Tiruvalla (MACFAST-Autonomous), Pathanamthitta, Kerala, India

R. Mohanraju, Shehin Sadaka M.C., Shibin Eranhottu, Fahmeeda Parveen P.S., Department of Ocean Studies and Marine Biology, Pondicherry University, Port Blair, Andaman and Nicobar Islands, India

 | https://doi.org/10.1515/9783112225509-005

1 Introduction

More and innovative electronic devices are produced as a result of the extensive usage of electronic technologies and the growth of the electrical and electronic industries [1].A significant amount of "e-waste" is generated when these goods approach the culmination of their convenient lives. Over 50 million metric tons of -waste were produced in 2019, and 74 million metric tons are expected to be produced by 2030 [2]. Nonetheless, 9.3 metric tons of e-waste generated account for 17.4% of the total quantity of produced e-waste [3]. The United Nations divided e-waste into six categories, depending on its metallic constitution, kind, genesis, stage of development, producer, and component parts: telecommunication equipment/ small information technology, temperature exchange equipment, lamps, monitors and screens, large equipment, and small equipment [1]. However, as they interconnect various electrical components, PCBs (printed circuit boards) are essential in numerous technologies. Depending on their alignment and arrangement, PCBs can be one-sided, double-sided, or multidimensional [4]. About 0.5 metric tons of PCBs are produced annually, and the average increase in PCB production worldwide is 4%. Similarly, a comparatively large percentage of precious metals are found in computer central processing units [1].

Polymeric substances, ferrous and nonferrous metals, flame retardants, glass fibers, and other hazardous and/or nonhazardous materials make up the highly diverse makeup of e-waste, which can be hazardous to people and the environment, if not handled properly [5]. Furthermore, compared to some ores, e-waste has higher concentrations of a number of metals, rare earth metals, base metals, and valuable metals [6]. These metals' exceptional chemical resistance and electrical conductivity make them useful in a wide range of industries, including aviation, dentistry, jewelry, automobile, computers, and electronics. According to certain figures, e-waste contains between 10 and 1,000 g of gold/ton, which is seventeen times more than that of mineral ore [7]. In addition to preventing environmental deterioration and supporting the preservation of primary resources (ore), repossession of metal from e-waste (secondary source) aids in the shift to a circular economy. Because e-waste contains valuable metals in higher quantities than primary ore, it has been classified as secondary resource in urban mining. Therefore, it is imperative to create a novel technology for metal recovery from e-waste to increase the value addition of waste materials. Metals are extracted from e-waste using various techniques, such as hydrometallurgical, pyrometallurgical, and mechanical separation [8]. However, these methods are marked by significant energy and chemical consumption, high operating expenses, high metal losses, the production of secondary byproducts, and environmental risks [9]. Using microbial strains for metal recycling from e-waste is an alternate approach to account the aforementioned concerns.

A sustainable and green method, bioleaching (also called bio-hydrometallurgy), employs the physiological mechanisms of microbial world (archaea, fungus, and bacteria) to bio-extract metals from e-waste, mineral concentrates, or low-grade ores [1, 3, 10]. To put it another way, it is a process whereby microorganisms transform non-soluble metal complexes found in solid matrix into solution versions [2, 6]. It is regarded as an easy, efficient, low-cost, environmentally friendly, and energy-efficient approach [1]. Additionally, bioleaching makes it easier to recover metals during the recycling operation and produces less secondary waste. However, substantial impediments to the extensive usage of bioleaching include low metal yields and long leaching times. Microbe-secreted metabolites (or lixiviants) help to solubilize and marshal insoluble metals from e-waste. The bacterial physiology and mineralogical composition of e-waste are the principal features that govern the efficiency of leaching [1]. Thus, the methods used by microbes for metals recovery from e-waste are the core theme of this chapter. Furthermore, methods for economic mobilization of metallic substances from e-waste are described in detail, along with bioprocess factors that affect the effectiveness of metal bioleaching.

2 Effects of e-waste on environment and human health

Human health and environment are seriously threatened when e-waste is improperly disposed of in dumpsites or landfills [11]. A variety of toxic inorganic and organic contaminants, including heavy metals, dioxins, brominated flame retardants, polychlorinated biphenyls, and polybrominated diphenyl ethers are found in e-waste and are a menace to ecological systems, plants, and animals. These materials are discharged either directly or indirectly as hazardous compounds, acids, and heavy metals. Exposition directly can happen by ingestion of contaminated dust, dermal contact with perilous materials, or inhalation of tiny and coarse particles [12]. As e-waste advances through the food chains and emanates with human contact, its concentration increases and can cause a number of wellness issues, like DNA damage, birth defects, heart failure, cancer, and skin rashes [12, 13]. Additionally, when e-waste is indecorously disposed of, contaminants (such as dioxins) or dust particles are released into the atmosphere, causing air contamination and adverse effects on respiratory system [1]. Furthermore, combusting e-waste releases harmful chemicals and fumes that contaminate the surrounding air. E-waste comprises unsafe constituents that seep into the earth and pollute surface and subsurface water, acidifying and poisoning it. This destroys freshwater and marine life, affects biodiversity, damages environmental systems, and alters the biological structure of soil [3, 12].

3 Metal bioleaching from e-waste

In a process called bioleaching, microorganisms are essential to the elimination of metal from e-waste [14] (Figure 1). Cyanogenic bacteria, fungi, iron and sulfur oxidizers are among the few microorganisms that participate in metal bioleaching from e-waste [15, 16]. The microbial contenders for the recovery of metals from e-waste typically fall into three categories: fungus, heterotrophic bacteria, and chemolithotrophic bacteria. Below are a detailed discussion of these bacteria and their leaching productiveness.

3.1 Bioleaching via heterotrophic microbes

Heterotrophic microorganisms, such as fungus and bacteria, use water and organic carbon sources as sources of energy and electrons [3]. This category of organisms can also be used to treat moderately alkaline waste products since they can withstand a wider pH range and complicated metals [17].

3.1.1 Fungal-mediated bioleaching

Metals from e-waste are mobilized by massive volumes of organic acids secreted by heterotrophic group of fungi, including lactic acid, gluconic acid, carboxylic acid, and citric acid [1]. This is accomplished through the processes of bioaccumulation, redoxolysis, acidolysis, chelation, and complex formation that occur when organic acids react with metals [18]. Furthermore, the reduction of extremely oxidized metallic composites in the context of specific biologically degradative extracellular enzymes may possibly be the cause of metal solubilization by fungus [1]. Fungi are a superior option for mineral bioleaching, owing to their capacity to acclimatize and resist ecological stressors like contamination by metals. Considering they have bioconversion reactions in their metabolic machinery, fungal taxa such *Phanerochaete*, *Saccharomyces*, *Candida*, *Trichoderma*, *Aspergillus*, and *Penicillium* are well known for their capability to mobilize metals from e-waste [3, 16, 18]. Patel and Lakshmi [19] examined the capacity of *Aspergillus fumigatus* A2DS to mobilize metals from PCBs of mobile phones. The two variable species of *Aspergillus nigrum* and *Rhizopus oligosporus* reported leaching efficiencies of 46.92% (copper) and 30.63% (lead) to 8.53% (copper) and 19.61% (lead), respectively, from PCB garbage from a recycling outlet [20]. The species of *Saccharomyces cerevisiae* recovered 98% gold (III) ion from a central processing unit aqua regia leachate in 10 min [21]. Liu et al. [22] documented leaching efficiency (60.96%) of *Phanerochaete chrysosporium* in the case of copper metal from waste PCBs at 30 °C (150 rpm) for 2 weeks, 2% pulp density, and pH 5. Arshadi et al. [23] studied metal bioleaching from PCBs by using *Penicillium simplicissimum*, with recovery efficiencies of nickel (89%) and copper (90%) at pH values of 2 and 7, 150 rpm, 30°C, 10^6 and 10^7 spore concentrations, respectively. Cui et al. [24] demonstrated that inoculation of *Aspergillus niger* in an optimal growth medium (pH 7.0), amended with indium tin oxide (1% w/v) for 2 weeks at 30 °C (200 rpm), led to the retrieval rate of 100% indium from waste liquid crystal display panels. The maximal leaching capabilities for lithium (100%), manganese (72%), nickel (45%), cobalt (38%), copper (94%), and aluminum (62%) were observed at 30 °C for the period of a month in the case of bioleaching efficacy from wasted lithium-ion cell phone batteries by *Aspergillus niger* [25].

A collective of heterotrophic fungus improves the secretion of organic acid to mobilize metals from e-waste [18]. The microbial combination of *Aspergillus tubingensis* and *Aspergillus niger* was used to recover 72% nickel, 77% aluminum, 95% manganese, 95% lithium, and 52% cobalt from mobile phone batteries at 140 rpm, 30 °C, and pulp density of 10 g/L [26]. The metal bioleaching efficacy was found to be >95% for the metals of nickel, lead, zinc, and copper under the cocktail of *Penicillium simplicissimum* and *Aspergillus niger* at conditions of 30 °C, 150 rpm, pH 6, and 1–10% (w/v) pulp density [27]. Metals were biorecovered from nickel-cadmium batteries by a consortium of *Aspergillus flavipes*, *A. fumigatus*, *A. japonicus*, *A. versicolor*, *A. tubingensis*, and *A. niger*, with leaching efficiencies of 90% (for cobalt), and 80% (each for zinc, cadmium, and nickel) at particle size <200 μm, 150 rpm, 27 °C, pH 5, and 1 g/L pulp density [28]. Madrigal-Arias et al. [29] reported leaching efficiency of 87% (gold) at 14 days by employing a combination of *A. niger* MX7 and *A. niger* MXPE6 in metal repossession from cell phone PCBs.

3.1.2 Bacteria-mediated bioleaching

Heterotrophic bacteria synthesize byproducts that aids in mobilization of metals from e-waste. Cyanogenic bacteria produce HCN (hydrogen cyanide) to extract metals of value from e-waste by using the process of cyanogenesis [1]. These types of organisms use oxidative decarboxylation to produce cyanide under alkaline circumstances using glycine as a precursor between exponential and stationary phases, facilitated by HCN synthase enzyme. The metabolite extracts valuable metals via the foundation of soluble metal-cyanide composite [1]. However, e-waste toxicity to cyanogenic bacteria and low cyanide outputs from organisms hinder extensive recovery. Also, dissolved oxygen content, pH, temperature, and glycine all impact microbial ability to produce cyanide [1]. Cyanogenic bacteria effectively extract valuable metals, including titanium palladium, platinum, silver, and gold from e-waste. Base metals are collected from e-waste before precious metals are bioleached using heterotrophic organisms. These metals form complexes with metabolites, reducing the mobilization of valuable metals. The cyanide-producing strains of *Pseudomonas* sp., *Bacillus megaterium*, and *Chromobacterium violaceum* are well-known players in the process of metal bioleaching.

Li et al. [30] experimented bioleaching efficiency (70.6% gold) of *C. violaceum* ATCC 12471 from a pretreated waste PCB. Chi et al. [31] employed *C. violaceum* in the investigation of metal bioleaching from waste mobile PCBs, inferring improved recovery of copper (4.9–11.4%) and gold (7.78–10.8%) from solid matrix under conditions of increased pH (8–11), 150 rpm, pulp density of 15 g/L, and 30 °C within the time interval of 8 days. Furthermore, the strain of *Pseudomonas balearica* SAE1, isolated from an e-waste recycling facility in India, reported recovery efficiencies of 33.8% (silver) and 68.5% (gold) at the optimal conditions of glycine (5 g/L), pulp density (10 g/L), pH 9, and 30 °C [20]. Similarly, the strain of *Pseudomonas putida* WSC361, isolated from e-waste shredding dust, reported 48% (gold) recovery efficiency under adjusted conditions of 150 rpm, 10 g/L pulp density, 10 g/L glycine, and pH 7.2–8.4 within 3 h [32]. The mutated strain of *Pseudomonas fluorescens* CICC21620 bioleached gold with recovery efficiency of 54% from waste PCB under standardized settings of (3.33% v/v) bacterial density, 1:10 glycine–methionine ratio, 0.133 mol/L glycine, 0.33% (w/v) pulp density, and pH 9 [9]. The strain of *B. megaterium* PTCC 1656 exhibited maximal percentage retrieval of rhenium (98%) and platinum (15.7%) under conditions of pulp densities of 4% (w/v) and 12.8 g/L, respectively, and optimal concentration of glycine [33]. The consortium of cyanogenic bacteria improves overall efficiency of metal bioleaching. The combined microbial action of *P. fluorescens* and *C. violaceum* reported recovery efficiencies of 49% (zinc), 73% (gold), 13% (iron), 8% (silver), and 83% (copper) from computer e-waste, under optimized conditions of 150 rpm, 30 °C, and pH 7.2 [34]. The co-culture of *B. megaterium* and *P. putida* (cyanide-producing bacterial strains) mobilized 83.59% gold from waste PCBs under 34 h leaching time, 5 g/L pulp density, and pH 10 [35]. The combination of *P. fluorescens* and *C. violaceum* reported gold bioleaching retrieval proficiency of 8% from e-waste under 2% and 4% (w/v) pulp densities [36].

3.2 Bioleaching via chemolithotrophic microbes

These microbial strains are also known as iron- and sulfur-oxidizing bacteria, or acidophilic bacteria. Metal bioleaching involves the oxidation of ferrous ions to ferric ions and elemental sulfur to sulfuric acid [37]. Figure 1 explains the bioleaching process. Biogenic sulfuric acid and ferric iron act as lixiviants (oxidizing agents), mobilizing base metals from the solid matrix through redoxolysis and acidolysis bioleaching mechanistic pathways [1]. Microbes use inorganic molecules like elemental sulfur, reduced sulfurous compounds, Fe^{2+} and S_8, $S_2O_3^{2-}$, H_2S, or polysulfide as energy source, and CO_2 from the atmosphere as carbon source. This assembly of microbes stands out from others in evaluating the ability to remove metals from e-waste via bioleaching and bio-oxidation [38]. Chemolithotrophic bacteria can be classified as extremely thermophilic (60–80 °C), moderately thermophilic (40–60 °C), and mesophilic (28–37 °C), depending on their preferable temperature: *Thermoplasma acidophilum, Sulfobacillus thermosulfidooxidans*, *Leptospirillum ferriphilum*, *Thiobacillus thiooxidans*, *Acidithiobacillus ferrooxidans*, and *Acidithiobacillus thiooxidans* [9, 40]. Most of these bacteria can endure higher metal concentrations (molybdenum, uranium, and silver) and lower pH levels [1]. However, the representatives of genus *Acidithiobacillus* are notable extensively researched organisms for metal bioleaching, primarily because of their low nutritional needs for metal solubilization and their exceptional tolerance to heavy metal toxicity [39]. The strains of *Acidiphilum acidophilum* and *A. thiooxidans* reported bioleaching efficiency of 75–100% (copper) from computer PCBs in 2.5 h and 9 days, respectively [40]. The recovery efficiencies of 54% nickel and 100% iron and copper were reported from the metal biorecovery of electronic waste combination under standardized conditions of 130 rpm agitation speed, 30°C, 10% v/v inoculum size, and 15 g/L pulp density [23]. The metal bioleaching process from discarded mobile phone PCBs extracted 82% (nickel) and 98% (copper) in the presence of *A. thiooxidans* [41]. At optimal conditions of incubation period (6 days), initial pH (0.5), and $FeSO_4 \cdot 7H_2O$ (60 g/L), the strain of *Acidithiobacillus ferrooxidans* bioleached 72% zinc, 74% nickel, and 76% copper from waste PCB sludge [42]. Arshadi and Yaghmaei [43] studied the effectiveness of *Acidithiobacillus ferrooxidans* in mobilizing metals like nickel (88%) and copper (90%) from waste PCBs obtained from a recycling center, Pars Charkhesh Asia, Tehran, Iran.

Utilizing the consortia of chemolithotrophic organisms or mixed cultures have reported to deliver more beneficial outcomes than monocultures [17]. The joint bioleaching activity of *A. acidophilum* and *Acidithiobacillus ferrooxidans* on waste computers PCBs resulted in the biorecovery of 74.5% lead, 75% nickel, 94.5% zinc, and 96% copper under the leaching time of 18 days and 7.5 g/L pulp density [44]. The adapted co-cultures of acidophilic strains of *Acidithiobacillus ferrooxidans* and *A. thiooxidans* demonstrated 50.4% cobalt, 89.4% nickel, and 99.2% lithium retrieval efficiencies from spent lithium-ion laptop batteries, under conditions of 36.7 g/L $FeSO_4$, 5 g/L sulfur, pH 1.5, 40 g/L pulp density, and <75 m particle size [45]. The consortium of *Ferroplasma* sp., *Sulfobacillus* sp., *L. ferriphilum*, and *Acidithiobacillus caldus* experimented recovery efficiencies of 84% lithium, 99% copper and nickel at 130 rpm, and 45°C from lithium-ion batteries [46]. The mixture of *S. thermosulfidooxidans* and *L. ferriphilum* recovered 94% copper by bioleaching 100 g/L PCB concentrates from waste PCBs in the duration of 9 days [1]. The combination blend of *A. thiooxidans* and *Acidithiobacillus ferrooxidans*, under conditions of 180 rpm agitation speed, 125 µm particle size, 30 ± 2 °C, leaching time 10 days, initial pH 1.8, 10% pulp density, and 10% (v/v) inoculum volume, reported recovery efficiencies of 59% aluminum, 88% zinc, 89% nickel, and 94% copper from discarded PCBs [47].

Figure 1: Examples of bioleaching applications (reproduced from [6]).

4 Effective strategies for metal bioleaching from e-waste

A wide variety of microorganisms use different methods to remove metals from solid waste. Most microorganisms, including fungi and bacteria, employ the following processes to regulate their actions.

4.1 Bioaccumulation

It entails the metal absorption from a bioleaching media via microbial outer membranes [16]. Metal ions build up in the cell vacuole as a result of the intracellular mobilization of soluble metals across the cell membrane in this microbe–metal interaction [18]. Metal dissolution from a solid matrix is aided by the attendance of explicit functional moieties on the cell wall, such as hydroxyl, carboxyl, amine, phosphate, etc. The metal ions attach to the functional groups and act as a cation exchanger. Fungal mycelium uses both passive adsorption and active biosynthetic processes to solubilize metal ions [18].

4.2 Biosorption

Microorganisms accumulate metals from the bioleaching media by biosorption [18]. It is a metabolically independent physico-chemical adsorption process that enables removal of metals from e-waste by ion-exchange, coordination, chelation, and complexation between metals using explicit types of inactive, lifeless, or living microbial biomass [17]. Numerous functional groups found in the microbial cell wall, such as phosphate, sulfonate, phosphodiester, and carbonyl groups, are necessary in the metal sequestration found in e-waste [48]. Heavy metals, valuable metals, and rare earth metals have been retrieved from e-waste, employing the biomasses of many algal, bacterial, and fungal species as biosorbents [1]. Because of huge surface-area-to-volume ratio of fungal hypha, quick development under regulated settings, and adaptability to environmental conditions, microorganisms are thought to be superior metal-absorbing agents [49]. Di Piazza et al. [50] experimented metal concentrations of lanthanum (390 ppm) and terbium (1,520 ppm) post treatment of waste electronic and electrical apparatus by *Penicillium expansum*. Metals from PCBs were bioleached using the cell-free supernatant of *S. thermosulfidooxidans* and *L. ferriphilum*, reporting copper retrieval proficiency of 100% after 2 h [1].

4.3 Complexolysis (or chelation)

The target metals from a solid matrix associate with ligands to create siderophores, organic acids, and cyanides, which aid in metal utilization from waste. Metals, including palladium, platinum, iron, gold, and silver, can be extracted by complexolysis, which proceeds after acidolysis [51]. Microbes secrete iron-chelating chemicals, which act as a solubilizing agent to retrieve Fe^{3+}. Furthermore, during metal mobilization, fungal organic acids increase the complexing capacity and generate protons [18]. *Aspergillus niger, Pseudomonas aeruginosa, B. megaterium, P. fluorescens, C. violaceum*, and other heterotrophic bacteria and fungi share this process [1].

4.4 Redoxolysis

Chemolithotrophic microbes such as *A. thiooxidans, Acidithiobacillus ferrooxidans*, and *Leptospirillum ferrooxidans* frequently use redoxolysis, an oxidation–reduction method, enabling the translation of insoluble metallic form to conforming soluble metals [1]. Fe^{3+} acts as an oxidizing agent during the bioconversion process, undergoing reduction to Fe^{2+} for metal mobilization, including copper, zinc, lead, and other metals, for recovery from e-waste [1].

4.5 Acidolysis

Acidolysis is the process of mobilizing metals from a solid matrix, mediated by protonation of oxygen atoms in metal oxides. Oxygen atoms form bonds with water molecules to extract metals from their source and incorporate them into a bioleaching solution [3]. Pyruvic acid, formic acid, gluconic acid, and other organic acids facilitate the protonation of oxygen atoms. The organic acids are necessary to maintain the low pH needed for an improved process of bioleaching [18]. Additionally, the acid increases solubility of metal ions from e-waste by decreasing the accessibility of anions to cations in the reaction mixture. Copper, nickel, zinc, and lead have all been extracted using this quick and common process found in heterotrophic fungi and bacteria [1].

5 Techniques for metal biorecovery from e-waste

When microorganisms are present, metals are extracted from e-waste using the following techniques:

5.1 One-step bioleaching

In this method, the e-waste and a freshly cultivated microbial culture are introduced in a sterilized bioleaching mixture under sterile environments and by shaking [1]. The bioleaching procedure is conducted under ideal bioprocess conditions that support organism growth, biolixiviant secretion, and metal biorecovery from waste materials. However, the toxic effect of e-waste is a significant obstacle to the biorecovery process since it hinders organism growth and lowers metabolite yields, which in turn decreases efficiency of metal recovery [4].

5.2 Two-step bioleaching

This method involves cultivating microbes aseptically, without the use of e-waste, under stirring, and ideal and suitable conditions of bioprocessing and fresh culture medium until the organism reaches maximal cellular progression and production of biolixiviant [1]. Following this, the stage 2 involves adding sterilized e-waste to the culture medium at appropriate particle size and pulp density, followed by additional incubation for a specific amount of time [52]. This method increases the organisms' ability to leach metals by reducing the e-waste-mediated inhibitory effects on metabolite secretion and microbial growth [1]. It can also be carried out at an increased pulp density and is quick. Several researches have documented the applicability of two-step bioleaching as a favored technique in the precious-metal retrieval from e-waste [1].

5.3 Spent-medium bioleaching

Using this technique, an organism is cultivated in a culture medium, under shaking without e-waste, until it reaches its maximum density of cells and production of biolixiviant. After that, e-waste is introduced in a cell-free supernatant that was generated by filtering or centrifuging the cultured broth, and it is then incubated for a longer time period [1]. The metabolites in the supernatant make it easier to recover metals from e-waste. In this method, the organisms are not in adjacent proximity to the e-waste, in contrast to single- and two-step bioleaching. Since it permits larger volumes of e-waste at higher pH without taking into account the harmful effects of waste on microorganisms, this process has been identified as the most successful method [53]. Furthermore, it prevents adverse reactions between metabolites and organisms [1].

6 Impact of bioprocess variables on metal bioleaching

The physiological characteristics of the organisms and the accessibility of appropriate physical and chemical characteristics at the ideal circumstances, necessary to enable microorganism development and metabolite secretion, are key factors in the metal biorecovery from e-waste [32]. The efficiency of bioleaching techniques – spent-medium, two-step, or one-step bioleaching – as well as the modus operandi – shake-flask, bioreactor, heap, or column operation – are additional variables that impact metal bioleaching [6, 123]. Both individually and collectively, these factors affected the bioleaching efficiency [124, 125]. Below is a discussion of some of the distinct variables that influence bioleaching:

6.1 Leaching substrate

E-waste's mineralogical and elemental makeup are important determinants of bioleaching effectiveness. This establishes the kinds and concentrations of metals in the solid matrices [16]. Pretreatment for the elimination of copper is essential in the valuable metal biorecovery from e-waste by cyanogenic organisms since it can form a stable complex with cyanide, lowering the yield of free cyanide and ultimately preventing the bioretrieval of precious metals [1]. *Acidithiobacillus ferrooxidans*, hydrogen peroxide, or nitric acid is used in this procedure. Furthermore, the particle size of e-waste influences the microbial growth and biorecovery of metals from the solid matrix and is often measured in millimeters or micrometers. As particle size decreases, bioleaching rises.

6.2 Microorganisms: type and physiology

Metals from e-waste are recovered using different microbes (fungi and bacteria) or their metabolic products. These organisms are divided into chemolithotrophic and heterotrophic categories, according to their energy and carbon needs. Depending on their temperature requirements, the former are further classified as extreme thermophiles, moderate thermophiles, and mesophiles [3]. Additionally, the microorganisms can be used in the bioleaching process as a consortium, mixed cultures, or pure cultures. The success of these organisms in the process of bioleaching depends on their ideal metal tolerance, inoculum volume, and adaption to the metallic environment and toxic nature of e-waste [54].

6.3 Oxygen availability

During metal bioleaching, microbial growth and metabolic activity depend on the availability of oxygen. Microbes (aerobes) receive oxygen through stirring/shaking that is measured in rpm (revolutions per minute) [55]. For a bioleaching experiment, the optimal shaking speed is typically between 130 and 170 rpm [13]. Nevertheless, excessive agitation causes microbial cell friction, abrasion, or rupture, which lowers the efficiency of metal biorecovery [54]. The oxidation of elemental sulfur and ferrous ions can be delayed by a low oxygen supply, which lowers the efficacy of bioleaching [56].

6.4 Culture media

The proficiency of leaching is determined by the chemistry of culture medium employed to cultivate microorganisms, since it affects the metabolic processes of the organisms. Microbes require inorganic and organic nutrients in this medium to thrive and produce metabolites. To recover valuable metals from e-waste, cyanogenic microorganisms are grown in a growth medium that contains organic materials like amino acids, glycine, yeast extract, peptone, etc. [57]. Chemolithotrophic organisms thrive in a chemically defined media that contains elements such as elemental sulfur, iron sulfate, magnesium sulfate, potassium hydrogen phosphate, and ammonium sulfate.

These nutrient compounds serve as a cofactor for the biosynthetic processes that are catalyzed by enzymes and result in the production of metabolites that simplify the extraction of metals from electronic waste [1].

6.5 Liquid-to-solid ratio

The liquid-to-solid ratio is another name for pulp density. It is the quantity of an e-waste-containing solid matrix in a solution. A low pulp density is indicated by a larger liquid-to-solid ratio [58]. The metabolic and growth activities of organisms are impacted by high pulp density, which adds to the cytotoxicity of e-waste. To put it another way, a higher pulp density lowers the leaching efficiency, restricts oxygenic mass-transfer rate, and prevents metabolite production and microbial growth [39]. The species of *B. megaterium*, *Roseovarius tolerans* DSM11457, and *C. violaceum* have reported optimal pulp densities of 0.5%, 1%, and 4%, respectively, in the efficient biorecovery of valuable metals from e-waste [33].

6.6 Temperature

Temperature affects the bioleaching process's performance since every organism has a preferred temperature for the development and production of metabolites [1]. For example, the ideal temperature ranges for several acidophilic and cyanogenic bacteria and fungi are 25–35 °C, respectively [1, 18]. On the other hand, thermophilic bacteria can be used for bioleaching at high temperature between 50 and 80 °C [1].

6.7 pH

The pH of bioleaching medium has a significant impact on the recovery of metals from e-waste because it influences the generation of biolixiviant and the development of microorganisms. However, the organisms' development, metabolic processes, and leaching efficiency may be negatively impacted by a variation from their ideal pH requirement [58]. Various groups of microorganisms have various pH requirements. For example, bacteria that oxidize iron and sulfur prefer pH values between 2.0 and 2.5, where the concentration of protons is abundant [1]. An alkaline pH is ideal for the development of cyanogenic organisms.

7 Limitations in metal bioleaching from e-waste

Microorganisms are an effective and promising method of metal bioleaching from e-waste. However, there are certain restrictions associated with this technology; hence, commercialization has not yet been developed. Among them are the following:

1. One major barrier to the industrial metal recovery of from e-waste is the effectual and affordable metal retrieval and recuperation from bioleaching medium.
2. One significant disadvantage that impacts the commercial large-scale metal retrieval from e-waste is the lengthy bioleaching processes.
3. The bioleaching process faces significant challenges due to the noxiousness of e-waste, which hinders the development of organisms that produce metabolites. Traditionally, microorganisms with a high tolerance to hazardous metal ions are chosen by sophisticated identification of microbial strains. Additionally, before waste materials are added to the bioleaching medium, the microorganisms are permitted to reach specific growth phase in which optimal biolixiviant production and maximum cell density are reached. Additionally, autochthonous organisms found in waste can be used to counteract the noxiousness of e-waste to microbial cells. Since these microorganisms are thought to function better in their home settings than in exotic ones, investigating indigenous organisms may be very important for metal bioleaching.
4. As bio-lixiviant quantities are consequently quite low, metal recovery from e-waste utilizing microorganisms is less effective than chemical leaching. To improve metal dissolving, leaching parameters and processing conditions must be changed.
5. The generation of metabolites (such as cyanide and organic acids) by microbial strains may be restricted because various organisms require different pH levels for growth and lixiviant production. To put it another way, a successful metal bioleaching experiment is hampered by different pH requirements for metabolite production and microbial growth. As a result, the pH must be optimized for the generation of metabolites, while yet allowing the organisms to proliferate. Which can be accomplished through the usage of metabolically modified organisms that have high pH tolerance or bioreactors, statistical experimental designs (ANN or RSM).

8 Conclusion and futuristic perspectives

The irresponsible management of e-waste, particularly after the items have reached the culmination of their convenient lives, has resulted in environmental and health risks due to the rapid global expansion of the electrical and electronic sectors as well as the constant need for electronic and electrical equipment. Moreover, e-waste is an intricate substance that contains a lot of metallic components. A sustainable and viable method of valuing e-waste is bioleaching, which recovers precious metals for usage in a diversity of industrial applications. The existence of ideal and suitable bioprocesses variables for the massive metal extractions from e-waste governs the microbial performance. For handling of e-waste, innovative and sustainable technologies are required to improve the metal bioleaching efficacy. The following are some of these creative methods for further research:

1. For the effective metal recovery during bioleaching progression, microbial cells can be immobilized on synthetic (polyurethane, polyvinyl, or polypropylene) or natural (alginate agar, or cellulose,) support materials. Additionally, immobilizing microorganisms on an appropriate carrier makes them more stable across a wide variety of pH and temperature conditions and allows them to be reused with lower operating expenses.
2. The invention of unique physico-mechanical methodology for the dichotomy of potentially dangerous aspects of waste components from benign elements can lower the noxious level of e-waste preceding the bio-extraction processes.
3. Further research should focus on the use of hybrid technology for effective and improved metal recovery from e-waste due to the complexities of e-waste, which contains a range of metals. In order to effectually recover metals from e-waste, an assortment of leaching machineries, including chemical, biological, and physical processes, must be cohesive and unified.
4. Organic sources of carbon, for example, agro-based manufacturing wastes like mango peels, straw, rice bran, corncobs, etc., should be encouraged as an affordable and ecologically advantageous medium for microbial cell growth and improved metabolite secretions in the biorecovery of metal from e-waste.
5. For bioleaching of metal from e-waste, further researches are needed to properly understand the mechanistic steps of action of consortium/mixed microbial strains, synergistic interactions, and community distribution.
6. A healthier option for improved recovery of metal from e-waste may be to investigate the leaching capacity of unknown or genomically altered microbes, such as thermophilic fungus, archaea, and bacteria, under various bioprocessing settings. In comparison to wild-type strains, the genetically modified organisms are extremely effective at recovering metals and quickly adjust to changes in their surroundings. It entails increasing the gene expression that code for the synthesis of biolixiviants.
7. An effective and cost-effective scheme for fashioning a successful comprehensive bioleaching process is to use appropriate concentrations of nonmetal species (like quartz and activated carbon) or metal ions (such as Ag^{+}, Cu^{2+}, Hg^{2+}, Co^{2+}, and Bi^{3+}) catalysts in the bioleaching medium to improve the slow kinetics of recovery of metal from e-waste. Furthermore, the cocktail application of these catalysts in metal bioleaching from e-waste need to be promoted since it may lead to increased catalysis, which would improve microbe-mineral interaction and ultimately increase metal recoveries.

References

[1] Adetunji AI, Oberholster PJ, Erasmus M. Bioleaching of metals from e-waste using microorganisms: a review. Minerals. 2023;13(6):828. https://doi.org/10.3390/min13060828.

[2] Al-Salem SM, Leeke GA, El-Eskandarany MS, Van Haute M, Constantinou A, Dewil R, Baeyens J. On the implementation of the circular economy route for E-waste management: A critical review and an analysis for the case of the state of Kuwait. Journal of Environmental Management. 2022;323:116181.

[3] Dutta D, Rautela R, Gujjala LK, Kundu D, Sharma P, Tembhare M, Kumar S. A review on recovery processes of metals from E-waste: A green perspective. Science of the Total Environment. 2023;859:160391.

[4] Magoda K, Mekuto L. Biohydrometallurgical recovery of metals from waste electronic equipment: current status and proposed process. Recycling. 2022;7(5):67.

[5] Dutta D, Goel S, Kumar S. Health risk assessment for exposure to heavy metals in soils in and around E-waste dumping site. Journal of Environmental Chemical Engineering. 2022;10(2):107269.

[6] Zhang X, Shi H, Tan N, Zhu M, Tan W, Daramola D, Gu T. Advances in bioleaching of waste lithium batteries under metal ion stress. Bioresources and bioprocessing. 2023;10(1):19. https://doi.org/10.1186/s40643-023-00636-5.

[7] Van Yken J, Boxall NJ, Cheng KY, Nikoloski AN, Moheimani NR, Kaksonen AH. E-waste recycling and resource recovery: A review on technologies, barriers and enablers with a focus on oceania. Metals. 2021;11(8):1313.

[8] Meng L, Wang Z, Zhong Y, Guo L, Gao J, Chen K, Cheng H, Guo Z. Supergravity separation for recovering metals from waste printed circuit boards. Chemical Engineering Journal. 2017;326:540–550.

[9] Li J, Wen J, Guo Y, An N, Liang C, Ge Z. Bioleaching of gold from waste printed circuit boards by alkali-tolerant *Pseudomonas fluorescens*. Hydrometallurgy. 2020;194:105260.

[10] Abdel Azim A, Bellini R, Vizzarro A, Bassani I, Pirri CF, Menin B. Highlighting the role of archaea in urban mine waste exploitation and valorisation. Recycling. 2023;8(1):20.

[11] Adetunji AI, Olaniran AO. Treatment of industrial oily wastewater by advanced technologies: a review. Applied Water Science. 2021;11(6):98.

[12] Li W, Achal V. Environmental and health impacts due to e-waste disposal in China – A review. Science of the Total Environment. 2020;737:139745.

[13] Rautela R, Arya S, Vishwakarma S, Lee J, Kim KH, Kumar S. E-waste management and its effects on the environment and human health. Science of the Total Environment. 2021;773:145623.

[14] Nasiri T, Mokhtari M, Teimouri F, Abouee E. Remediation of metals and plastic from e-waste by iron mine indigenous acidophilic bacteria. Waste Management & Research. 2023;41:894–902.

[15] Argumedo-Delira R, Díaz-Martínez ME, Gómez-Martínez MJ. Microorganisms and plants in the recovery of metals from printed circuit boards of computers and cell phones: A mini review. Metals. 2020;10:1120.

[16] Trivedi A, Vishwakarma A, Saawarn B, Mahanty B, Hait S. Fungal biotechnology for urban mining of metals from waste printed circuit boards: A review. Journal of Environmental Management. 2022;323:116133.

[17] Ji X, Yang M, Wan A, Yu S, Yao Z. Bioleaching of typical electronic waste-printed circuit boards (WPCBs): A short review. International Journal of Environmental Research and Public Health. 2022;19:7508.

[18] Dusengemungu L, Kasali G, Gwanama C, Mubemba B. Overview of fungal bioleaching of metals. Environmental Advances. 2021;5:100083.

[19] Patel F, Lakshmi B. Bioleaching of copper and nickel from mobile phone printed circuit board using *Aspergillus fumigatus* A2DS. Brazilian Journal of Microbiology. 2021;52(3):1475–1487.

[20] Netpae T, Suckley S. Bioleaching of Cu and Pb from printed circuit boards by *Rhizopus oligosporus* and *Aspergillus niger*. Environmental & Experimental Biology. 2019;17(2).

[21] Saitoh N, Fujimori R, Nakatani M, Yoshihara D, Nomura T, Konishi Y. Microbial recovery of gold from neutral and acidic solutions by the baker's yeast *Saccharomyces cerevisiae*. Hydrometallurgy. 2018;181:29–34.

[22] Liu Q, Bai JF, Gu WH, Peng SJ, Wang LC, Wang JW, Li HX. Leaching of copper from waste printed circuit boards using *Phanerochaete chrysosporium* fungi. Hydrometallurgy. 2020;196:105427.

[23] Arshadi M, Yaghmaei S, Mousavi SM. Study of plastics elimination in bioleaching of electronic waste using *Acidithiobacillus ferrooxidans*. International Journal of Environmental Science and Technology. 2019;16(11):7113–7126.

[24] Cui J, Zhu N, Mao F, Wu P, Dang Z. Bioleaching of indium from waste LCD panels by *Aspergillus niger*: Method optimization and mechanism analysis. Science of the Total Environment. 2021;790:148151.

[25] Bahaloo-Horeh N, Mousavi SM, Baniasadi M. Use of adapted metal tolerant *Aspergillus niger* to enhance bioleaching efficiency of valuable metals from spent lithium-ion mobile phone batteries. Journal of Cleaner Production. 2018;197:1546–1557.

[26] Alavi N, Partovi K, Majlessi M, Rashidi M, Alimohammadi M. Bioleaching of metals from cellphones batteries by a co-fungus medium in presence of carbon materials. Bioresource Technology Reports. 2021;15:100768.

[27] Brandl H, Bosshard R, Wegmann M. Computer-munching microbes: Metal leaching from electronic scrap by bacteria and fungi. Hydrometallurgy. 2001;59:319–326.

[28] Kim MJ, Seo JY, Choi YS, Kim GH. Bioleaching of spent Zn-Mn or Ni-Cd batteries by *Aspergillus* species. Waste Management. 2016;51:168–173.

[29] Madrigal-Arias JE, Argumedo-Delira R, Alarcón A, Mendoza-López MR, García-Barradas O, Cruz-Sánchez JS, Ferrera-Cerrato R, Jiménez-Fernández M. Bioleaching of gold, copper and nickel from waste cellular phone PCBs and computer goldfinger motherboards by two *Aspergillus niger* strains. Brazilian Journal of Microbiology. 2015;46(3):707–713.

[30] Li J, Liang C, Ma C. Bioleaching of gold from waste printed circuit boards by *Chromobacterium violaceum*. Journal of Material Cycles and Waste Management. 2015;17:529–539.

[31] Chi TD, Lee JC, Pandey BD, Yoo K, Jeong J. Bioleaching of gold and copper from waste mobile phone PCBs by using a cyanogenic bacterium. Minerals Engineering. 2011;24:1219–1222.

[32] Marra A, Cesaro A, Rene ER, Belgiorno V, Lens PNL. Bioleaching of metals from WEEE shredding dust. Journal of Environmental Management. 2018;210:180–190.

[33] Motaghed M, Mousavi SM, Rastegar SO, Shojaosadati SA. Platinum and rhenium extraction from a spent refinery catalyst using *Bacillus megaterium* as a cyanogenic bacterium: Statistical modeling and process optimization. Bioresource Technology. 2014;171:401–409.

[34] Pradhan JK, Kumar S. Metals bioleaching from electronic waste by *Chromobacterium violaceum* and Pseudomonads sp. Waste Management & Research. 2012;30:1151–1159.

[35] Zhou G, Zhang H, Yang W, Wu Z, Liu W, Yang C. Bioleaching assisted foam fractionation for recovery of gold from the printed circuit boards of discarded cellphone. Waste Management. 2020;101:200–209.

[36] Pharm VA, Ting YP. Gold bioleaching of electronic waste by cyanogenic bacteria and its enhancement with bio-oxidation. Advanced Materials Research. 2009;71–73,661–664.

[37] Vardanyan A, Khachatryan A, Castro L, Willscher S, Gaydardzhiev S, Zhang R, Vardanyan N. Bioleaching of sulfide minerals by *Leptospirillum ferriphilum* CC from polymetallic mine (Armenia). Minerals. 2023;13:243.

[38] Vera M, Schippers A, Hedrich S, Sand W. Progress in bioleaching: Fundamentals and mechanisms of microbial metal sulfide oxidation–part A. Applied Microbiology and Biotechnology. 2022;106:6933–6952.

[39] Sarkodie EK, Jiang L, Li K, Yang J, Guo Z, Shi J, Deng Y, Liu H, Jiang H, Liang Y, Yin H. A review on the bioleaching of toxic metal(loid)s from contaminated soil: Insight into the mechanism of action and the role of influencing factors. Frontiers in microbiology. 2022;13:1049277.

[40] Chandane P, Jori C, Chaudhari H, Bhapkar S, Deshmukh S, Jadhav U. Bioleaching of copper from large printed circuit boards for synthesis of organic-inorganic hybrid. Environmental Science and Pollution Research. 2020;27(6):5797–5808.

[41] Kadivar S, Pourhossein F, Mousavi SM. Recovery of valuable metals from spent mobile phone printed circuit boards using biochar in indirect bioleaching. Journal of Environmental Management. 2021;280:111642.

[42] Bai J, Gu W, Liao C, Yuan W, Zhang C, Wang J, Dong B, Shih K. Bioleaching for Extracting Heavy Metals from Electronic Waste Sludge. Amsterdam, The Netherlands: Elsevier; 2019, pp. 525–551.

[43] Arshadi M, Yaghmaei S. Advances in bioleaching of copper and nickel from electronic waste using *Acidithiobacillus ferrooxidans*: Evaluating daily pH adjustment. Chemical Papers. 2020;74:2211–2227.

[44] Priya A, Hait S. Biometallurgical recovery of metals from waste printed circuit boards using pure and mixed strains of *Acidithiobacillus ferrooxidans* and *Acidiphilium acidophilum*. Process Safety and Environmental Protection. 2020;143:262–272.

[45] Heydarian A, Mousavi SM, Vakilchap F, Baniasadi M. Application of a mixed culture of adapted acidophilic bacteria in two-step bioleaching of spent lithium-ion laptop batteries. Journal of Power Sources. 2018;378:19–30.

[46] Ghassa S, Farzanegan A, Gharabaghi M, Abdollahi H. Novel bioleaching of waste lithium-ion batteries by mixed moderate thermophilic microorganisms, using iron scrap as energy source and reducing energy. Hydrometallurgy. 2020;197:105465.

[47] Arslan V. Bacterial leaching of copper, zinc, nickel and aluminium from discarded printed circuit boards using acidophilic bacteria. Journal of Material Cycles and Waste Management. 2021;23:2005–2015.

[48] Kaur P, Sharma S, Albarakaty FM, Kalia A, Hassan MM, Abd-Elsalam KA. Biosorption and bioleaching of heavy metals from electronic waste varied with microbial genera. Sustainability. 2022;14(2):935.

[49] Xia L, Xu X, Zhu W, Huang Q, Chen W. A comparative study on the biosorption of Cd^{2+} onto *Paecilomyces lilacinus* XLA and *Mucoromycote* sp. XLC. International Journal of Molecular Sciences. 2015;16:15670–15687.

[50] Di Piazza S, Cecchi G, Cardinale AM, Carbone C, Mariotti MG, Giovine M, Zotti M. *Penicillium expansum* link strain for a biometallurgical method to recover REEs from WEEE. Waste Management. 2017;60:596–600.

[51] Sethurajan M, Gaydardzhiev S. Bioprocessing of spent lithium-ion batteries for critical metals recovery – A review. Resources, Conservation and Recycling. 2021;165:105225.

[52] Dave SR, Sodha AB, Tipre DR. Microbial technology for metal recovery from e-waste printed circuit boards. Journal of Bacteriology and Mycology. 2018;6:241–247.

[53] Reed DW, Fujita Y, Daubaras DL, Jiao Y, Thompson VS. Bioleaching of rare earth elements from waste phosphors and cracking catalysts. Hydrometallurgy. 2016;166:34–40.

[54] Yaashikaa PR, Priyanka B, Kumar S, Karishma S, Jeevanantham S, Indraganti S. A review on recent advancements in recovery of valuable and toxic metals from e-waste using bioleaching approach. Chemosphere. 2022;287:132230.

[55] Amiri F, Yaghmaei S, Mousavi SM. Bioleaching of tungsten-rich spent hydrocarbon catalyst using *Penicillium simplicissimum*. Bioresource Technology. 2011;102:1567–1573.

[56] Guezennec A-G, Joulian C, Jacob J, Archane A, Ibarra D, De buyer R, Bodénan F, d'Hugues P. Influence of dissolved oxygen on the bioleaching efficiency under enriched atmosphere. Minerals Engineering. 2017;106:64–70.

[57] Faramarzi MA, Moghrabi-Manzari M, Brandl H. Bioleaching of metals from wastes and low-grade sources by HCN-forming microorganisms. Hydrometallurgy. 2020;191:105228.

[58] Vardanyan A, Vardanyan N, Aâtach M, Malavasi P, Gaydardzhiev S. Bio-assisted leaching of non-ferrous metals from waste printed circuit boards- importance of process parameters. Metals. 2022;12:2092.

Ramlath Beegam L., Riya G. Roshan, Treesa Varghese*

Microbial upcycling of e-wastes: a renewable technology

Abstract: The increasing accumulation of electronic waste (e-waste), filled with toxic metals, plastics, and toxics, has become a significant world environmental problem. Mainstream management approaches like landfilling, combustion, and chemical treatment are often unsustainable, expensive, and low-efficiency options, necessitating environmentally benign substitutes. Microbial degradation is an emerging green technology in which bacteria, fungi, and actinomycetes play a prominent role in the bioleaching of valuable metallic components, such as copper, gold, zinc, and nickel, as well as the degradation of nonbiodegradable organic pollutants, including plastics and flame retardants through specific enzymatic mechanisms. Various driven forces including pH, temperature, nutrient availability, and microbial interactions may critically influence these processes with varying degrees of efficiency. Green technology using microbial approaches has better influences than mainstream technologies in terms of cost, sustainability, and circularity of the bioeconomy, though drawbacks like low degradation rates, toxicity intolerance, and scalability remain a hindrance. Recent breakthroughs in genetic manipulation, microbial consortia, and integrated approaches with nanotechnology have improved degradation rates and metal recovery potential. This chapter introduces e-waste composition, shortcomings of current practices, microbial mechanisms of degradation, merits, drawbacks, recent breakthroughs, and directions of research, highlighting microbial bioremediation as a sustainable route of efficient e-waste management and metal and resource recovery.

Keywords: Green technology, electronic waste, bioremediation, metal recovery, microbial solution

1 Introduction

The ongoing rise in worldwide demand for electronic devices means that efficient electronic waste management has become one of the most significant challenges in

*__Corresponding author: Treesa Varghese__, School of Biosciences, Mar Athanasios College for Advanced Studies, Tiruvalla (MACFAST), Pathanamthitta, Kerala, India, e-mail: treesav@gmail.com
Ramlath Beegam L., School of Biosciences, Mar Athanasios College for Advanced Studies, Tiruvalla (MACFAST), Pathanamthitta, Kerala, India; Department of Microbiology and Biochemistry, St. Berchmans College (Autonomus), Changanacherry, Kottayam, Kerala, India
Riya G. Roshan, School of Biosciences, Mar Athanasios College for Advanced Studies, Tiruvalla (MACFAST), Pathanamthitta, Kerala, India

 | https://doi.org/10.1515/9783112225509-006

current solid waste management practices [1]. The challenge of managing electronic waste is a complicated, interdependent global issue that affects industrialized, developing, and emerging economies alike. While the international trade of electronic products is an increasingly important factor in the global economy, these same products become hazardous waste at the end of their service life. This waste contains toxic elements such as heavy metals, poisonous chemicals, and persistent, nondegradable plastics, which resist degradation environmental pollution and significant health risks [2]. A complex mixture of metallic and nonmetallic parts comprises e-waste with many toxics to the environment and living organisms. Metals frequently used in electrical and electronic equipment, such as lead, mercury, cadmium, nickel, copper, and zinc, are deemed toxic because they persist in the environment and biological systems for a long time. Discarded electronic products contribute various types of plastics to the environment like PET, polystyrene, PVC, and polypropylene from different components. These materials are also found alongside ceramics, printed circuit boards, wood composites, and other synthetic materials [3]. E-waste is laden with huge quantities of toxic organic pollutants, including PAHs, PCDDs, PBDDs, DP, and PCBs. These contaminants persist in the environment and contaminate natural resources such as soil, air, and water. These wastes are highly injurious to health and ecologically hazardous since they are carcinogenic, resistant to biodegradation, and capable of bioaccumulating within food chains. Up to 75% of electronic waste is still stored in houses and industries, and when disposed of badly such as by burning or shredding, they emit fumes that give rise to respiratory and skin issues [4]. Microorganisms have an unavoidable role in managing e-waste, particularly fungi and cyanogenic bacteria, and can assist in bioremediation by generating organic acids and hydrogen cyanide to enhance the bioleaching of toxic substances. Improving microbial applications offers a feasible strategy for e-waste management. This review aims to provide an overview of the role of microbes and their enzymes in the degradation and detoxification of toxic elements of e-waste, including heavy metals, PAHs, and azo dye [5].

2 Need for sustainable e-waste management

The fast growth of technology and people's increasing use of electronic devices have greatly added to the amount of electronic waste (e-waste) produced around the world. Modernization increases the rate of e-waste dumping every year and only a small fraction is recycled or properly managed. Urbanization is the route course and is higher in developed countries than in developing or underdeveloped countries, leading to an increase in global e-waste problem [6].

E-waste accumulation and their underprocessing are considered as the most life-threatening route course in the environment. A majority of electronic gadgets are built using harmful metals including cadmium, mercury, and lead (Table 1). Informal recycling or improper disposal methods are equally harmful to both ecosystems and biodiversity. These toxic pollutants may build up and accumulate in water, soil, and air and can enter into the food chain that leads to many disease conditions. Nonbiodegradable plastics and chemical residues also stay in the environment for a long time, adding to the problem [7]. To solve these issues, sustainable ways to manage e-waste are essential. This includes reducing waste at the source, encouraging recycling and reuse, and ensuring safe disposal methods. Governments and companies can help by creating policies like extended producer responsibility (EPR), designing eco-friendly products, and raising public awareness. New methods such as green technology and using microbes to clean up toxic materials can also make waste management safer and more eco-friendly. With these sustainable solutions, societies can reduce pollution, save resources, and protect the environment for the future [8].

Table 1: Components of electronic wastes and their environmental concerns.

Category/ component	Typical composition (%)	Major elements/ materials	Common sources	Environmental concerns	References
Alloy	50–60	Copper, iron, aluminum, nickel, tin, zinc, and lead	Different parts of an electronic assembly	Heavy metal drainage into the soil and groundwater causes contamination	[9]
Rare earth elements	0.1–2	Gold, silver, palladium, platinum, cobalt, and tantalum	Printed wiring board, cell phones, and motherboards	Processing elements produce harmful fumes and metal remnants	[9]
Resins (synthetic and nonsynthetic materials)	15–25	Acrylonitrile butadiene, styrene, polyvinyl chloride, polycarbonate, and polyethylene	Basic parts of computers and cables	Air pollution; by the release of dioxins and furans due to combustion	[10]
Glass and ceramics	10–15	SiO_2, Al_2O_3, and BaO	Screens, monitors, bulbs, and CRTs	Lead-containing glass is nonbiodegradable and hazardous	[11]
Printed circuit boards	30–35% metals, 25–30% plastics, 30–35% ceramics (by PCB weight)	Cu, Sn, Fe, Pb, Au, Ag, and Pd	Present in almost all electronic devices	Major source of heavy metals and brominated organic compounds	[12]

Table 1 (continued)

Category/ component	Typical composition (%)	Major elements/ materials	Common sources	Environmental concerns	References
Harmful ingredients	<1	Pb, Cd, Hg, Cr(VI), BFRs, and PFAS	Older TVs, cables, and plastic parts	Persistent, bioaccumulative, and toxic substances in ecosystems	[13]
Large appliances	60–70% metals and 20–30% plastics	Fe, Cu, and Al	Washing machines, refrigerators, and industrial appliances	Recycling is energy-intensive and may emit pollutants	[14]
Consumer electronics	45–55% glass, 20–30% plastics, and 10–15% metals	Pb, Sn, Cu, and Si	TVs, display panels, and monitors	Lead exposure from discarded glass poses serious health risks	[15]

3 Toxicological and environmental risks

E-waste constitutes a significant toxicological and environmental concern owing to its complex composition of heavy metals and persistent organic pollutants (POPs). Unscientific recycling and disposal methods induce the releases of the various hazardous chemical compounds into the surroundings. They include polycyclic aromatic hydrocarbons (PAHs), polychlorinated biphenyls (PCBs), brominated flame retardants (BFRs), and heavy metals like lead, cadmium, mercury, chromium, and nickel. In animals, these pollutants may accumulate in the body through injection, inhalation, or absorption process and induce various long-term toxic effects [16]. Scientific studies reported that chronic exposure to e-waste can have many persistence, bioaccumulative, and synergistic toxic influences. They may interfere with key biological events including induction of oxidative stress, mutation induced carcinogenesis, inflammatory responses, dysfunction of liver and kidney, neurotoxicity, endocrine disruption. Additionally, e-waste-derived chemicals interact with xenobiotic receptors and activate cellular toxicity pathways, resulting in apoptotic or necrotic cell death [17].

4 Traditional e-waste management approaches

4.1 Regular landfilling

The escalating global generation of e-waste presents a significant environmental challenge, with a substantial portion of this complex and hazardous material stream still managed through conventional landfilling practices. E-waste contains high levels of harmful substances such as toxic metals, PAHs, PCBs, polybrominated diphenyl ethers, polyvinyl chloride phthalates, and other similar pollutants [18]. The concentration of hazardous heavy metals and POPs within e-waste poses a significant threat within landfill environments; once released, these toxins infiltrate nearby ecosystems and move up the food chain via bioaccumulation and biomagnification, ultimately threatening human health through the consumption of contaminated food sources. The diverse and complex composition of e-waste, containing hundreds of different substances, makes it difficult to predict chemical interactions within a landfill environment. These interactions can accelerate corrosion and the release of unpredictable cocktails of toxins. Therefore, more sustainable waste management techniques should supplant landfilling, which is considered an inefficient method.

4.2 Incineration and airborne emission hazards

The incineration of e-waste, despite its effectiveness in volume reduction, introduces significant environmental and health hazards through the release of toxic airborne emissions. Toxic chemicals, including dioxins, along with fine dust particles, enter the environment through informal e-waste disposal practices, such as dismantling or melting components, resulting in air pollution and respiratory illness [19]. Within India's informal e-waste sector, the city of Moradabad serves as a primary center where a substantial volume of printed circuit boards is processed, largely through environmentally hazardous methods such as open-air burning [20]. Informal e-waste recycling methods – including open burning, manual dismantling, rudimentary incineration, ash washing, and acid leaching – are known to release numerous toxic and hazardous substances into the environment. These pollutants often consist of heavy metals (such as lead, cadmium, nickel, copper, and zinc), BFRs, various neurotoxins, and volatile compounds containing nitrogen and chlorine [21]. Research indicates that the fine particulates generated during e-waste burning processes can contribute to pulmonary and cardiovascular health issues [22]. Pyrometallurgy involves using high-heat processes like incineration, smelting, and melting to eliminate plastic and organic materials from e-waste, thus concentrating the remaining metals [23]. The extant literature consistently and unequivocally indicates that incineration facilities are significant contributors to the burden of hazardous air pollutants. Consequently, these hazards are directly implicated in a broad spectrum of negative public health outcomes, underscoring the necessity for stringent regulatory controls, and the exploration of safer waste management alternatives.

4.3 Chemical and mechanical recycling techniques

The escalating global volume of e-waste, a complex and rapidly growing solid waste stream, presents significant environmental hazards due to toxic components and an opportunity for urban mining due to its high content of valuable and finite resources. Many electronic devices and appliances contain plastic, nearly a third of their total weight. In thermosetting plastics, their complex structure provides intrinsic protection against external damage, such as exposure to chemicals and heat. The use of these materials provides better mechanical strength and durability of the device [24]. In recycling, the common methods used can directly affect the purity and quality of materials [1]. A suitable way to handle this waste stream and material recycling can achieve an advanced recycling approach. In traditional e-waste recycling, material separation is done in two stages: (i) manual dismantling and (ii) mechanical separation. Dismantling is the process of disassembling the product using hand tools, because each electronic component has a different structure, so automated separation is difficult [25]. Mechanical separation of e-waste involves crushing and dismantling, which alter the physical properties of the materials [26]. For the recovery of high-grade metals, various processes such as hydrometallurgy, extraction, replacement, and ion exchange can be used, in which the crushed e-waste is placed in an acidic or alkaline solution [27]. For instance, copper is recovered using an acid bath mechanism, such as treating circuit boards with sulfuric acid for 10–15 h, until the metal is fully dissolved. Copper sulfate is precipitated with chemicals and then washed off under a steam bath to ensure the complete removal of copper from the acid. Other metals like lead, silver, and gold also adopt the same approaches for acid treatments [28].

5 Microbial degradation as a green alternative

The escalating global production of e-waste poses a significant environmental and human health challenge due to its complex and hazardous composition, which includes heavy metals and POPs. Traditional management methods such as landfilling, incineration, and conventional pyrometallurgical and hydrometallurgical processes are often associated with high costs, energy consumption, and the release of secondary pollutants. In response to these limitations, the concept of microbial bioremediation has emerged as a promising, cost-effective, and environmentally sustainable alternative, harnessing the inherent metabolic capabilities of specific microorganisms to break down, transform, or recover valuable and toxic components from e-waste. Microorganisms often produce biosurfactants to begin the breakdown of PAHs. These biosurfactants, which are surfactants secreted outside the cell by specific microbes, help speed up the biodegradation process [29]. Even though the traditional e-waste recovery process like hydrometallurgical and pyrometallurgical extraction methods are much more effective, their impact on cost, and environmentally damaging, generating toxic emissions, and secondary waste are hectic [4]. The pursuit of eco-friendly solutions for the growing e-waste crisis has brought microbial biotechnology to the forefront, utilizing specific types of bacteria, fungi, and archaea to detoxify and recover valuable resources from discarded electronics. It was reported that the PCBs are the richest e-waste sources for gold and copper, often exhibiting concentrations higher than those found in natural ores [3].

5.1 Biochemistry of bioleaching

Bioleaching is a microorganism-assisted technology to remove metals from sulfide mineral ore, the richest source of numerous valuable metals including copper, silver and gold, zinc, nickel, iron, and palladium [1, 2, 29]. Bioleaching is a promising low input and cost-effective solution to the problem where traditional extraction methods are not economically viable, by exploiting microbial metabolisms to break metal ores down. Even though the microbial colonization is a simple process, the biochemical mechanism behind it is a little bit complicated, involving a series of complex enzymes. It was reported that in this process, microorganisms get energy through oxidation of iron and sulfur ions in the sulfide minerals; eventually, the oxidants then act on the sulfide minerals and release target metals [30].

Microorganisms involved in bioleaching are typically chemolithoautotrophic acidophiles (Table 2). On the other hand, chemolithotrophic bacteria, mainly *Thiobacillus thiooxidans, and T. ferrooxidans* induce the recovery of metals from sulfide minerals by converting insoluble metal sulfides into soluble metal sulfates, whereas heterotrophic bacteria and fungus can be used to treat non-sulfide minerals and ores [31]. Acidophilic chemolithoautotrophic organisms live in acidic environments; they receive their carbon through CO_2 fixation process, and exploit inorganic electron sources (e.g., iron and sulfur) as their energy source. Whereas the heterotrophs metabolize and act on organic compounds and can inhibit chemolithotrophic activity, exhibiting an indirect role in mineral dissolution [32]. Bacterial species with these peculiar properties include many of the acidophiles, such as *Acidithiobacillus ferrooxidans, Acidithiobacillus thiooxidans, Leptospirillum ferriphilum*, and *Sulfobacillus thermosulfidooxidans*, as well as heterotrophs like *Pseudomonas putida* and *Chromobacterium violaceum*. Additionally, fungi such as *Aspergillus niger, Penicillium simplicissimum*, and *Saccharomyces cerevisiae* may also take part in this process [30].

Recently, studies established the use of recombineering techniques to modify *Pseudomonas putida* strain 06,909 to produce the high-affinity metal-binding peptide EC20 with the ability to both enhance cadmium sequestration and mitigate the metal's toxicity in the cell [31–33]. Solubilization of metals from e-waste involves biological processes through which microorganisms convert insoluble metal compounds into soluble and recoverable forms. They solubilize metals through oxidation-reduction reactions by secreting organic acids such as citric, oxalic, and gluconic acids that chelate and mobilize metal ions [5, 6]. This transformation mainly occurs via two distinct pathways: direct (contact) and indirect (noncontact) mechanisms. In the direct mechanism, bacteria (*Acidithiobacillus ferrooxidans* and *Leptospirillum ferrooxidans*) and fungus (*Aspergillus niger* and *Penicillium simplicissimum*) electrostatically attach to metal surfaces and oxidize metal sulfides using enzymatic machineries, to release the metal ions into the surrounding medium. In contrast, the indirect mechanism involves the microbial oxidation of ferrous ions (Fe^{2+}) or reduced sulfur compounds to ferric ions (Fe^{3+}) and sulfuric acid respectively, which then chemically dissolve metals from solid matrices [33]. More precisely, the exopolymers present in *Acidithiobacillus ferrooxidans*, contain iron(III) ions, complexed with two uronic acid residues. The positive charge of this complex structure facilitates the interaction with negatively charged pyrite. Additionally, electrons from the metal sulfides are then used to reduce molecular oxygen using complex redox chains present in the cytoplasmic membrane, periplasmic space, and outer membrane of these microbes [34].

Table 2: Microorganisms and metal recovery.

Microorganism	Acidophiles/ heterotrophs	Types of e-waste	Metal leached
Chromobacterium violaceum	Heterotrophs	Printed circuit boards	Gold and copper
Pseudomonas putida	Heterotrophs	Plastics, PCBs, ceramics, glass, etc.	Gold, silver, iron, aluminum, lead, mercury, cadmium, nickel, etc.
Sulfobacillus thermosulfidooxidans	Acidophiles	PCBs, mobile phones, and computers	Aluminum, silver, zinc, and nickel
Acidithiobacillus thiooxidans	Acidophiles	Spent lithium-ion batteries, PCBs	Lithium, nickel, cobalt, lead, zinc, and copper
Acidithiobacillus ferrooxidans	Acidophiles	PCBs	Lead, gallium, zinc, tin, aluminum, nickel, and copper
Leptospirillum ferriphilum	Acidophiles	Cell phones, PCBs, spent lithium-ion batteries	Copper, cobalt, lead, iron, zinc, and nickel

5.1.1 Factors influencing bioleaching

How effectively and safely e-waste can be managed has been determined by many physicochemical factors essential for their optimal growth and survival of a particular microorganism used along with their peculiar physiological nature [7]. Setting up an optimum temperature is crucial for better bioleaching efficiency. Interestingly, thermophiles leaching bacteria function at a temperature between 40 and 45°C, whereas the majority of the acidophiles exhibit their optimum temperature between 25 and 30°C [35]. Studies have demonstrated that maintaining an optimum pH of the medium (between 2 and 2.5) is essential for better metal recovery, whereas a high pH can lead to metal loss through the formation of precipitates. [36]. Sufficient aeration and agitation also have a vital role in determining bioleaching ability of an organism. Many of the microorganisms used for the bioleaching process are typically aerobics in nature. Additionally, insufficient agitation may result in the deprivation of essential nutrients followed by accumulation of toxic by-products compounds [37–39]. Maximum availability of the essential macro and micronutrients will determine the net productivity of an organism along with the continued microbial growth during biological treatment processes, serving as energy and electron sources for microorganisms. The efficiency of microbial processing depends not only upon nutrient availability but also on nutrient balance [42]. However high concentration of metals can also limit their efficiency and retard their growth performance. Microbial synergism or combined action of mixed microbial flora for e-waste recovery is a better strategy to enhance the recovery performance of metals including copper and zinc, gold, and silver [8–10]. There have been reports of synergistic associations between metal-resistant and acidophilic bacteria that could aid in degrading electronic waste substances more effectively [42].

5.2 Biodegradation of persistent organic pollutants

POPs, including PCBs, PAHs, BFRs, and dioxins, are highly toxic compounds frequently generated during the improper recycling and disposal of electronic waste. These materials are chemically stable and resist natural breakdown, leading to their buildup in food chains. This presents significant long-term dangers to ecosystems and people. In contrast, biodegradation presents an eco-conscious, lasting approach to lessen contamination from these pollutants linked to electronic waste [43]. Both aerobic and anaerobic microorganisms including bacterial species of *Pseudomonas, Sphingomonas, and Rhodococcus,* as well as fungal strains like *Phanerochaete chrysosporium* possess the enzymatic capacity to degrade these pollutants to less toxic molecules such as carbon dioxide and water [41].

5.2.1 Microbial Enzyme in bioremediation

Enzymatic pathways provide a sustainable and green option for traditional chemical recycling, encompassing both biological activities and the detoxification of e-waste (Figure 1). Microorganisms, particularly bacteria and fungi, can decompose organic pollutants, oxidize metals, and degrade plastic components in e-waste. Microbial techniques such as bioleaching, biosorption, and biodegradation have attracted interest, as they not only support the recovery of valuable metals, but also aid in neutralizing the environmental and health risks of hazardous chemicals [10, 35, 40].

Several enzymes play a significant role in the management of e-waste via biological methods, breaking down harmful materials and restoring valuable metals in an eco-friendly manner. Among these, enzymes such as oxidoreductases – including laccases, peroxidases, and dehydrogenases – help oxidize toxic organic pollutants BFRs, PCBs, and PAHs. Using this process, these toxic compounds are converted into less toxic and more stable compounds. Hydrolases are enzymes used for the decomposition of plastic envelopes and polymeric materials found in electronic devices. In bioleaching, specific enzymes such as oxidases and sulfur-oxidizing proteins (rusticyanin and sulfur dioxygenase) are used to help the solubilization of metal sulfides into ionized metals and extract them efficiently [44, 45]. All these enzymes use mechanisms to address the detoxification of e-waste and improve material recovery for further use. When we use these enzymes in modern recycling systems, they give more positive results in reducing pollution and enhancing metal recovery, thus providing a way to improve sustainable, green e-waste management technologies [46, 47].

6 Bioengineering approaches for e-waste management

Two methods, genetic engineering and synthetic biology, enable innovative and sustainable management of biowaste through greater efficiency in biological degradation and conversion of waste streams into value-added products. The application of microorganisms in e-waste management is a synthetic biology application that represents a large new research area. Synthetic biology builds on engineering principles and molecular biology to provide tools to improve enzyme activity, change substrate preferences, and increase tolerance to toxic metals, thereby positively contributing to the expansion of novel or improved potentials that are instrumental in proceeding bioremediation and recovering valuable materials [48]. Traditional microbial remediation has many limitations and has led to a multitude of research into genetically engineered microorganisms (GEMs) for various bioremediation processes. This research

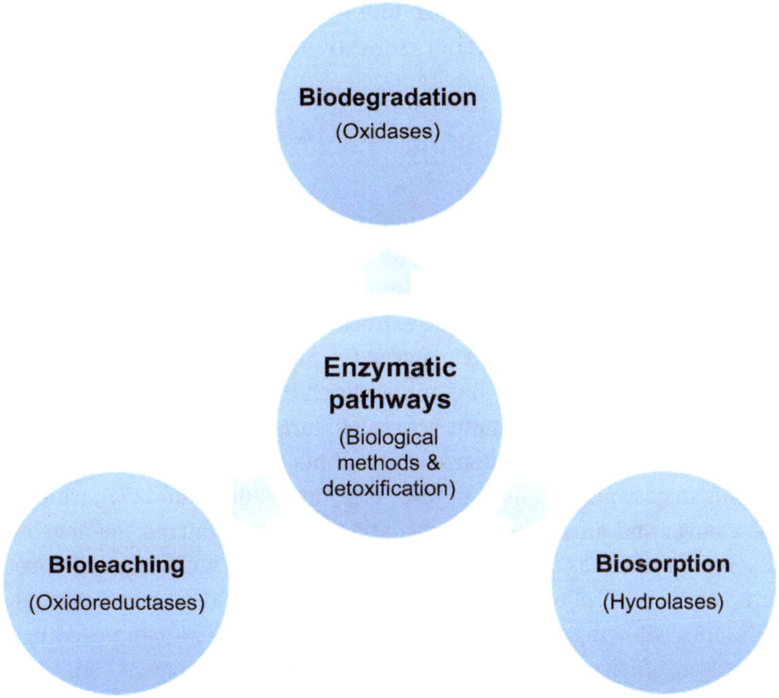

Figure 1: Microbial enzymes in bioremediation process.

specifically targets the various mechanisms of metal capture and removal, including detection, adsorption-chelation, absorption, and bioconversion, as detailed in comprehensive reviews by Diep and Sharma et al. [49, 50].

6.1 Development of microbial consortia

The proliferation of electronic and electrical industries has led to a significant increase in discarded devices, known as e-waste, which poses a serious threat to the environment and public health if not properly managed. Traditional e-waste management strategies often have setbacks in terms of safety and environmental impact, driving the need for eco-friendly and sustainable alternatives. To address this challenge, the development and application of specific microbial consortia have emerged as a promising and sustainable biotechnology approach for the effective bioremediation and valuable material recovery from complex e-waste streams. Most contemporary studies are centered on engineering communication among different *E.coli* strains within consortia, specifically by leveraging quorum-sensing systems [51, 52]. The pioneering research involved the development of synthetic acyl homoserine lactones – receiver devices intended to enable fine-tuned gene expression control in microbial consortia, supplemented by a software tool specifically designed to identify orthogonal communication channels for better consortium designs [53].

6.2 Integration with nanotechnology

Nanotechnology improves the extraction of precious metals (e.g., gold, silver, and palladium) from complex components like printed circuit boards. Advanced methods, such as laser ablation, facilitate the highly pure and economically viable recovery of metal nanoparticles. Nanoscale techniques can facilitate the extraction of valuable metals from electronic waste, a resource-rich but challenging material for conventional recycling. A wide array of natural airborne particles, including mineral-rich dust, has been employed as interfaces to eliminate atmospheric pollutants generated during e-waste disposal and industrial processes. Traditional e-waste management techniques often involve fossil fuel combustion, resulting in gaseous emissions of toxic organic pollutants and CO_2, which is linked to climate change and atmospheric degradation. This has spurred the development of hybrid solutions, such as integrated bioreactors coupled with recyclable iron oxide nano-/microparticle adsorption interfaces. These innovations use natural particles to eliminate pollutants, allowing for the integration of novel detoxification features while preserving process efficiency and effectiveness in e-waste management [54].

7 Challenges and limitations

7.1 Slow degradation rates, toxicity tolerance, and microbial survival

The primary issue with e-waste in natural environments is not slow degradation, but that it is largely nonbiodegradable and persists for an extremely long time. While organic waste can decompose in months, plastic components in e-waste can take up to a million years to break down, and metals anywhere from 50 to 500 years. E-waste is a complex mix of engineered materials (plastics, metals, glass) that natural biological processes struggle to break down effectively. The presence of hazardous chemicals such as lead, mercury, cadmium, and BFRs inhibits microorganisms that could otherwise assist in breaking down organic materials. Modern electronics feature intricate designs with mixed materials (e.g., glue holding parts together and embedded components) that are difficult for both natural processes and traditional mechanical recycling methods to separate. Plastics are often treated with flame retardants, making them more durable and resistant to environmental factors like UV light and microbial action. Atmospheric discharge of dangerous particles, specifically those containing flame retardants and heavy metals is a consequence of certain recycling methods [55].

7.2 Scale-up and economic feasibility

The e-waste sector presents a duality of significant hurdles and promising paths forward. Primary challenges involve ensuring worker safety, safeguarding the environment, and addressing the critical lack of processing infrastructure in developing nations, a situation often worsened by waste exports from wealthier countries. Despite these issues, the problem offers chances for meaningful improvement. Progress relies on deploying advanced recycling technology, strengthening regulatory enforcement, and embracing cutting-edge innovations to enhance the industry's efficiency and safety standards. Specialized attention is required for data security concerns, particularly for entities managing sensitive information like government bodies, as improperly handled e-waste is vulnerable to breaches. While economic constraints persist, targeted governmental policies, such as financial incentives and subsidies, can stimulate the development of safer, more effective e-waste systems globally. The continuous expansion of the digital landscape ensures a growing need for sustainable management, offering a fertile ground for innovative enterprises to create lasting positive change [56].

7.3 Regulatory and safety concerns

E-waste management is governed by a patchwork of international, national, and regional regulations that aim to mitigate severe safety risks posed by hazardous materials. The primary concerns center on the exposure to toxic substances and the challenges in enforcing environmentally sound management practices, particularly within the large informal sector. International policies and legislation are essential for e-waste management, as they define the necessary standards and mandatory rules to regulate the behavior of all parties involved in handling electronic waste across government and private industries [57]. The Basel Convention requires its member nations to ensure the proper monitoring and environmentally sound disposal of hazardous and other wastes moved across international borders [58]. Ultimately, the broad adoption of national or shared policies by 71% of the global population demonstrates significant progress in establishing a foundational framework for worldwide governance [59].

8 Future prospects and research directions

The future of effective e-waste management depends on strategies that improve metal recovery, incorporate advanced hybrid recycling technologies, and function within strong, sustainability-focused policy frameworks. Enhancing metal extraction efficiency through better mechanical processing, optimized hydrometallurgy, and improved bioleaching techniques allows for greater resource recovery while reducing environmental harm and limiting reliance on newly mined minerals. At the same time, hybrid methods that integrate biological, chemical, and physical processes provide higher precision, lower energy consumption, and improved recovery rates, making them ideal for supporting a circular bioeconomy. These innovations enable the conversion of discarded electronics into new materials, bio-based products, and industrial resources, helping maintain a closed-loop system centered on resource conservation. However, widespread implementation of these technologies requires strong policy support. Measures that improve collection systems, formalize recycling sectors, enforce extended producer responsibility (EPR), and promote eco-friendly technologies are key to long-term advancement. International collaboration, investments in formal recycling facilities, and partnerships among stakeholders further strengthen responsible e-waste management. Collectively, these developments form a sustainable, efficient, and economically resilient approach to e-waste handling that supports global efforts toward a circular and resource-efficient future.

References

[1] Babu BR, Parande AK, Basha CA. Electrical and electronic waste: a global environmental problem. Waste Management & Research. 2007;25:307–318.

[2] Osibanjo O, Nnorom IC, Adie GU, Ogundiran MB, Adeyi AA. Global management of electronic wastes: challenges facing developing and economy-in-transition countries in:John Wiley & Sons, Ltd. Metal Sustainability: Global Challenges, Consequences, and Prospects. First Edition. 2016;52–84.

[3] Estrada-Ruiz RH, Flores-Campos R, Gámez-Altamirano HA, Velarde-Sánchez EJ. Separation of the metallic and non-metallic fraction from printed circuit boards employing green technology. Journal of Hazardous Materials. 2016;311:91–99.

[4] Tang X, Shen C, Shi D, Cheema SA, Khan MI, Zhang C, Chen Y. Heavy metal and persistent organic compound contamination in soil from Wenling: an emerging e-waste recycling city in Taizhou area, China. Journal of Hazardous Materials. 2010;173:653–660.

[5] Chaudhary P, Pant M, Sinha S, Pant K. Importance of microorganisms in metal recovery from E-waste. In Microbial technology for sustainable e-waste management 2023 (pp. 145–155). Cham: Springer International Publishing.

[6] Vishwakarma A, Kanaujia K, Hait S. Global scenario of E-waste generation: trends and future predictions. In Global e-waste management strategies and future implications 2023 (pp. 13–30). Elsevier.

[7] He Y, Kiehbadroudinezhad M, Hosseinzadeh-Bandbafha H, Gupta VK, Peng W, Lam SS, Tabatabaei M, Aghbashlo M. Driving sustainable circular economy in electronics: A comprehensive review on environmental life cycle assessment of e-waste recycling. Environ Pollut. 2023;342:123081.

[8] Naik S, Eswari JS. Electrical waste management: Recent advances, challenges and future outlook. Science of the Total Environment. 2022;1:100002.

[9] Fornalczyk A, Willner J, Francuz K, Cebulski J. E-waste as a source of valuable metals. Archives of Materials Science and Engineering. 2013;63:87–92.

[10] Lee S, Huang M, Lee J, Choi H, Jo IY, Na H, Lee Y, Youk JH, Yoon MH, Shim BS. Eco-Friendly Materials for a Zero E-Waste Society: Challenges and Opportunities in Engineering Plastics. Advanced Sustainable Systems. 2024;8:2300428.

[11] Lekshmi S, Lekshmi AA, Chandran SS, Achuthavarier D. Sustainable Ways to Manage Electronic Waste Contaminated Soil Using Phytoremediation Techniques. e-waste. 2025;211–222.

[12] Bouraoui H, Bourzama G, Iratni N, Sayad L, Bensefia N, Deutscher J, Milohanic E. Biodegradation of Printed Circuit Boards by Pseudomonas aeruginosa and Achromobacter sp. Isolated from E-Waste Contaminated Soil. Microbiol. 2024;93:939–953.

[13] Bhardwaj LK, Rath P, Jain H, Choudhury M. Exploring the effects of e-waste on soil, water quality and human health. Discover Civil Engineering. 2025;2:12.

[14] Roy SK, Rao AS, Singh SK, Somasundaram R, Dash S. Harnessing Bio-Based products for sustainable E-Waste management in biomanufacturing: green solutions. In Environmental Applications of Carbon-Based Materials 2024 (pp. 351–379). IGI Global.

[15] Bhattacharjee J, De A, Kamila B, Mandal A. Recent scenario of e-waste recycling: chemical engineering. International Journal of Chemical Reactor Engineering. 2025;23:675–700.

[16] Owonikoko WM, Alimba CG. Systematic literature review of heavy metal contamination of the Nigerian environment from e-waste management: Associated health and carcinogenic risk assessment. Toxicol. 2024;505:153811.

[17] Eze C, Vinken M. E waste: mechanisms of toxicity and safety testing. FEBS Open bio. 2024;14:1420–1440.

[18] Wong MH, Wu SC, Deng WJ, Yu XZ, Luo Q, Leung AO, Wong CS, Luksemburg WJ, Wong AS. Export of toxic chemicals – A review of the case of uncontrolled electronic-waste recycling. Environ Pollut. 2007;149:131–140.

[19] Mundada MN, Kumar S, Shekdar AV. E waste: a new challenge for waste management in India. International Journal of Environmental Studies. 2004;61:265–279.

[20] Gangwar C, Singh A, Pal R, Kumar A, Sharma S, Tripathi A. Environmental susceptibility: soil contamination of heavy metals in the territory of e -waste recycling area. International Journal of Plant and Environment. 2017;3:25–31.

[21] Kiddee P, Naidu R, Wong MH. Metals and polybrominated diphenyl ethers leaching from electronic waste in simulated landfills. Journal of Hazardous Materials. 2013;252:243–249.

[22] McAllister L. The human and environmental effects of e-waste. Population Reference Bureau. 2013.

[23] Ebin B, Isik MI. Pyrometallurgical processes for the recovery of metals from WEEE. In WEEE recycling: Research, development, and policies Chagnes A, Cote G, Ekberg C, Nilsson M, Retegan T (eds). 2016 (pp. 107–137).

[24] Jung H, Shin G, Kwak H, Hao LT, Jegal J, Kim HJ, Jeon H, Park J, Oh DX. Review of polymer technologies for improving the recycling and upcycling efficiency of plastic waste. Chemosphere. 2023;311:138089.

[25] Ostroukh A, Surkova N, Varlamov O, Chernenky V, Baldin A. Automated process control system of mobile crushing and screening plant. Journal of Applied Engineering Science. 2018;16:343–348.

[26] Li J, Lu HZ, Guo J, Xu ZM, Zhou YH. Recycle technology for recovering resources and products from waste printed circuit boards. Environmental Science & Technology. 2007;41:1995–2000.

[27] Ashiq A, Kulkarni J, Vithanage M. Hydrometallurgical recovery of metals from e-waste. In Electronic waste management and treatment technology 2029 (pp. 225–246). Butterworth-Heinemann, Elsevier.

[28] Pingle BP, Godse JS, Gadodiya UV, Shisodia SU, Pawar RP. A review on Study of disposal of E-Waste and its Impact on Human Health. International Journal of Scientific Research in Science and Technology. 2023;10:257.

[29] Zhao H, Zhang Y, Zhang X, Qian L, Sun M, Yang Y. The dissolution and passivation mechanism of chalcopyrite in bioleaching: An overview. Minerals Engineering. 2019;136:140–154.

[30] Jones S, Santini JM. Mechanisms of bioleaching: iron and sulfur oxidation by acidophilic microorganisms. Essays in Biochemistry. 2023;67:685–699.
[31] Bosecker K. Bioleaching: metal solubilization by microorganisms. FEMS Microbiology Reviews. 1997;20:591–604.
[32] Shiers DW, Collinson DM, Watling HR. Life in heaps: a review of microbial responses to variable acidity in sulfide mineral bioleaching heaps for metal extraction. Research in Microbiology. 2016;167:576–586.
[33] Borthakur A, Singh P. Microbial biotechnology for sustainable electronic waste management. In Conference Proceedings – Asia-Pacific Conference on Biotechnology for Waste Conversion Asia-Pacific Conference on Biotechnology for Waste Conversion 2016, Hong Kong SAR, P.R 2016 China. 5–8 December 2016.
[34] Rohwerder T, Gehrke T, Kinzler K, Sand W. Bioleaching review part A: progress in bioleaching: fundamentals and mechanisms of bacterial metal sulfide oxidation. Applied Microbiology and Biotechnology. 2003;63:239–248.
[35] Marra A, Cesaro A, Rene ER, Belgiorno V, Lens PN. Bioleaching of metals from WEEE shredding dust. Journal of Environmental Management. 2018;210:180–190.
[36] Bleiwas DI, Kelly TD. Obsolete Computers, Gold Mine, Or High-Tech Trash?: Resource Recovery from Recycling. In US Geological Survey 2001 Denver, USA.
[37] Abhilash ST, Ghosh A, Meshram P, Hullebusch ED. Microbial processing of waste shredded PCBs for copper extraction cum separation – comparing the efficacy of bacterial and fungal leaching kinetics and yields. Metals (Basel). 2021;11:317.
[38] Arshadi M, Nili S, Yaghmaei S. Ni and Cu recovery by bioleaching from the printed circuit boards of mobile phones in non-conventional medium. Journal of Environmental Management. 2019;250:109502.
[39] Natarajan G, Ting YP. Gold recovery with improved strategy through spent medium leaching with pH modification. Chemosphere. 2015;136:232–238.
[40] Wu CH et al. Engineering plant microbe symbiosis for rhizoremediation of heavy metals. Applied and Environmental Microbiology. 2006;72:1129–1134.
[41] Li JY, Xu XL, Liu WQ. Thiourea leaching gold and silver from the printed circuit boards of waste mobile phones. Waste Management. 2012;32:1209–1212.
[42] Koyama K, Tanaka M, Lee J. Copper leaching behavior from waste printed circuit board in ammoniacal alkaline solution. Materials Transactions. 2006;47:1788–1792.
[43] Ju Y, Wang Z, Lin H, Hou R, Li H, Wang Z, Zhi R, Lu X, Tang Y, Chen F. Modulating the microenvironment of catalytic interface with functional groups for efficient photocatalytic degradation of persistent organic pollutants. Chemical Engineering Journal. 2024;479:147800.
[44] Soon CF, Yee SK, Nordin AN, Rahim RA, Ma NL, Hamed ISLA, Tee KS, Azmi NH, Sunar NM, Heng C. Advancements in biodegradable printed circuit boards: Review of material properties, fabrication methods, applications and challenges. International Journal of Precision Engineering and Manufacturing. 2024;25:1925–1954.
[45] Charitopoulou MA, Papadimitriou M, Papadopoulou L, Achilias DS. Extraction-based pretreatment of end-of-life plastics from waste electrical and electronic equipment for brominated flame retardant removal and subsequent valorization via pyrolysis. Processes. 2025;13:1458.
[46] Murkute S, Patil R, Patil NK, Jain A. Potential Applications of Enzymes in Waste Treatment. In Management of Waste to Control Environmental Pollutions: Sustainability and Economic Feasibility 2025 (pp. 9–21). Cham: Springer Nature Switzerland.
[47] Jain M, Kumar D, Chaudhary J, Kumar S, Sharma S, Verma AS. Review on E-waste management and its impact on the environment and society. Waste Management Bulletin. 2023;1:34–44.
[48] De Lorenzo V, Prather KL, Chen GQO, Day E, Von kameke C, Oyarzún DA, Hosta-Rigau L, Alsafar H, Cao C, Ji W, Okano H. The power of synthetic biology for bioproduction, remediation and pollution

control: the UN's Sustainable Development Goals will inevitably require the application of molecular biology and biotechnology on a global scale. EMBO reports. 2018;19:45658.
[49] Capeness MJ, Horsfall LE. Synthetic biology approaches towards the recycling of metals from the environment. Biochemical Society Transactions. 2020;48:1367–1378.
[50] Diep P, Mahadevan R, Yakunin AF. Heavy metal removal by bioaccumulation using genetically engineered microorganisms. Frontiers in Bioengineering and Biotechnology. 2018;6:157.
[51] Sharma P, Sirohi R, Tong YW, Kim SH, Pandey A. Metal and metal (loids) removal efficiency using genetically engineered microbes: Applications and challenges. Journal of Hazardous Materials. 2021;416:125855.
[52] Kylilis N, Tuza ZA, Stan GB, Polizzi KM. Tools for engineering coordinated system behaviour in synthetic microbial consortia. Nature Communication. 2018;9:2677.
[53] Alnahhas RN, Sadeghpour M, Chen Y, Frey AA, Ott W, Josić K, Bennett MR. Majority sensing in synthetic microbial consortia. Nature Communication. 2020;11:3659.
[54] Karra Geetha SA, Kiran RS, Rao TR. An Overview Of Nano Technology In Waste Management. International Journal of Novel Research and Development. 2023;8:281–289.
[55] Kiddee P, Naidu R, Wong MH. Electronic waste management approaches: an overview. Waste Management. 2013;33:1237–1250.
[56] Quinto S, Law N, Fletcher C, Le J, Jose SA, Menezes PL. Exploring the e-waste crisis: Strategies for sustainable recycling and circular economy integration. Recycling. 2025;10:72.
[57] Baldé CP, Van den brink S, Forti V, Van der schalk A, Hopstaken F. The Dutch WEEE flows 2020: What happened between 2010 and 2018. United Nations University (UNU)/United Nations Institute for Training and Research (UNITAR); 2020.
[58] Hackett DP. Assessment of the Basel Convention on the control of Transboundary Movements of Hazardous Wastes and Their Disposal. American University Journal of International Law and Policy. 1989;5:31.
[59] Kaur RR, Sahni S, Kaur G, Singh RI Environmental Governance and Key Challenges at Local Level in Indian Context.

Tijo Cherian*, Sijo Asokan, N.Sharmila Devi, Smitha Vijayan, Beena Cherian, Sini Kurien

Green e-revolution: microbial magic for e-waste degradation and environment protection

Abstract: Globally, millions of tons of e-waste are disposed of every year. The production of materials for EEE (electrical and electronic equipment) is rising in tandem with the rapid expansion of the industrial and technological sectors. Modern lifestyle changes and the population's capacity to afford new technologies have improved the EEE sector. A number of raw materials and components are used in the manufacturing processes. The eventual destinations of produced items are altered by shipping and exporting. This business is placing more strain on poor and underdeveloped nations, who also mostly receive low-quality EEE items. The establishment of factories in these nations was boosted by the low cost of labor, resulting in tons of e-waste from the production of a variety of e-products that were subsequently exported to other countries. Notwithstanding the information revolution's numerous beneficial features, the drawbacks of accumulated hazardous materials are becoming more pronounced. With their high concentrations of hazardous metals and contaminants, the EEE business has the potential to pose a serious harm to human health and the environment, according to recent studies that examined its associated risks and concerns. Public health awareness has lately grown, leading to the identification of potential hazards and the development of environmentally appropriate solutions for accumulated e-waste. This chapter explains the prospective benefits of bioremediation techniques and the ability of microorganisms to break down and remove various kinds of discarded e-waste, as well as the associated dangers and technological constraints.

Keywords: Electronic waste, environmental sustainability, ecological restoration, microbial degradation, organic pollutants

***Corresponding author: Tijo Cherian,** Post Graduate Department of Food and Industrial Microbiology, Bishop Kurialacherry College for Women, Amalagiri 686561, Kottayam, Kerala, India, e-mail: tvarghese891@gmail.com

Sijo Asokan, School of Biosciences, Mar Athanasios College for Advanced Studies Tiruvalla (MACFAST-Autonomous), Kerala, India

N. Sharmila Devi, Post Graduate Department of Food and Industrial Microbiology, Bishop Kurialacherry College for Women, Amalagiri 686561, Kottayam, Kerala, India

Smitha Vijayan, Beena Cherian, Sini Kurien, Department of Food Technology and Quality Assurance, Mar Athanasios College for Advanced Studies Tiruvalla (MACFAST – Autonomous), Kerala, India

 | https://doi.org/10.1515/9783112225509-007

1 Introduction

Effective treatment of electronic trash, or "e-waste," has become a crucial issue in the field of solid waste management due to the increasing need for electronic goods worldwide [1]. This issue affects established, developing, and transitioning countries alike, creating a complex web of interrelated problems. (Figure 1) Global trade depends on the transportation of numerous electronic devices over international boundaries, yet once their useful lives are over, they become hazardous waste made up of dangerous materials including nonbiodegradable polymers, heavy metals, and toxic chemicals. Pollution and the emergence of serious health problems are the outcomes of this change [1]. The environment and its ecosystems are seriously threatened by the complex mixture of metallic materials and heavy metals found in e-waste, all of which are lethal [2]. Zinc, copper, nickel, cadmium, mercury, lead, and other metallic substances commonly present in electrical devices are regarded as dangerous [3]. Additionally, disposing of e-waste releases a range of plastic materials into the environment, including plywood, printed circuit boards, ceramics, polypropylene, polyvinyl chloride (PVC), polystyrene, and polyethylene terephthalate esters [4]. Polychlorinated biphenyls (PCBs), dechlorane plus, polybrominated dibenzo-*p*-dioxins, polychlorinated dibenzo-*p*-dioxins (PCDDs), and polycyclic aromatic hydrocarbons (PAHs), which are equally hazardous to the environment, are among the organic materials present in e-waste [1].

E-waste undoubtedly contaminates the environment and its natural resources, including land, water, air, and soil surfaces, either directly or indirectly. Because they are primarily carcinogenic and contain nonbiodegradable polymers, acids, and heavy metals, these wastes pose a threat to the health of both plants and animals [5]. The proper administration from collection to disposal is required due to their potential to biologically magnify the food chain [6]. Approximately 75% of this waste is kept in residences, workplaces, or businesses where it will eventually be dumped. When non-recyclable garbage is disassembled, shredded, or even burned, large amounts of hazardous smoke containing heavy carcinogenic substances are released. Skin and respiratory problems are caused by these emissions, which worsen health. For example, microbes have demonstrated remarkable efficacy in addressing environmental pollutants. Organotrophs include both fungus and cyanogenic microorganisms. Organic acids are produced by fungi, and when organic carbon is available, cyanobacteria synthesize hydrogen cyanide. The bioleaching process is aided by the interaction between hydrogen cyanide and organic acids [7]. Enhancing the utilization of microbes as biological remediation drivers is essential to advancing the goal of ecological sustainability. The importance of different bacteria in the remediation of the main biodegradable components found in e-waste, such as azo dye, PAHs, and heavy metals, their mode of action, and the obstacles involved in the process are all highlighted in this chapter. It also gives a quick overview of how different microbial enzymes contribute to the biotransformation of e-waste constituents.

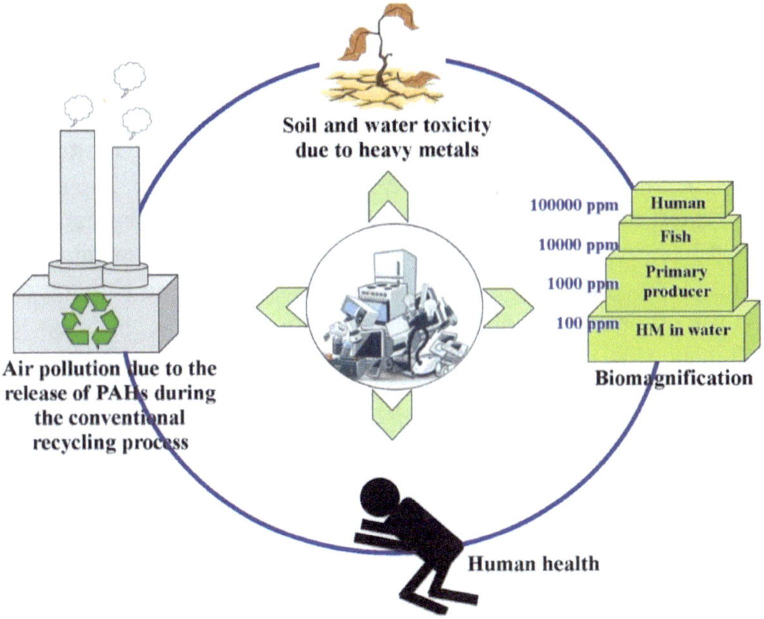

Figure 1: Schematic illustration of the comprehensive impact of e-waste on human health and the environment (reproduced from [1]).

2 Components and main sources of e-waste

With an exponential rise in volume, e-waste is currently the waste source with the fastest rate of growth [2]. About billions metric tons of e-waste are produced worldwide each year, and this amount is predicted to increase by 4–5% annually [8]. Industrialization, urbanization and our reliance on electrical and electronic components are some of the key causes of this astounding growth. The increase in demand for a variety of electronic devices has been attributed to both local consumption and foreign exports [9]. Notably, one of the most rapidly expanding industries in India is the electronics sector [10]. The content of e-waste varies depending on the manufacturer, type of equipment, and age, including an assortment of >1,000 distinct compounds. Iron and steel make up more than half of the ferrous metal portion in e-waste, which is composed of roughly 16.5% nonferrous metals, 26% plastics, and 38% ferrous metals. E-waste frequently contains metals in elementary forms or as composites of several elements [1]. Modern technologies include an incredible diversity of up to 60 parts, making them more complex in a time of growing invention. The number of metals with alloying, conductive, and luminous properties increases with increasing complexity [1].

Different electrical and electronic gadgets contain a wide range of metals in different quantities and combinations. For the production of components like printed circuit boards, which are widely used in cell phones, personal computers, laptops, and other devices of a similar nature, exact measurements of elements are necessary. Hazardous elements including copper, nickel, lead, zinc, and chromium, whether in their elemental or alloyed form, may be included in these components. Because of their greatest magnetization capacity and low iron loss, electrical steels are frequently used in electronics. TV monitors frequently use display technologies such light-emitting diodes (LEDs), liquid crystal displays (LCDs), and cathode ray tubes (CRTs), because of their accessibility, high resolution, and persistent magnetization. Rare earth metals like lanthanum (La) and lithium cobalt oxides, and lithium oxides are components of rechargeable batteries (lithium/lithium-ion polymer), which are widely used in mobile phones and laptops [11]. Furthermore, arsenic, heavy metals, and europium (rare earth elements) are crucial components of electrical components [2]. Additionally, some electronic appliances like air conditioners and refrigerators contain halogenated substances like brominated flame retardants (BFRs), polybrominated biphenyls (PBBs), polybrominated diphenyl ethers (PBDEs), PCBs, and CFCs.

3 Effects of human health and environment

The concerns about the distribution and existence of organic pollutants, such as heavy metals, in the natural environment have grown in recent years [12]. The ecosystem and human health are severely impacted by any manner of disposing of waste, whether it is in bodies of water or landfills [13]. Heavy metals, which have detrimental effects on the skin, lungs, blood, kidneys, and central nervous system, are found in the majority of e-waste. Furthermore, elements including dibenzofurans, beryllium, and barium can lead to a number of skin and lung conditions as well as cancer. Direct exposure to e-waste at recycling facilities, ingestion of heavy metals via water, or breathing of contaminated air following combustion can all have negative health impacts [14].

Typically, these substances fall into one of three categories:

1. Heavy metals and halogenated substances such as zinc, nickel, barium, cadmium, and, lead are considered primary pollutants [1].
2. Chemicals including PHAHs, PAHs, and dioxins are found in secondary pollutants, which include by-products of improper recycling processes [15].
3. Reagents employed in hydro- and pyrometallurgical processes are examples of tertiary pollutants [1].

Heavy metals are released during recycling processes such ion exchange, membrane separation, chemical extraction, adsorption, heat treatment, and chlorination. These metals directly penetrate the surface of soil and endanger soil ecology. As a result, this garbage may pollute water supplies, endangering marine life [16]. The process of biomagnification can be initiated, for example, when aquatic species are negatively impacted by cadmium contamination in systems of groundwater that exceeds the usual threshold. When this water is utilized for irrigation, plants store and absorb heavy metals in their tissues, putting human and plant health at risk if consumed. The BFRs in plastics have the potential to cause serious medical problems such as thyroid gland dysfunction. Furthermore, heavy metals like mercury, which are frequently present in electronic parts like switches, LED screens, and fluorescent tubes, have detrimental impacts on health, including memory loss, dermatitis, and sensory impairment [17]. A common insulator in electrical wires, PVC has a tendency of bioaccumulation.

4 Management techniques for e-waste in India

E-waste management is similar in India, where urban families gather, classify, repair, and disassemble old electronics in order to identify work. However, there is no custom in India of consumers voluntarily donating unwanted electronics to reputable electronic waste processing facilities, unlike in developed countries. The NRDC (National Research Development Corporation) states that recycling facilities are primarily found in African and Asian nations [18].

4.1 Microbe-mediated degradation of e-waste

The term "biodegradation" describes the aerobic or anaerobic chemical breakdown of compounds by biological organisms. The degradation of organic molecules is greatly impacted by this mechanism [1]. The majority of microorganisms emit biosurfactants to start the PAH breakdown process. Certain microbes release extracellular surfactants called biosurfactants, which speed up the process of biodegradation. Because of their surface activity, multifunctionality, potent emulsifying powers, low toxicity, ionic characteristics, and environmental compatibility, biosurfactants are seen as potential choices for bioremediation. Furthermore, biosurfactants have a wide range of binding capability and metal selectivity, as well as a variety of chemical structures, which increases their capacity to eliminate impurities.

PAHs, PCBs, and heavy metals make up the majority of e-waste. A variety of substances, including metals, pesticides, medicines, and hydrocarbons like PAHs, PCBs, and oil can be broken down, transformed, or accumulated by some microorganisms due to their broad catabolic capacity [19]. Numerous microbes, including *Sulfobacillus* sp., *A. caldus, Alicyclobacillus* sp., *Leptospirillum ferriphilum, Saccharomyces cerevisiae, Penicillium simplicissimum, B. megaterium, Chromobacterium violaceum, A. tubingensis, A. versicolor, A. niger, A. japonicus, A. flavipes, A. fumigatus, Pseudomonas fluorescens, Pseudomonas aeruginosa, Pseudomonas putida, S. thermosulfidooxidans, A. thiooxidans,* and *A. ferrooxidans,* through a variety of mechanisms can absorb different heavy metals and other components like PCBs and PAHs. Heavy metals may be able to change from one chemical state to another, which would lessen their environmental hazard even though they are not biodegradable [20]. During the mineralization process, microbes help convert pollutants into final products like water and carbon dioxide as well as other intermediary physiological compounds. The process of transforming substances into a state that renders them unavailable in the environment is known as immobilization. It has been discovered that *B. cereus* and *E. asburiae* have a role in absorbing heavy metals [20]. Both in situ and ex situ immobilization procedures are possible. The ex situ procedure involves moving contaminated soils from the contaminated site to a different location where the metal ions causing the contamination are immobilized using a microbiological technique. The in situ method, on the other hand, involves addressing contamination at its point of origin [21]. Designing microbial methodologies that consider elements like responses to changing environmental conditions, metabolic capabilities, microbial growth dynamics in contaminated areas, and, regulatory mechanisms are some effective ways to increase the efficacy of bioremediation processes in particular locations. Microbes have evolved defense systems, despite the fact that exposure to specific organic solvents can cause cell membrane rupture [22]. These include the development of solvent or hydrophobic efflux pumps that protect the outermost membrane of cells [22]. The main ways that bacteria aid in the e-waste bioremediation include bio-oxidation, reduction, biomineralization, biosorption, biotransformation, bioaccumulation, and bioleaching.

4.2 Biological degradation of azo dye constituents of e-waste

About 80% of all dyes produced worldwide are azo dyes, making them the most extensively produced form of dye [1]. These dyes, which are made by a simple coupling and diazotization method, are essential to the market for dyeing and printing [23]. Functional dyes designed for high-tech applications, like imaging technologies (thermal printing and electrophotography), electronic materials (organic semiconductors), and optoelectronics (liquid crystal displays, dye-sensitized solar cells, and photochromic materials) have become more popular lately. These dyes are used in a variety of electrical equipment, such as fuel cells, nonlinear optical devices, lasers, and thermal transfer printers [23]. Additionally, the possibility of high molar absorptivity new azo-cyanine dyes as cyanine photosensitizers in the creation of innovative photodynamic treatment agents has been studied [1]. However, because azo dyes pose health concerns and have serious environmental repercussions, there is rising concern regarding their use in these industries. Numerous research support bioremediation methods for azo dye remediation. The capacity of microorganisms, especially bacteria, to effectively break down a variety of pigments in both aerobic and anaerobic environments has drawn attention from all over the world [1]. For instance, microorganisms such as *Phoma tropica* and *Dichotomomyces cejpii* MRCH 1–2 show a 95% degradation rate for regularly used dyes like Congo Red in industries like printing [23]. However, as there is little research on azo dyes in the field of electronics, more research is necessary to understand the degradation routes of azo dyes used in devices and their particular environmental impact.

4.3 Biological degradation of PAHs

E-wastes contain a variety of volatile organic compounds, PAHs, and PCBs. Biosurfactants, which lower the surface tension of these oil-like compounds and break them down into smaller particles so that the cells may absorb them for further metabolism, can be released by a wide variety of bacteria (Figure 2). Complex organic pollutants known as PAHs are mostly created by partial combustion reactions [1]. These pollutants spread throughout the world via water and air currents after being released into the ecosystem by both natural and human processes. As they ascend the food chain, they accumulate in organisms and pollute food, plants, and the air. Concerns over possible health dangers like cancer are heightened by rising PAH levels, particularly from inappropriate e-waste disposal. Asthma and cardiovascular disorders are additionally rendered much more probable by prolonged exposure to PAHs [24]. The capacity of some microorganisms to break down high molecular weight PAHs has been well demonstrated. Microbial cells of *Achromobacter*, *Pseudomonas*, *Rhodococcus*, *Mycobacterium*, and *Bacillus* are important bacterial genera involved in PAH breakdown [1]. As a class of actinomycetes, mycobacteria are naturally resistant to harsh environments and are especially skilled at breaking down different PAHs, polychloro derivatives of phenols, and heavy metals. Usually, bacteria use enzymatic processes like peroxidases and oxygenases to break down PAHs. Cytochrome P450 from yeast species including *C. tropicalis* and *C. maltosa*, naphthalene monooxygenase from *P. putida*, and AlkB from *Pseudomonas* species are a few examples [25].

Numerous fungi have also shown the ability to degrade PAHs, including white-rot fungus, deuteromycetes, and basidiomycetes. In contrast to bacteria, fungi use PAHs in addition to other sources of carbon to produce products of oxidization, such as CO_2 (carbon dioxide) [26]. Since they produce extracellular ligninolytic enzymes like laccase (Lac), manganese peroxidase (MnP), and lignin peroxidase, white-rot fungus species like *Phanerochaete chrysosporium* are very effective at eliminating PAH compounds [27]. There are impediments even if bacteria and fungi have the ability to break down PAHs. The simultaneous breakdown of many PAH types is difficult be-

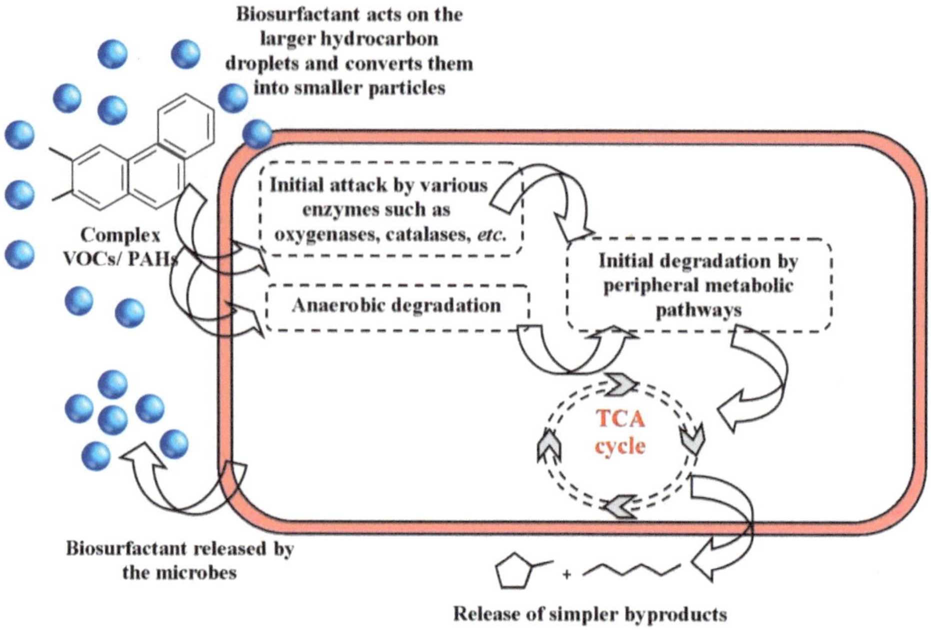

Figure 2: Diagrammatic representation of the initial degradation of PAHs and VOCs by microbial biosurfactants (reproduced from [1]).

cause of factors like metabolic interactions and bioavailability, and some bacteria display challenges efficiently digesting PAHs. However, co-metabolism is essential for the synergistic breakdown of PAHs, which enables certain bacteria to break down a greater variety of PAHs, especially ones with high molecular mass. Understanding the dynamics of PAHs in marine and soil ecosystems is another major obstacle to PAH bioremediation. The majority of released PAHs are retained by black-clayish carbon particles and coal tar, which drastically lowers their bioavailability [26].

5 Strategies employed in the biological degradation of e-waste

Below is a discussion of the specific processes by which bacteria contribute to the breakdown of e-waste (Figure 3):

a) **Enzyme-mediated degradation of e-waste:** Bioremediation is an environmentally benign biotechnological technique that uses biological agents including bacteria, plants, and their enzymes to change dangerous contaminants into non-harmful/less toxic compounds through a variety of metabolic processes. Numerous enzymes derived from microorganisms (fungi and bacteria) and plants have been found to be crucial for the bioremediation of contaminants [28]. Many studies have been conducted on the enzymatic activities of significant enzymes, including hydrolases, dioxygenases, and oxidoreductases. Numerous heavy metals and PAHs can be changed into less hazardous forms by microbial enzymes like dioxygenases, dismutases, catalase, carboxylesterases, esterases, laccases, and reductases. In several phases of bioremediation, microbe-mediated enzymatic mechanisms are crucial because they interact with harmful pollutants and transform them into innocuous compounds [29]. Benefits of enzymes include reduced reliance on hazardous chemicals, ease of handling and storage, independence from microbial growth rates, substrate specificity, and consistency. Different enzymes have different capacities for breaking down heavy metal contaminants in e-waste [28]. In order to break down various classes of pollutants, enzymes such glyoxal oxidase, glucose oxidase, MnP (manganese peroxidase), proteases, laccases, amidases, esterases, dihydrodiol dehydrogenase, nitrilases, and cytochrome p450 are crucial [29, 30]. As they are more affordable, natural enzymes are typically chosen. However, there are encouraging prospects to create more effective enzymes thanks to cutting-edge technologies including genetic engineering, recombinant methods, and nanotechnology [28]. This is due to the fact that these technologies can be modified to alter the amino acid sequences of enzymes in order to attain particular metabolic characteristics, stress resistance, temperature tolerance, and pH required for heavy metal bioremediation [29]. However, by transferring coding genes for expression, genetic engineering can improve the output of enzymes. Additionally, it is anticipated that recombinant enzymes would be excellent in breaking down pollutants in natural settings [31].

b) **Biomineralization:** In order to benefit the biological system, biomineralization involves the microbial production of particular inorganic compounds using substrate molecules. This mechanism involves bacteria collecting anions or ligands, which subsequently bind to harmful metal pollutants and precipitate. It is a commonly used technique for degrading or precipitating polymers toxic and heavy metals found in e-waste [32]. Consequently, organic molecules break into less dangerous and more stable states, whereas polluting metals change into somewhat stable forms. Biologically controlled mineralization (BCM) and biologically induced mineralization (BIM) are the two forms of biomineralization. BIM can occasionally result in the formation of minerals on cell surfaces or inside cells [33]. However, BCM involves the synthesis of extracellular minerals as a result of microbial metabolic capacities [4]. *Cupriavidus metallidurans*, a metallophilic bacterium, is a possible option for collecting Au(III) since it can help with cellular detoxification [34]. Furthermore, *Cupriavidus metallidurans*, *Sporosarcina ginsengisoli*, and *Bacillus fusiformis* are well-known for their part in the process of biomineralization, which successfully removes heavy metals including lead, arsenic, and cadmium. Another study by Achal [35] highlighted how well *Sporosarcina ginsengisoli* can biomineralize arsenic (As(III)). The level of contamina-

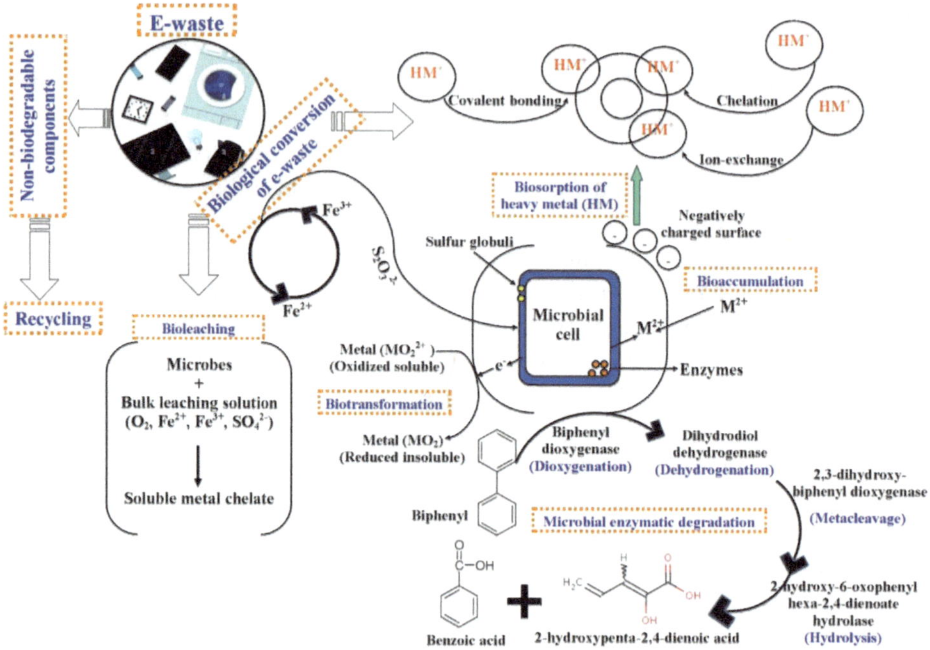

Figure 3: Various microbial mechanisms involved in the biodegradation of e-waste (reproduced from [1]).

tion in the affected area and the efficacy of the bacteria to be used are some of the aspects that also play a large part in biomineralization, despite the fact that the process has gotten a lot of attention lately.

c) **Bioaccumulation:** Through a detoxifying mechanism and an active process, bioaccumulation utilizes microbial capacity to absorb harmful metals and accumulate them inside their cell vacuoles. Energy is needed for the detoxification of vacuoles and absorption of metals. Remediation results from the removal of metals from the environment and their sequestration inside living cells. When utilized for bioremediation of metal-contaminated areas, plants and microbes are effective at removing metals through accumulation. This strategy produces better heavy metal removal when combined with methods like phytodegradation [36]. Bioaccumulation is the process by which metals are integrated into biological biomass. Several native strains of *Bacillus sphaericus* were tested for tolerance to metals like chromium (Cr), iron (Fe), cobalt (Co), mercury (Hg), and arsenic (As). Additionally, bioaccumulation in live biomass was evaluated and it was reported that both living and dead cells demonstrated an enormous capacity for bioaccumulation of metals. Zinc (Zn) and cadmium (Cd) bioaccumulation has been accomplished using gram-positive bacteria like *Tsukamurella paurometabola* and gram-negative bacteria like *Pseudomonas aeruginosa* [36]. Additionally, the study examined how *Fusarium flocciferum*, *Pseudomonas putida*, and *S.*

cerevisiae, removed mercury (Hg), arsenic (As), cadmium (Cd), and lead (Pb) [36, 37]. *Aspergillus niger* was found to be highly sensitive to all chromium(VI) concentrations compared to the bioaccumulation of chromium(VI), lead(II), and copper(II), inferring *A. niger* as a useful living biosorbent for removing heavy metal ions [38]. Further, researchers have also documented the role of recombinant *E. coli* in cadmium bioaccumulation through the cytosolic expression of metallothionein (MT) [39]. The growth rate and capacity of the microbes used in this process to bioaccumulate heavy metals are key factors in this technology. Furthermore, compared to in vitro results, the accuracy rates for field experiments continue to lag significantly. The cost-effectiveness of this approach, however, cannot be completely discounted out.

d) **Biotransformation:** In the context of e-waste remediation, biotransformation refers to the chemical modification of metals by microorganisms or transformations in their oxidative states brought about by the removal or addition of electrons by microbial agents. This process is crucial in converting chemical pollutants into compounds that are more environmentally friendly. High metal-binding capacity has been shown by a number of fungi, including *Stachybotrys* sp., *Phlebia* sp., *Klebsiella oxytoca*, *Botryosphaeria rhodina*, and *Allescheriella* sp. [40]. Furthermore, strains of gram-positive bacteria, like *Cellulosimicrobium* sp., have demonstrated resistance to heavy metals like lead, chromium, mercury, cadmium and xenobiotics. A strain of bacteria called *Bacillus* sp. SFC 500 was reported to convert chromium into its less harmful form by biotransformation [41]. Additionally, studies have demonstrated the effectiveness of fungi like *Rhodobacter sphaeroides* in removing hydrophobic hazardous metals from soil, such as cadmium and zinc [1]. There are two types of metal biotransformation: indirect and direct. Harsh multivalent metals are reduced through direct biotransformation, sometimes referred to as enzymatic biotransformation, which uses microbial enzymes to alter oxidation states [1]. On the other hand, in sedimentary and subterranean environments, metal-reducing bacteria stabilize multivalent dangerous metals by immobilizing them.

e) **Biosorption:** The biosorption mechanism is based on the binding and absorption of ionized harmful metals onto the cell surface [1]. Metabolism-dependent biosorption takes place in the presence of ATP through mechanisms like chelation, a special process in which molecules and ions bind to metal ions by creating di/poly coordinate bonds between a single central atom and a polydentate ligand. A film of the adsorbate is also produced on the surface of the adsorbent via a surface phenomenon called physical adsorption [1]. The phenomenon of chemiosmotic gradient potential controls processes like covalent bonding, ion exchange, and adsorption, by which biosorption takes place in the absence of ATP. Biosorption routes can be categorized as either metabolism-dependent or metabolism-independent based on the requirements of cell metabolism and the kind of metal contamination. Functional groups on bacterial cell surfaces interact physicochemically with metals through metabolism-independent processes that include ion exchange, chemical sorption, and physical adsorption [42].

In microbial cells, metal-absorbing groups like carboxyl, amino, sulfate, and phosphate groups are found in proteins, lipids, and carbohydrates. Microbes are known as biosorbents due to their ability to form bonds with e-waste in aqueous solutions [43]. Analyzing the characteristics of biosorbents, such as maximal sorption capacity, regeneration, sorption kinetics, and recovery of related metals, is crucial for assessing their microbial stability. In the biosorption of heavy metals, dead biomass, living cells, or polymers produced by their metabolic activities are used as biosorbent materials [44]. Streptomycetes, *Penicillium, Mucor, Rhizopus, Aspergillus,* and Actinomycetes are filamentous fungi that have a high potential to absorb metals. By absorbing metal ions in their mycelium and spores, fungi may both survive and detoxify them. It is even known that *Cephalosporium aphidicola* and *Aspergillus parasitica* can clean up regions contaminated with lead [45].

Due to their functional groups, microbes exhibit different biosorption capacities. Metal ions are tightly bound by negatively charged functional groups including carboxyl (CO^-), phosphate (PO_4^{3-}), and hydroxyl (OH^-) [1]. Bacterial functional groups like the sulfate (SO_4^{2-}) and carboxyl (COO^-) groups of uronic acid can also promote ion exchange. The cell walls of gram-negative bacteria, which contain enzymes, phospholipids, lipopolysaccharides, glycoproteins, and lipoproteins and gram-positive bacteria, which contain teichoic acid, meso-diaminopimelic acid, glutamic acid, and alanine function as ligands for actively binding metal ions. As their cell walls contain polysaccharides, yeasts are also regarded as desirable biosorbents. Heavy metals can be adsorbed by *Streptomyces longwoodensis, Saccharomyces cerevisiae,* and *Candida tropicalis* [1]. *Saccharomyces cerevisiae* is a yeast strain that has shown ease in holding onto metal ions like copper and cobalt [46]. Through mechanisms like ion exchange, yeasts like *S. cerevisiae* can also function as bioremediation agents. Because they may thrive in a range of environmental conditions, bacteria and fungi are appealing biosorbents for e-waste treatment. Because of their large biomass, algae have a tremendous ability of biosorption via ion exchange mechanism that makes them extremely effective when compared to other microorganisms [47]. With functional groups like NH_2, SH, SO_3^-, and COO^-, brown sea algae (like *Fucus vesiculosus*) efficiently remove metals including lead, nickel, and cadmium. Due to its adaptability to various environmental circumstances, biosorption is frequently employed as a biological instrument for the accumulation of environmentally dangerous heavy metals through physicochemical pathways of uptake. However, the biosorbent materials and related expenses determine its efficacy [46].

f) **Bioleaching:** Utilizing acidophilic microorganisms to help dissolve solid heavy metals within a sediment matrix is known as bioleaching [48]. This technique works especially well for pollutants like sulfur or iron [49]. There are two types of bioleaching processes: “direct” and “indirect.” Electron transport from the metal sulfide to the cell attached to the mineral surface takes place immediately in direct leaching. Conversely, metal ions like iron (III) ions cause indirect leaching. Iron (II)-oxidizing microbes, which can be adhered to the mineral surface or float freely, generate these ions and act as agents that oxidize metal sulfides. Certain organisms are frequently used in the field of bioleaching due to their capacity to remove metals. Fungi like *Penicillium simplicissimum* and *Aspergillus niger*, as well as bacteria like *Acidithiobacillus ferrooxidans, Leptospirillum ferriphilum, T. ferrooxidans*, and *Thiobacillus thiooxidans*, have been widely used in the extraction of metals from electronic waste materials [50]. The three main types of microorganisms that actively participate in the bioleaching of metals are heterotrophic fungi (like *Penicillium* and *Aspergillus*), heterotrophic bacteria (like *Bacillus* and *Pseudomonas*), and autotrophic bacteria (like *Sulfobacillus benefaciens* and *Thiobacillus*) [48]. These microbes are able to extract metals from mineral concentrates and ores that contain iron or sulfide. Among these, *Aspergillus niger* is notable for its capacity to generate organic acids, including malic, oxalic, gluconic, and citric acids. By acting as potent chelating agents during the bioleaching process, these organic acids enable the recovery of metals from materials like lithium-ion batteries. *Aspergillus niger* produces gluconic acid, which may chelate and dissolve significant amounts of several metals. Additionally, *Penicillium* sp. is frequently employed in gold recovery bioleaching techniques while *Aspergillus niger* may leach zinc oxide [51]. *Aspergillus niger* has effectively extracted metals from fly ash [52].

6 Prospective applications for genetically modified organisms (GMOs) in the breakdown of components of e-waste

PAHs, which are sometimes difficult to treat using traditional bioremediation techniques, are among the heavy metals and pollutants that genetic engineering offers intriguing prospects to mitigate. Compared to physiochemical approaches, which are less environmentally friendly and may endanger human life, genetically modified organisms (GMOs) provide substantial advantages for bioremediation because of their reduced health risks and environmentally benign nature. For example, *E. coli* JM109 transformed with plasmids pGPMT (M4) and pCLG2 (M5) showed improved Cd^{2+} absorption as M4 expressed the cadmium transport system and metallothionein (MT), thereby doubling the initial absorption capability of the strain [53]. Huang et al. [54] used the CmarsM gene from the heat-resistant algae *Cyanidioschyzon merolae* to methylate and volatilize arsenic (As) using a genetically engineered *Bacillus subtilis* strain 168, which may help clean up compost contaminated with As. In order to improve the bioreduction of heavy metals like chromium, Li et al. [55] used a novel technique called STAR, which uses CRISPR-ddAsCpf1 to increase the electron transfer capacity of *Shewanella oneidensis* MR-1. Additionally, some enzymes have the ability to change HMs (heavy metals) into less hazardous forms. For instance, *E. coli* BL308 can withstand higher quantities of mercury(II) and transform it into less harmful mercury(0) when the mercury resistance gene merA from *Deinococcus radiodurans* is expressed in the bacterium [56]. HMs are captured by metal-binding peptides, such as copM for copper (Cu)(II), pbrT and pbrD for lead (Pb)(II), and cadB for zinc (Zn)(II) and cadmium (Cd)(II), while metallothioneins with sulfhydryl groups and cysteine are used for HM binding [57, 58]. GMOs have demonstrated efficacy in breaking down PAHs in addition to heavy metals. GEMs (genetically modified microorganisms) use certain enzymes such dehydrogenase, hydroaldolase, monooxygenase, and dioxygenase to break down PAHs [59]. Variations in the enzymes produced by functional genes are frequently responsible for changes in degradation routes and efficiency [59]. These functional genes are often used to create GEMs that break down PAHs. For instance, when Mohtashami et al. [60] introduced *Pleurotus ostreatus* laccase gene (poxa1b) into *E. coli* BL21, benzo[α]pyrene was oxidized by 17%. By co-expressing pdoAB with plasmid pBRCD, they were able to oxidize benzo[α]pyrene, anthracene, pyrene, and phenanthrene with the help of electron-transferring factors from plasmid pBRCD. To the best of our knowledge, however, no literature demonstrates the effective field application of GMOs in the degradation of e-waste components.

7 Challenges relating to biological remediation

Bioremediation has a number of limitations despite scientific advancements and economy when compared to more traditional methods like biomass combustion or landfilling. Strongly aromatic hydrocarbons and chlorinated organic compounds are examples of e-waste components that are resistant to bacterial breakdown [61]. The efficiency of bioremediation is influenced by the kind and quantity of pollutants, adsorption by soil particles, geographical location, and soil texture. The bioremediation selectivity requires the employment of specific microorganisms, appropriate conditions for growth, and adequate nourishment accessibility. Bioremediation is affected by moisture, oxygen levels, pH, temperature, post-inoculation decrease, microbial competition, and other ecological factors [62]. As the temperature rises, pollutants become more soluble. Another roadblock is a lack of thorough understanding of gene expression, microbial ecology, physiology, and site-specific factors. It is still difficult to develop cutting-edge bioremediation systems appropriate for extensively contaminated sites with a variety of hazardous contaminants. The approval criteria for bioremediation are in turmoil, and there isn't a commonly recognized explanation or method of treatment [63]. Because the suppression of microorganisms by harmful heavy metals depends on variables such as redox potential, metal ion concentration, and ecological circumstances, evaluating bioremediation capacity is complicated. Microorganism cell wall structure and sorption sites are two factors that affect how well metal-microbe complex stabilization works. In general, the substrate being treated and the particular environmental conditions at hand dictate how effective bioremediation is.

8 Futuristic perspectives

It is impossible to overestimate the significance of managing the present rate and volume of e-waste in addition to its consequences on the environment. The current situation demonstrates how poorly maintained e-waste reuse procedures lead to the discharge of persistent hazardous materials, such as PCDDs and PBDEs, into the atmosphere, water, soil, airborne particles, residual ash, and the immediate surroundings. Additionally, as demonstrated by Miller et al. [64], these dangerous substances inevitably reach terrestrial and oceanic ecosystems, initiating a process of biomagnification and bioaccumulation. The accessibility of obtainable elements decreases as these dangerous compounds continue to build up. In order to prevent catastrophic consequences, scientists have created environmentally safe and suitable techniques for recovering and recycling hazardous materials from garbage. These actions have major immediate and long-term consequences on the environment in addition to improving human health [65]. Furthermore, the use of bioremediation techniques to clean up groundwater reservoirs and landfills has grown significantly. Even within the scope of improvement, there is still a lack of appropriate application of the variety of waste management solutions accessible, both in developing and developed nations. However, it is impossible to overstate the urgent need for clear rules, upkeep procedures, and all-encompassing policies to keep an eye on environmental and health problems caused by hazardous metals. This urgent requirement is still a top priority going forward and is especially pertinent in the current situation.

9 Conclusion

The fast rise in e-waste is a serious issue that needs to be addressed promptly. This issue has global repercussions, affecting areas all over the globe with a variety of electronic waste management challenges. In order to have a sustainable future, it is imperative that we address this issue. New technologies, each with pros and cons of its own, are being developed as science advances to address this looming catastrophe. Any innovation, however, is only useful if it can benefit humanity in a more ecologically friendly and clean way. Through a variety of processes, microorganisms present a promising answer to this problem. Microorganisms have been shown to be capable of efficiently and economically remediating a variety of e-waste, comprising hydrocarbons such as PAHs, oil, medicines, and PCBs. Additionally, it has been noted that some bacteria aid in the leaching process, which could lead to new developments in metallurgy and the extraction of metals from ores. It is crucial to remember that, in comparison to physical and chemical approaches, microbial breakdown processes can take longer. However, there is a great deal of promise for enhancing microbial degradative mechanisms in the future by contemporary biotechnological advancements.

References

[1] Khan S, Alam FS, Borah D. Microbial cell factories in the remediation of e-wastes: an insight. Biotechnology for the Environment. 2024;1(1):4. https://doi.org/10.1186/s44314-024-00003-4.

[2] Kumar A, Saini HS, Kumar S. Bioleaching of gold and silver from waste printed circuit boards by *Pseudomonas balearica* SAE1 isolated from an e-waste recycling facility. Current Microbiology. 2018;75:194–201.

[3] Chakraborty SC, Qamruzzaman M, Zaman MW, Alam MM, Hossain MD, Pramanik BK, Nguyen LN, Nghiem LD, Ahmed MF, Zhou JL, Ibrahim M, Mondal H, Hossain MA, Johir MAH, Ahmed MB, Sithi JA, Zargar M, Moni MA. Metals in e-waste: occurrence, fate, impacts and remediation technologies. PSEP. 2022;162:230–252.

[4] Mor RS, Sangwan KS, Singh S, Singh A, Kharub M. E-waste management for environmental sustainability: an exploratory study. Procedia CIRP. 2021;98:193–198.

[5] Mitra S, Chakraborty AJ, Tareq AM, Emran TB, Nainu F, Khusro A, Idris AM, Khandekar MU, Osman H, Alhumaydhi FA, Simal-Gandara J. Impact of heavy metals on the environment and human health: Novel therapeutic insights to counter the toxicity. Journal of King Saud University – Science. 2022;34 (3): 101865.

[6] Ankit S, Kumar L, Tiwari V, Sweta J, Rawat S, Singh J, Bauddh K. Electronic waste and their leachates impact on human health and environment: global ecological threat and management. Environmental Technology and Innovation. 2021. https://doi.org/10.1016/j.eti.2021.102049.

[7] Srichandan H, Mohapatra RK, Parhi PK, Mishra S. Bioleaching approach for extraction of metal values from secondary solid wastes: a critical review. Hydrometallurgy. 2019;189:105122.

[8] UNU-United Nations University (2020): Global e-waste surging: up 21% in 5 years; https://unu.edu/press-release/global-e-waste-surging-21-5-years

[9] India's Electronics Manufacturing and Export Market Report (2023): https://www.ibef.org/research/case-study/india-s-electronics-manufacturing-and-export-market

[10] Pulicherla KK, Adapa V, Ghosh M, Ingle P. Current efforts on sustainable green growth in the manufacturing sector to complement "make in India" for making "self-reliant India. Environmental Research. 2022;206:112263.

[11] Ahsan Z, Cai Z, Wang S, Wang H, Ma Y, Song G, Zhang S, Yang W, Imran M, Wen C. Enhanced stability and electrochemical properties of lanthanum and cerium co-modified $LiVOPO_4$ cathode materials for Li-ion batteries. Journal of Rare Earths. 2022;41(10): 1002–1721.

[12] Briffa J, Sinagra E, Blundell R. Heavy metal pollution in the environment and their toxicological effects on humans. Heliyon. 2020;6(9): 2020.

[13] Siddiqua A, Hahladakis JN, Al-Attiya WAKA. An overview of the environ mental pollution and health effects associated with waste landfilling and open dumping. Environmental Science and Pollution Research. 2022;29(39): 58514–58536.

[14] Okereafor U, Makhatha M, Mekuto L, Uche-Okereafor N, Sebola T, Mavumengwana V. Toxic metal implications on agricultural soils, plants, animals, aquatic life and human health. International Journal of Environmental Research and Public Health. 2020;17(7): 2204.

[15] Yang L, Zhang H, Zhang X, Xing W, Wang Y, Bai P, Zhang L, Hayakawa K, Toriba A, Tang N. Exposure to atmospheric particulate matter-bound polycyclic aromatic hydrocarbons and their health effects: a review. International Journal of Environmental Research and Public Health. 2021;18(4): 2177.

[16] Aziz KHH, Mustafa FS, Omer KM, Hama S, Hamarawf RF, Rahman KO. Heavy metal pollution in the aquatic environment: efficient and low cost removal approaches to eliminate their toxicity: a review. RSC Advances. 2023;13(26): 17595–17610.

[17] Bakulski KM, Seo YA, Hickman RC, Brandt D, Vadari HS, Hu H, Park SK. Heavy metals exposure and Alzheimer's disease and related dementias. Journal of Alzheimer's Disease. 2020;76(4): 1215–1242.

[18] National Research Development Corporation (NRDC 2020): Single Use Plastics 101; https://www.nrdc.org/stories/single-use-plastics-101#pollution

[19] Miglani R, Parveen N, Kumar A, Ansari MA, Khanna S, Rawat G, Panda AK, Bisht SS, Upadhyay J, Ansari MN. Degradation of xenobiotic pollutants: an environmentally sustainable approach. Metabolites. 2022;12(9): 818.

[20] Fashola MO, Ngole-Jeme VM, Babalola OO. Heavy metal immobilization potential of indigenous bacteria isolated from gold mine tailings. International Journal of Environmental Research. 2020;14:71–86.

[21] Cao HL, Liu C, Cai FY, Qiao XX, Dichiara AB, Tian C, Lř J. In situ immobilization of ultra-fine AgNPs onto magnetic Ag@RF@Fe_3O_4 core-satellite nanocomposites for the rapid catalytic reduction of nitrophenols. Water Research. 2020;179:115882.

[22] Huang L, Wu C, Gao H, Xu C, Dai M, Huang L, Hao H, Wang X, Cheng G. Bacterial multidrug efflux pumps at the frontline of antimicrobial resistance: an overview. Antibiotics. 2022;11(4): 520.

[23] Benkhaya S, M'rabet S, El Harfi A. Classifications, properties, recent synthesis and applications of azo dyes. Heliyon. 2020;6(1): e03271.

[24] Godfred S-A, Francis A, Ismaila E, Francis GO. A review of the sources, distribution sequences, and health risks associated with exposure to atmospheric polycyclic aromatic hydrocarbons. Cogent Engineering. 2023;10:1.

[25] Elufisan TO, Rodríguez-Luna IC, Oyedara OO, Sánchez-Varela A, Hernán dez-mendoza A, Dantán Gonzalez E, Paz-González AD, Muhammad K, Rivera G, Villalobos-Lopez MA, Guo X. Pemsol isolated from Mexico. PeerJ. 2020;8:e8102.

[26] Patel AB, Shaikh S, Jain KR, Desai C, Madamwar D. Polycyclic aromatic hydrocarbons: sources, toxicity, and remediation approaches. Front Microbiol. 2020;11:562813.

[27] Abo-State MAM, Osman ME, Khattab OH, El-Kelani TA, Abdel-Rahman ZM. Degradative pathways of polycyclic aromatic hydrocarbons (PAHs) by *Phanerochaete chrysosporium* under optimum conditions. JRRAS. 2021;14(1): 507–520.

[28] Mousavi SM, Hashemi SA, Iman Moezzi S,M, Ravan N, Gholami A, Lai CW, Chiang WH, Omidifar N, Yousefi K, Behbudi G. Recent advances in enzymes for the bioremediation of pollutants. Biochemistry Research International. 2021;2021:5599204.

[29] Bhandari S, Poudel DK, Marahatha R, Dawadi S, Khadayat K, Phuyal S, Shrestha S, Gaire S, Basnet K, Khadka U, Parajuli N. Microbial enzymes used in bioremediation. Journal of Chemistry. 2021;2021:1–17.

[30] Khalid F, Hashmi MZ, Jamil N, Qadir A, Ali MI. Microbial and enzymatic degradation of PCBs from e-waste-contaminated sites: a review. Environmental Science and Pollution Research. 2021;28:10474–10487.

[31] Mital S, Christie G, Dikicioglu D. Recombinant expression of insoluble enzymes in *Escherichia coli*: a systematic review of experimental design and its manufacturing implications. Microbial Cell Factories. 2021;20:1–20.

[32] Zhang Y, Hu X, Wang Y, Jiang N. A critical review of biomineralization in environmental geotechnics: applications, trends, and perspectives. Biogeotechnics. 2023;100003.

[33] Dong H, Huang L, Zhao L, Zeng Q, Liu X, Sheng Y, Shi L, Wu G, Jiang H, Li F, Zhang L, Guo D, Li G, Hou W, Chen H. A critical review of mineral microbe interaction and co-evolution: mechanisms and applications. National Science Review. 2022;9(10): nwac128.

[34] Reith F, Etschmann B, Grosse C, Moors H, Benotmane MA, Monsieurs P, Grass G, Doonan C, Vogt S, Lai B, Martinez-Criado G, George GN, Nies DH, Mergeay M, Pring A, Southam G, Brugger J. Mechanisms of gold biomineralization in the bacterium *Cupriavidus metallidurans*. Proceedings of the National Academy of Sciences. 2009;106(42): 17757–17762.

[35] Achal V, Pan X, Fu Q, Zhang D. Biomineralization based remediation of As (III) contaminated soil by *Sporosarcina ginsengisoli*. Journal of hazardous materials. 2012;201:178–184.

[36] Nnaji ND, Onyeaka H, Miri T, Ugwa C. Bioaccumulation for heavy metal removal: a review. SN Applied Sciences. 2023;5(5): 125.

[37] Massoud R, Hadiani MR, Hamzehlou P, Khosravi-Darani K. Bioremediation of heavy metals in food industry: application of Saccharomyces cerevisiae. Electronic Journal of Biotechnology. 2019;37:56–60.

[38] Dursun AY, Uslu G, Cuci Y, Aksu Z. Bioaccumulation of copper (II), lead (II) and chromium (VI) by growing *Aspergillus niger*. Process Biochemistry. 2003;38(12): 1647–1651.

[39] Lu CW, Ho HC, Yao CL, Tseng TY, Kao CM, Chen SC. Bioremediation potential of cadmium by recombinant *Escherichia coli* surface expressing metallothionein MTT5 from *Tetrahymena thermophila*. Chemosphere. 2023;310:136850.

[40] D'Annibale A, Leonardi V, Federici E, Baldi F, Zecchini F, Petruccioli M. Leaching and microbial treatment of a soil contaminated by sulphide ore ashes and aromatic hydrocarbons. Applied Microbiology and Biotechnology. 2017;74:1135–1144.

[41] Thatoi H, Das S, Mishra J, Rath BP, Das N. Bacterial chromate reductase, a potential enzyme for bioremediation of hexavalent chromium: a review. Journal of Environmental Management. 2014;146:383–399.

[42] Priya AK, Gnanasekaran L, Dutta K, Rajendran S, Balakrishnan D, Soto Moscoso M. Biosorption of heavy metals by microorganisms: Evaluation of different underlying mechanisms. Chemosphere. 2022;307:135957.

[43] Ali Redha A. Removal of heavy metals from aqueous media by biosorption. Arab Journal of Basic and Applied Sciences. 2020;27(1): 183–193.

[44] Concórdio-Reis P, Reis MA, Freitas F. Biosorption of heavy metals by the bacterial exopolysaccharide FucoPol. Applied Sciences. 2020;10(19): 6708.

[45] Akar T, Tunali S, Cabuk A. Study on the characterization of lead (II) biosorption by fungus *Aspergillus parasiticus*. Applied Biochemistry and Biotechnology. 2017;136:389–406.

[46] Savastru E, Bulgariu D, Zamfir C-I, Bulgariu L. Application of *Saccharomyces cerevisiae* in the biosorption of Co(II), Zn(II) and Cu(II) ions from aqueous media. Water. 2022;14(6): 976.

[47] Ankit S, Kumar L, Tiwari V, Sweta J, Rawat S, Singh J, Bauddh K Electronic waste and their leachates impact on human health and environment: global ecological threat and management. Environ Technol Innov. 2021. https://doi.org/10.1016/j.eti.2021.102049

[48] Sarkodie EK, Jiang L, Li K, Yang J, Guo Z, Shi J, Deng Y, Liu H, Jiang H, Liang Y, Yin H, Liu X. A review on the bioleaching of toxic metal(loid)s from contaminated soil: insight into the mechanism of action and the role of influencing factors. Front Microbiol. 2022;13:1049277.

[49] Bhandari G, Gupta S, Chaudhary P, Chaudhary S, Gangola S. Bioleaching: a sustainable resource recovery strategy for urban mining of E-waste. Microb technol Sustaina E-waste Manag. Cham: Springer; 2023, Vol. 157 175. https://doi.org/10.1007/978-3-031-25678-3_10.

[50] Adetunji AI, Oberholster PJ, Erasmus M. Bioleaching of metals from e-waste using microorganisms: a review. Minerals. 2023;13(6): 828.

[51] Saitoh N, Fujimori R, Nakatani M, Yoshihara D, Nomura T, Konishi Y. Microbial recovery of gold from neutral and acidic solutions by the baker's yeast *Saccharomyces cerevisiae*. Hydrometallurgy. 2018;181:29–34.

[52] Annamalai M, Gurumurthy K. Microbiological leaching of metals and its recovery from waste electrical and electronic equipment: a review. World Review of Science, Technology and Sustainable Development. 2019;15(1): 1–6.

[53] Deng X, Yi XE, Liu G. Cadmium removal from aqueous solution by gene modified *Escherichia coli* JM109. Journal of Hazardous Materials. 2007;139(2): 340–344.

[54] Huang K, Chen C, Shen Q, Rosen BP, Zhao FJ. Genetically engineering *Bacillus subtilis* with a heat-resistant arsenite methyltransferase for bioremediation of arsenic-contaminated organic waste. Applied and Environmental Microbiology. 2015;81(19): 6718–6724.

[55] Liang Y, Zhao CZ, Yuan H, Chen Y, Zhang W, Huang JQ, Yu D, Liu Y, Titirici M, Chueh Y, Yu H, Zhang Q. A review of rechargeable batteries for portable electronic devices. InfoMat. 2019;1(1): 6–32.

[56] Brim H, Mcfarlan SC, Fredrickson JK, Minton KW, Zhai M, Wackett LP, Daly MJ. Engineering *Deinococcus radiodurans* for metal remediation in radioactive mixed waste environments. Nature Biotechnology. 2000;18(1): 85–90.

[57] Giner-Lamia J, López-Maury L, Florencio FJ. CopM is a novel copper binding protein involved in copper resistance in *Synechocystis* sp PCC 6803. Microbiologyopen. 2015;4(1): 167–185.

[58] Xavier JC, Costa PES, Hissa DC, Melo VMM, Falcão RM, Balbino VQ, Mendonça LAR, Lima MGS, Coutinho HDM, Verde LCL. Evaluation of the microbial diversity and heavy metal resistance genes of a microbial community on contaminated environment. Applied Geochemistry. 2019;105:1–6.

[59] WU C, Li F, Yi S, Ge F. Genetically engineered microbial remediation of soils co-contaminated by heavy metals and polycyclic aromatic hydrocarbons: advances and ecological risk assessment. Journal of Environmental Management. 2021;296:113185.

[60] Mohtashami M, Fooladi J, Haddad-Mashadrizeh A, Housaindokht M. Molecular cloning, expression and characterization of poxa1b gene from *Pleurotus ostreatus*. Molecular Biology Reports. 2019;46:981–990.

[61] Vishwakarma A, Hait S. E-Waste Valorization and Resource Recovery. Management of Electronic Waste: Resource Recovery, Technology and Regulation. 2024;202–233.

[62] Han P, Teo WZ, Yew WS. Biologically engineered microbes for bioremediation of electronic waste: wayposts, challenges and future directions. Engineering Biology. 2022;6(1): 23–34.

[63] Sharma S, Sharma S. Microbial Biotechnology for Circular Economy in Wastewater Treatment: Potentials, Technologies, and Challenges. In Advanced and Innovative Approaches of Environmental Biotechnology in Industrial Wastewater Treatment 2023 Aug 1 (pp. 1–21). Singapore: Springer Nature Singapore.

[64] Miller ME, Hamann M, Kroon FJ. Bioaccumulation and biomagnification of microplastics in marine organisms: A review and meta-analysis of current data. PloS one. 2020;15(10): e0240792.

[65] Seif R, Salem FZ, Allam NK. E-waste recycled materials as efficient catalysts for renewable energy technologies and better environmental sustainability. Environment, Development and Sustainability. 2023;1–36.

Harlina Ahmad

Phytoremediation of e-waste-contaminated sites: a sustainable approach

Abstract: The escalation of electronic waste, commonly referred to as "e-waste," constitutes a significant threat to ecosystems and human health. Hazardous substances, such as heavy metals and persistent organic pollutants, present in e-waste, pose risks to both humans and the environment. Conventional disposal techniques, including incineration and landfilling, are associated with high costs and have the potential to exacerbate environmental issues. Phytoremediation offers a safe and effective alternative for addressing e-waste contamination. This method employs specific plant species to absorb, stabilize, or decompose pollutants in contaminated soil and water. Notably, it is effective even in resource-constrained regions. Furthermore, this chapter delineates the various stages of phytoremediation, including rhizofiltration, phytoextraction, phytostabilization, and phytodegradation. It also examines how interactions between plants and microorganisms enhance the efficiency of environmental remediation. The careful selection of plant species, based on their tolerance to pollutants, biomass yield, rooting characteristics, and environmental adaptability, is likely to be highly effective for phytoremediation. Several practical applications of phytoremediation in areas impacted by e-waste are also included. Remediation strategies, such as integrating phytoremediation with constructed wetlands and nano-phytotechnology, are described. Although phytoremediation is an attractive approach, it has several limitations, including lengthy cleanup periods, challenges in treating complex pollutants, and potential environmental impacts. Overcoming these challenges necessitates ongoing research, technological advancements, and interdisciplinary collaboration.

Keywords: Phytoremediation, e-waste management, heavy metals, sustainable remediation, plant-microbe interactions

1 Introduction

The production and consumption of electrical and electronic devices have increased substantially in recent years, resulting in a significant rise in electronic waste. Currently, with approximately 60.0 million metric tons of e-waste generated worldwide in 2023, the yearly increase in e-waste was 3–4% [1, 2]. Rising consumption, short life-

*Corresponding author: Harlina Ahmad, Environmental Technology Division, School of Industrial Technology, Universiti Sains Malaysia, 11800 Penang, Malaysia, e-mail: harlinaa@usm.my

 | https://doi.org/10.1515/9783112225509-008

spans, and limited repair have led to increased e-waste generation. Poor or insufficient management jeopardizes the environment and human health, as e-waste contains hazardous substances. In 2023, only 17.4% of e-waste generated was recovered and recycled into a valuable economic asset [1]. Considering the worldwide rise in electronic waste, innovative and sustainable consumption and production methods are essential to safeguard the environment.

E-waste poses significant risks to human health and environmental integrity, particularly in the absence of regulated recycling procedures. It incorporates toxic substances such as heavy metals, brominated flame retardants, and polycyclic aromatic hydrocarbons, which frequently contaminate the air, water, and soil in the vicinity. Chronic exposure to these chemicals can lead to carcinogenesis, neurotoxicity, and respiratory disorders [3]. Unregulated or unlawful disposal methods amplify these health and environmental dangers.

Electronic waste management primarily uses methods such as acid leaching, thermal processing, incineration, landfilling, pyrometallurgy, and hydrometallurgy. These methods can cause environmental and health hazards. When disposal and recycling are unregulated, hazardous substances like heavy metals, persistent organic pollutants (POPs), and other toxic chemicals often contaminate soil and water. Traditional, unregulated practices, driven by economic and social factors, underpin e-waste management in many developing countries [4]. These methods are often ineffective, harm the environment, and consume substantial resources. Improper disposal can expose individuals to toxic substances and safety hazards that may cause health issues such as cancer and skin problems.

Phytoremediation is an inexpensive and environmentally friendly method for removing pollutants from contaminated sites. The sources of heavy metals and organic matter in industrial, agronomic, and municipal wastewater will be reduced in a more environmentally friendly manner. Phytoremediation increases economic value by producing biofuels, essential oils, and biomass. Aquatic plants effectively treat wastewater by absorbing and degrading contaminants, including nitrates, phosphates, and heavy metals [5]. Conventional approaches to remediate e-waste-contaminated sites, such as physical and chemical treatments, are both costly and environmentally harmful. In contrast, phytoremediation is regarded by some as an attractive approach for in situ remediation of trace metals and organic contaminants, as it is of low capital and is greener [6]. Nano-phytoremediation, a hybrid technique combining phytotechnology and phytogenetics, has been shown to enhance the capacity of phytoremediation to address e-waste pollution, suggesting it could be further developed as a new treatment approach. [7]

Phytoremediation can eliminate contaminants from soil and water sources. Nevertheless, it has limitations, including prolonged remediation times and the need for continuous maintenance. The efficacy of this technique depends on the bioaccumulation potential of the selected plant species, the pollutant concentrations in the environment, and the specific characteristics of the plants used. More research is needed to enable large-scale application of phytoremediation and to elucidate the mechanisms of pollutant removal and the treatment methods. Phytoremediation comprises several methods, including phytoextraction, phytostabilization, phytodegradation, phytovolatilization, and phytofiltration, in which plants can be used to accumulate, immobilize, or degrade contaminants from air, soil, and water. The use of plants, in conjunction with plant-associated bacteria, in this approach, demonstrates greater efficacy in bioremediation, encompassing not only organic compounds but also inorganic pollutants. This chapter provides an overview of the potential of phytoremediation as an environmentally friendly, sustainable, and practical approach to e-waste treatment and recycling, with particular reference to current waste management practices.

2 E-waste composition and environmental pathway

2.1 E-waste composition

Lead, cadmium, mercury, arsenic, and chromium are among the heavy metals that pose risks to soil, water, and human health. POPs, including PCBs, PBDEs, and TBBPA, are characterized by their environmental persistence and health hazards. Inadequate plastic waste management and disposal also contribute to the release of toxic chemicals that harm air, soil, and water. Table 1 details the constituents of e-waste and their environmental impacts.

Table 1: Comprehensive overview of the constituents of e-waste and their associated environmental impacts.

Constituent	Environmental impact	References
Heavy metals (e.g., lead, cadmium, mercury, arsenic, and chromium)	Contamination degrades soil quality, harms plants, microbes, and human health. Poses risks, including neurodevelopmental issues, mutagenic effects, and bioaccumulation in the food chain.	[8–10]
Polychlorinated biphenyls (PCBs)	Persistent in the environment and may cause cancer and other health problems in humans.	[11]
Polybrominated diphenyl ethers (PBDEs)	Persist and accumulate in the environment and organisms. They may act as an endocrine disruptor and a neurotoxin in humans.	[12]
Tetrabromobisphenol A (TBBPA)	Persistent, and may threaten the environment and health.	[13]

Table 1 (continued)

Constituent	Environmental impact	References
Polyvinyl chloride (PVC)	Dioxins and furans are toxic substances released during improper disposal, which can cause cancer and health issues.	[14]
Batteries (e.g., lead-acid and lithium ion)	Release of heavy metals and toxins contaminates soil and water.	[15]
Cathode ray tubes (CRTs)	Contains lead and heavy metals that can leak into the environment.	[16]
Printed circuit boards (PCBs)	Contaminants, including heavy metals and toxins, pose health risks during recycling.	[17]
Plastics	Burning or improper disposal releases toxic chemicals, polluting the air and soil.	[18]

2.2 Release and transport mechanisms of e-waste contaminants

The release and transport of e-waste contaminants encompass various mechanisms and factors. Polyhalogenated compounds, such as polybrominated diphenyl ethers (PBDEs), are mainly released through thermal treatment and open burning. Their transport occurs primarily via airborne particles and soil. Recycling processes emit heavy metals that are relatively immobile due to their low vapor pressure. Antimony, originating from e-waste microplastics, is released via surface dissolution, a process influenced by environmental parameters such as pH and agitation rate. Microplastics also release decabromodiphenyl ether (BDE-209), which diffuses during transport. Landfills and processing facilities discharge per- and polyfluoroalkyl substances, which eventually infiltrate soil and water. Hazardous wastes move through environments via adsorption, convection, and diffusion. Understanding these processes helps establish effective management strategies to mitigate the environmental and health impacts of electronic waste contaminants. Table 2 summarizes the release mechanisms and transport pathways of these contaminants into environmental compartments.

Table 2: Summary of the release and transport mechanisms of e-waste contaminants.

Contaminant	Release mechanism	Transport mechanism	References
Polyhalogenated compounds (e.g., PBDEs and PCBs)	Thermal treatment and open burning release PBDEs; diffusion from plastic matrices occurs.	Airborne particles, residual ashes, and gases; short-range soil transport.	[19, 20]

Table 2 (continued)

Contaminant	Release mechanism	Transport mechanism	References
Heavy metals (e.g., Cd, Cu, and Pb)	Released during e-waste recycling.	Low vapor pressure limits transport near sources.	[21]
Antimony (Sb)	Surface dissolution and diffusion of e-waste microplastics depend on pH, agitation, and salinity.	Dissolution in water depends on particle size and environmental conditions.	[22]
Decabromodiphenyl ether (BDE-209)	Diffusion from e-waste microplastics.	Internal diffusion, interfacial mass transfer, and environmental diffusion.	[23]
Per-/poly-fluoroalkyl substances (PFAS)	Released from e-waste sites and landfills.	Diffusion into water and soil causes ecotoxicity.	[24]

2.3 Ecotoxicological impacts of e-waste

E-waste contaminants pose complex and significant risks. These pollutants harm human health and the environment. Heavy metals such as lead and cadmium induce neurodevelopmental issues, respiratory illnesses, and reproductive disorders. POPs, including PBDEs and polychlorinated biphenyls, interfere with endocrine function and increase cancer risk. E-waste exerts ecotoxicological effects through substances such as PBDEs, heavy metals, brominated flame retardants, polychlorinated dibenzofurans, polycyclic aromatic hydrocarbons, and dioxins, all of which threaten ecological systems. The improper disposal of hazardous substances, including lead, mercury, and cadmium, into soil, air, and water sources results in environmental pollution, destruction, and resource depletion.

E-waste endangers biodiversity and wildlife by containing toxic substances, including heavy metals, flame retardants, and POPs, which contaminate ecosystems. These hazardous materials disturb habitats and poison organisms at all trophic levels. Pollutants derived from e-waste bioaccumulate and magnify through food chains, leading to long-term ecological damage and adverse effects on living organisms. Improper disposal of e-waste results in toxic materials polluting ecosystems and poses threats to both human and animal health. Prolonged exposure causes respiratory issues, neurological impairments, and cancer, particularly in communities situated near informal recycling facilities. Exposure to e-waste also disrupts iron metabolism, blood glucose regulation, and the functioning of repair enzymes. It can impair antibody production, causing nutritional deficiencies and health complications within vulnerable populations [25].

Research on e-waste recycling has identified many findings that emphasize the need to address technical, financial, and policy factors on a large scale to reduce the ecotoxicological risks associated with e-waste, as well as the significance of economic incentives [26]. The ecotoxicological impacts of e-waste on the environment are substantial. They contribute to the accumulation of toxic compounds in ecosystems, thereby adversely affecting human health. While existing regulations help mitigate these risks, coordinated efforts among governments, industries, and communities are essential to improve e-waste management. Table 3 summarizes the potential risks, affected groups, and impacted populations.

Table 3: Summary of the potential risks of e-waste contaminants.

Contaminant	Potential risks	Affected groups	References
Lead (Pb)	Neurodevelopmental disorders, respiratory problems, reduced blood cells, increased blood lead in children, stillbirths, and preterm births	Children, pregnant women, the elderly, and landfill workers	[27–29]
Cadmium (Cd)	Respiratory issues, kidney damage, and endocrine disruption.	General population and children	[30]
Mercury (Hg)	Neurological damage, respiratory issues, and cancer	Communities near recycling sites	[31]
Polybrominated diphenyl ethers (PBDEs)	Endocrine disruption, reproductive toxicity, and cancer	E-waste workers and children	[32]
Polychlorinated biphenyls (PCBs)	Cancer, endocrine disruption, and respiratory issues	E-waste workers and general population	[11]
Polycyclic aromatic hydrocarbons (PAHs)	Cancer, DNA damage, and oxidative stress	E-waste workers and general population	[33]
Persistent organic pollutants (POPs)	Developmental neurotoxicity, cancer, and endocrine disruption	Children, pregnant women, and the general population	[34]
Liquid crystal monomers (LCMs)	Bioaccumulation, toxicity, and health risks to workers and families	E-waste workers and infants	[35]
Heavy metals (general)	Genetic damage, sperm abnormalities, and reduced sperm count	General population and E-waste workers	[36]
Microbial contaminants	Pathogenic infections, Exposure to mycotoxins	E-waste workers	[37]

3 Fundamentals of phytoremediation in treating e-waste contaminants

Phytoremediation is an eco-friendly and sustainable method that uses plants to remove, degrade, or immobilize contaminants from the environment, including soil, water, and air. Plants achieve this through mechanisms such as phytovolatilization, phytoaccumulation, phytostabilization, phytofiltration, and phytodegradation, targeting pollutants like heavy metals, organic compounds, and pathogens. The success of phytoremediation depends on factors such as contaminant type, plant species, environmental conditions, and soil properties.

3.1 Mechanisms of phytoremediation

Phytoremediation is very important for cleaning up pollution from e-waste, which often contains hazardous substances such as heavy metals. These are the most important pathways by which phytoremediation works and hence are important for e-waste management:

3.1.1 Phytoextraction or phytoaccumulation

Plants absorb contaminants through their roots and translocate them to their aboveground parts, where they accumulate. This is particularly effective for heavy metals such as lead (Pb), cadmium (Cd), and zinc (Zn) [38].

3.1.2 Phytostabilization

This process involves establishing a plant cover that reduces the mobility of contaminants by either root accumulation or rhizosphere immobilization. It helps control leaching and erosion, thereby creating a more stable environment for microbial activity that degrades organic contaminants [39].

3.1.3 Phytodegradation

Plants and their associated microorganisms break down organic pollutants into less harmful substances. This mechanism is useful for degrading organic compounds in e-waste. This specific mechanism involves the breakdown of pollutants by plant enzymes and associated microorganisms. It is part of the broader phytoremediation process and is effective for organic pollutants [40].

3.1.4 Phytovolatilization

In phytovolatilization, plants uptake volatile organic compounds and certain heavy metals, transforming them into less toxic forms that are then released into the atmosphere. This process not only reduces pollutant concentrations in the soil but also minimizes their bioavailability, thereby mitigating their harmful effects on the ecosystem [41].

3.1.5 Rhizofiltration

The process includes absorption, accumulation, and precipitation of contaminants in plant roots, which can then be harvested and disposed of safely. Plant roots absorb, concentrate, and precipitate water contaminants. This process helps treat contaminated water. Plant species, soil type, and environmental conditions influence the effectiveness of phytoremediation [42].

3.2 Plant-microbe interactions in e-waste treatment

Plants and microorganisms interact in ways that are essential for e-waste bioremediation. These interactions help mitigate adverse environmental impacts sustainably. Plants work with microbes to remove toxins from polluted areas. They stabilize, degrade, or extract pollutants, with microbial processes such as phytodegradation and rhizoremediation, facilitating these processes through rhizosphere enzymes. Rhizoremediation emphasizes the cooperation between plant roots and microorganisms in detoxifying pollutants, thereby supporting plant health and ecological recovery.

At e-waste sites, microbial communities have developed resistance to heavy metals and POPs. These microbes can mobilize or stabilize metals, aiding in their removal and recovery. They also degrade e-plastics by using their metabolic functions to dismantle complex polymer by-products. Bioleaching employs genetically engineered microbes and microbial consortia to extract valuable metals from e-waste through microbial activity [43]. Improving the efficacy of phytoremediation requires a better understanding of plants, microbes, soil types, and pollutants, with the microbiota playing a crucial role in enhancing efficiency.

In this context, researchers can remediate e-waste-contaminated soils by developing plants with enhanced phytoextraction capacity and by developing microbial solubilizers. Emerging best practices in machine learning and omics-based technologies can transform bioremediation by enabling better characterization of microbial interactions and their effects. Microbe-plant interactions offer a promising strategy to tackle e-waste issues. By applying these methods, we can lessen the harmful effects of e-waste and help restore regular resource cycles. Additionally, phytoremediation technologies need to improve in efficiency and scalability.

3.3 Criteria for selecting candidate plant species

When selecting plant species for phytoremediation, it is important to evaluate factors such as growth rate, biomass production, root system characteristics, tolerance to contaminated soils, and tolerance to heavy metals and other pollutants. The following criteria are employed to identify suitable species.

3.3.1 Tolerance to heavy metals and toxic substances

Plants such as *Dianthus spiculifolius* and *Dianthus chinensis* exhibit considerable tolerance to heavy metals, including cadmium (Cd) and lead (Pb), making them suitable candidates for heavy-metal phytoremediation. Ramie (*Boehmeria nivea*) also demonstrates a notable capacity to tolerate cadmium, lead, and zinc. Conversely, tomato and alfalfa plants have shown tolerance to DDTs, without exhibiting any discernible phytotoxic effects [44–46].

3.3.2 Growth rate and biomass production

Researchers have observed substantial biomass production in *Acacia ampliceps* and *Eucalyptus camaldulensis*, following inoculation with growth-promoting bacteria. Additionally, *Zea mays* (maize) produces substantial biomass, thereby contributing to phytoremediation initiatives. Furthermore, scholars acknowledge *Dianthus spiculifolius* for its rapid growth and significant biomass accumulation, making it a promising candidate for phytoremediation.

3.3.3 Root system characteristics

Deep-rooted ramie germplasms exhibit greater phytoremediation efficacy, owing to their extensive root systems. *Morus alba* (mulberry) also possesses a well-developed root system that facilitates the stabilization and accumulation of heavy metals. Plants such as *Festuca arundinacea* and *Medicago sativa* exhibit high root surface areas, thereby enhancing their capacity for heavy metal accumulation [47].

3.3.4 Ability to thrive in contaminated soils

Sesuvium portulacastrum exhibits remarkable tolerance to high salinity and elevated levels of heavy metals, making it suitable for phytoremediation in coastal areas. *Phyla canescens* demonstrates considerable resilience to various heavy metals and can thrive in diverse contaminated soils. The application of arbuscular mycorrhizal fungi and plant growth-promoting rhizobacteria can improve plant performance in contaminated environments by enhancing nutrient absorption and stress resilience [48].

3.4 Candidate plant species for e-waste site remediation

E-waste remediation through phytoremediation involves using specific plant species to absorb and detoxify heavy metals and other pollutants from contaminated soils. This overview presents potential plant candidates for this purpose, emphasizing hyperaccumulators, indigenous species, rapidly growing varieties, and the advantages of perennials over annuals.

3.4.1 Hyperaccumulator plant species

Hyperaccumulators are plants capable of absorbing and storing heavy metals at levels that exceed those of many plants. Such plants can be vegetatively rich in polluted soils without toxic symptoms. For instance, the *Pteris vittata* and *Thlaspi caerulescens* species have been reported to efficiently absorb and accumulate metals, including cadmium and arsenic [49]. Hyperaccumulators use specialized metal transporters and chelation mechanisms to retain metals within their tissues. For example, they overexpress genes, encoding metal transporters such as heavy-metal ATPases or metallothioneins. Certain plants concentrate notable metals: *Euphorbia stellata absorbs manganese, Calystegia hederacea* takes up zinc, and some willows or sunflowers accumulate heavy metals [50].

3.4.2 Native plant species

Native species are fundamental to bioremediation because they can adapt to local conditions and facilitate ecosystem restoration. These plants enhance soil quality and support biodiversity, thereby aiding the recovery of contaminated sites. For instance, species such as *Pueraria phaseoloides* and *Acacia spp.* have been shown to serve as bioindicators of cadmium in polluted soils [51]. This approach will help prevent and control the introduction of foreign species into the region and continuously improve the environment for native fauna, thereby offering a sustainable solution for remediation.

3.4.3 Fast-growing plant species

Species characterized by rapid proliferation possess considerable potential for environmental remediation due to their accelerated biomass accumulation. These crops can expeditiously occupy contaminated sites, thereby reducing erosion and enhancing soil stability. An illustrative example is bamboo, renowned for its high biomass productivity and capacity to absorb heavy metals [52]. Although not classified as hyperaccumulators, these fast-growing species may significantly contribute to the uptake and stabilization of contaminants.

3.4.4 Perennial plants

The selection between annual and perennial species can markedly influence the efficacy of e-waste remediation. These plants typically feature more extensive root systems. Improvements in soil structure and enhancements to the nutrient cycle are essential. Perennial plants offer stable, long-term biomass production, which is advantageous for sustained remediation efforts. While they grow rapidly and provide immediate shelter, they typically must be replanted each season, which can disrupt soil health and nutrient cycling. Nonetheless, they are well-suited to short-term remediation initiatives. When selecting plants for e-waste remediation, priority should be given to hyperaccumulators, for their metal-absorption capabilities; native species, for ecological harmony; and fast-growing plants, for biomass production. Perennial plants confer enduring benefits for soil health and sustainability. By leveraging these characteristics, feasible solutions to e-waste pollution challenges can be identified.

4 Applied case studies and integrative approaches in e-waste phytoremediation

Phytoremediation, an environmentally friendly and cost-effective approach, is emerging as an efficient solution for addressing e-waste contamination. It uses plants to accumulate, degrade, or stabilize pollutants, making it a practical method for pollution mitigation in natural settings. To enhance the efficacy of phytoremediation for treating e-waste-contaminated sites, researchers and industry stakeholders have proposed and implemented numerous innovative methods. These integrative approaches combine standard phytoremediation techniques with modern technologies and ecological systems to improve pollutant removal, minimize environmental impact, and facilitate large-scale deployment. The subsequent sections outline the main techniques utilized in the remediation of e-waste.

4.1 Constructed wetlands

Constructed wetlands are widely used in phytoremediation. They use natural plants in artificial ecosystems to remove pollutants. For instance, an integrated system using free-floating, vertical, and horizontal subsurface-flow wetlands with plants such as *Eichhornia crassipes*, *Typha augustifolia*, and *Catharanthus roseus* achieved 73.8% efficiency in removing chromium from contaminated water [53].

4.2 Nano-phytoremediation

Combining nanotechnology with phytoremediation, known as nano-phytoremediation, enhances pollutant removal efficiency. For example, using *Impatiens balsamina* and nanoscale zero-valent iron (nZVI) significantly improved the removal of lead (Pb) and polychlorinated biphenyls (PCBs) from e-waste-contaminated soil [54].

4.3 Phytoextraction and phytostabilization

These methods involve using plants to extract or stabilize contaminants. Various plant species have been studied for their potential in Cd phytoextraction, when combined with EDTA and ammonium sulfate. Mulberry, for example, has shown high bioconcentration factors for Cd, indicating its ability to accumulate Cd from MSW compost. Other plants, such as *Portulaca oleracea* and alfalfa, have also demonstrated increased Cd uptake with EDTA treatment. The combination of EDTA and ammonium sulfate has been effective in increasing Cd translocation from roots to shoots, which is crucial for removing Cd from contaminated sites [55]. Similarly, combining aquatic macrophytes with other plants can enhance the remediation of acid mine drainage.

4.4 Case studies and practical applications of e-waste phytoremediation

Relevant applications and case studies have been documented in practice, demonstrating the potential of phytoremediation to treat e-waste pollution. These illustrations highlight how phytoremediation processes can be modified and broadened to other pollution sources, such as heavy metals in soils and microplastics in electronic waste. By combining ecological principles with new technologies, they demonstrate the practicality, versatility, and additional benefits of phytoremediation.

4.4.1 Energy recovery from biomass

A study on the phytoremediation of lead-contaminated soil using *Festuca arundinacea* (tall fescue) emphasizes its efficacy in lead removal and the advantages of energy recovery from biomass. This approach not only mitigates environmental impacts but also provides a sustainable means of managing contaminated biomass. The plant accumulates substantial amounts of lead in its roots and shoots, making it suitable for remediation. Comparative analysis of biomass management strategies indicates that energy recovery through cogeneration is more environmentally friendly and sustainable than direct landfill disposal, significantly reducing environmental impacts by 30–100%, as demonstrated by life cycle assessments. Although energy recovery processes involving conditioning and cogeneration contribute to environmental impacts, these are nonetheless lower than those associated with direct disposal, thereby reaffirming energy recovery as the preferred method [56].

4.4.2 Full-scale applications

Engineered wetland systems have been effectively used at full scale to treat a range of pollutants, including metals and organic compounds, notably in Newfoundland, Canada. These systems are passive, solar-powered, and visually appealing, making them an attractive remediation option. They effectively remove pollutants, including metals, pesticides, crude oil, polycyclic hydrocarbons, and landfill leachate. A case study from Newfoundland demonstrates their practical implementation and success [57].

4.4.3 Biodegradation of e-waste plastics

The fungus *Penicillium brevicompactum* shows strong potential for biodegrading microplastics from printed circuit boards, reducing their mass by up to 75% in 14 days. Its degradation process involves breaking down the plastics into smaller particles and altering their chemical structure through chain scission and oxidation. This capability offers a promising, sustainable approach to mitigating the environmental impact of e-waste plastics, which are difficult to recycle with conventional methods [58].

5 Challenges and limitations in phytoremediation of e-waste

Phytoremediation is cost-effective; however, substantial effort is required to assess its economic feasibility on a large scale. Therefore, leveraging the value-added product manufacturing capabilities (e.g., biofuels and essential oils) of phytoremediation may enhance its financial feasibility. Regulatory frameworks and policies to promote the implementation of phytoremediation are warranted. Governments, the private sector, and local communities will need to collaborate to support more sustainable e-waste management practices. Phytoremediation offers an attractive integrated approach for treating e-waste-contaminated sites. It is an eco-friendly approach to one of the world's most critical environmental issues. It operates through natural methods and innovative technologies. Plants play a vital role in removing, degrading, or stabilizing environmental contaminants. Although phytoremediation offers many benefits, such as cost-effectiveness and environmental sustainability, it also faces challenges that limit its application. Here are some factors that limit the effectiveness and success of phytoremediation, especially in the context of e-waste.

5.1 Complexity and toxicity of the contaminants

Some hazardous substances, such as lead, cadmium, mercury, PBDEs, and other organic pollutants, are found in e-waste. This variety of contaminants can complicate cleanup efforts. Different plants may be required to remove these pollutants effectively, simultaneously. Many of these contaminants are highly toxic and adversely affect plant growth and survival. At high concentrations, heavy metals can cause phytotoxicity by inhibiting root development, reducing nutrient uptake, and impairing phytoremediation effectiveness [59].

5.2 Plant selection

Few plant species can absorb or tolerate heavy metals and organic pollutants. This complicates the identification and cultivation of hyperaccumulator plants capable of thriving in contaminated environments. Such limitations can hinder their growth and reduce their effectiveness in pollutant remediation [60].

5.3 Soil and environmental factors

Soil texture, organic content, and pH are key factors that affect contaminant availability and plant growth. Heavy clay soils can impede root penetration, whereas acidic soils increase metal solubility, thereby increasing toxicity. Soil pH is crucial in determining the forms and accessibility of trace elements, which can influence plant absorption and the effectiveness of phytoremediation. Organic matter content also significantly impacts the mobility and availability of pollutants in the soil. Environmental conditions, such as temperature, rainfall, and sunlight, are crucial for plant development and for the breakdown of contaminants. Healthy plant growth requires access to water, which is vital for transporting nutrients and other compounds essential to plant health. Temperature and humidity influence plant physiology and the success of phytoremediation processes. Elevated CO_2 levels can enhance plant growth and yield, mitigate nutrient deficiencies, and improve nutrient uptake and utilization [61].

5.4 Duration for remediation

Plants need several months to grow and mature. Therefore, contaminant remediation may take longer to establish the chosen species within phytoremediation systems. This slow process might not be suitable for sites that require immediate clean up or pose urgent health risks. Monitoring and managing phytoremediation sites involve ongoing efforts to protect plant health while effectively removing contaminants.

5.5 Economic considerations

Implementing phytoremediation systems generally requires initial investments in land preparation, plant selection, and maintenance. Additionally, the economic viability of phytoremediation depends on the market value of the cleaned land. The financial rationale for phytoremediation might become less convincing if the land has limited future use or low market value.

5.6 Regulatory and public perception issues

Uncertainty about the law or prevailing practice can create confusion for operators. Regulatory bodies may require extensive testing and documentation, which can delay a project timeline. Public acceptance of the effectiveness and safety of plants for cleaning hazardous waste can also influence community support and, at some point, may discourage the adoption of this alternative treatment.

6 Conclusion

The fast-paced rise in e-waste has created severe environmental and health issues. Conventional e-waste treatment and recycling methods are insufficient, allowing contaminants to leach into the environment. Phytoremediation, in which plants use natural processes to remove pollutants, can efficiently and sustainably remove these pollutants. It helps to lower the risk of toxic contaminants and contributes to biodiversity and soil health. However, the challenges stem from the characteristics of various contaminants, plant selections, and environmental conditions. Plants, in contrast, grow more slowly in nature and are therefore less amenable to immediate clean up. Consequently, the factors outlined significantly influence plant growth, the extent of contaminant uptake, and their efficacy in contaminant removal, thereby affecting the overall remediation process. To effectively implement phytoremediation for site cleanup, it is essential to comprehend these constraints alongside site-specific variables and factors. Phytoremediation, when combined with other methods, could potentially provide a novel approach to this process. The interaction of these two approaches will create entirely new opportunities for ecological protection and improving public health. It will take collaboration among researchers, policymakers, plant growers, and industry stakeholders to promote and grow its adoption.

References

[1] Jadoun S, Chinnam S, Jabin S, Upadhyay Y, Jangid NK, Zia J. Recovery of metals from e-waste: facts, methods, challenges, case studies, and sustainable solutions. Environmental Science & Technology Letters. 2024;12(1):8–24.

[2] Reijnders L. The global challenge of e-waste generation. In Management of Electronic Waste: Resource Recovery, Technology and Regulation 2024 (pp. 15–38). Wiley.

[3] Routray SK. Electronic waste and environmental sustainability. IEEE Potentials; 2025, Vol. 44, 2–7.

[4] Wirtu YD, Godana UA, Tucho GT. A review of public perception and awareness of e-waste management in developing countries: challenges and opportunities for sustainable solutions. Environmental Reviews. 2025;33:1–12.

[5] Ali S, Abbas Z, Rizwan M, Zaheer IE, Yavas I, Unay A et al. Application of floating aquatic plants in phytoremediation of heavy metals polluted water: a review. Sustainability. 2020;12(5):1927.

[6] Deng S, Zhang X, Zhu Y, Zhuo R. Recent advances in phyto-combined remediation of heavy metal pollution in soil. Biotechnology Advances. 2024;72:108337.

[7] Purushothaman B, Kannan S. Green-based nanomaterials and plants in nano-phytoremediation strategies. In Green Nanoremediation: Sustainable Management of Environmental Pollution 2023 (pp. 277–288). Cham: Springer.

[8] Sharma P, Bakshi P, Kour J, Singh AD, Dhiman S, Kumar P et al. PGPR and earthworm-assisted phytoremediation of heavy metals. In Earthworm-Assisted Remediation of Effluents and Wastes 2020 (pp. 227–245). Singapore: Springer.

[9] Popoola OJ, Ogundele OD, Ladapo EA, Senbore S. The impact of heavy metal contamination in soils on soil microbial communities and its potential health risks for humans. In Soil Microbiome in Green Technology Sustainability 2024 (pp. 351–375). Cham: Springer.

[10] Acharya S. Heavy metal contamination in food: sources, impact, and remedy. In Food Safety and Quality in the Global South 2024 (pp. 233–261). Singapore: Springer.
[11] Hashmi MZ, Shoukat A, Pongpiachan S, Kavil YN, Alelyani SS, Alkasbi MM et al. Polychlorinated biphenyls induced toxicities upon cell lines and stem cells: a review. Environmental Geochemistry and Health. 2025;47(2):56.
[12] Kostenko O, Del Pino LF, Jorge-Montalvo P, Visitación-Figueroa L. Management of waste containing polybrominated diphenyl ethers: a review. Heliyon. 2024;10(22):e40229.
[13] Malkoske T, Tang Y, Xu W, Yu S, Wang H. A review of the environmental distribution, fate, and control of tetrabromobisphenol A released from sources. Science of the Total Environment. 2016;569:1608–1617.
[14] Kumar P, Singh S, Gacem A, Yadav KK, Bhutto JK, Alreshidi MA et al. A review on e-waste contamination, toxicity, and sustainable clean-up approaches for its management. Toxicology. 2024;508:153904.
[15] Gao X, Zhou Y, Liu S, Bi Z, Jiang M, Zhang M, Cai H, Wang X. Risk assessment of heavy metal contamination in air, water, soil, and crops adjacent to a spent lead-acid battery storage facility. Journal of Material Cycles and Waste Management. 2025;27:1557–1569.
[16] Tian XM, Wu YF. Recent development of recycling lead from scrap CRTs: a technological review. Waste Management. 2016;57:176–186.
[17] Wang Y, Xu J, Liu G. Characteristics and health risk assessment of heavy metals in dust of a waste printed circuit board recycling workshop, China. RSC Advances. 2023;13(32):22216–22225.
[18] Petrlik J, Beeler B, Ismawati Y, Bell L. Toxic contamination caused by plastic waste in countries of the global south. In Plastic Waste Trade: A New Colonialist Means of Pollution Transfer 2024 (pp. 113–128). Cham: Springer.
[19] Li TY, Zhou JF, Wu CC, Bao LJ, Shi L, Zeng EY. Characteristics of polybrominated diphenyl ethers released from thermal treatment and open burning of e-waste. Environmental Science and Technology. 2018;52(8):4650–4657.
[20] Sun B, He Y, Chi H, Tao S. Kinetics of brominated flame retardant (BFR) releases from granules of waste plastics. Environmental Science and Technology. 2016;50(24):13419–13427.
[21] Li H, Bai J, Li Y, Cheng H, Zeng EY, You J. Short-range transport of contaminants released from the e-waste recycling site in South China. Journal of Environmental Monitoring. 2011;13(4):836–843.
[22] Zhong Z, Du X, Wang Z, Wu L, Tao X, Dang Z et al. Dissolution release of antimony from e-waste microplastics in aquatic solution: kinetics, mechanisms and surface-release-layer thickness. Water Research. 2025;124619.
[23] Wang Z, Du X, Wu X, Zhong Z, Liang J, Tao X et al. Release mechanisms of decabromodiphenyl ether from typical e-waste microplastics into water: insights from molecular dynamics simulations. Journal of hazardous materials. 2025;138507.
[24] Singh D, Dikshit AK, Dangi MB. Technological advances in per- and polyfluoroalkyl substances (PFAS) removal from landfill leachate: source identification and treatment options. Emerging Contaminants. 2025;11(2):100458.
[25] Shoukat A, Nurulain SM, Hussain S, Khan SA, Ahmed H, Muneer Z et al. Chronic exposure to electronic waste poses a risk to liver toxicity with molecular interaction of GSTM1, GSTT1 null variants, and GSTP1. Environmental Pollution. 2023;334:122204.
[26] Jaiswal SK, Mukti SK. Prioritizing factors affecting e-waste recycling in India: a framework for achieving a circular economy. Circular Economy and Sustainability. 2025;5(1):461–481.
[27] Zafar S, Fatima S, Asad F, Nazir MM, Batool S, Ashraf A. Combating lead (Pb) contamination: integrating biomonitoring, advanced detection, and remediation for environmental and public health. Water, Air, and Soil Pollution. 2025;236(1):8.

[28] Son Y, Jeong YJ, Kim HY, Lee SH, Choi Y, Choi HD et al. Behavioral and neurochemical changes in mice induced by low-level lead exposure: implications for ADHD and conduct disorders. Ecotoxicology and Environmental Safety. 2025;302:118735.
[29] Parithathvi A, Choudhari N, Dsouza HS. Prenatal and early life lead exposure induced neurotoxicity. Human & Experimental Toxicology. 2024;43:09603271241285523.
[30] Schoeters G, Hond ED, Zuurbier M, Naginiene R, Van Den Hazel P, Stilianakis N et al. Cadmium and children: exposure and health effects. Acta Paediatrica. 2006;95:50–54.
[31] Niżnik Ł, Świdniak A, Toporowska-Kaźmierak J, Jurowski K. Spilled liquid mercury: case studies, proposition of safety procedures with assessment of commercial cleanup kit and recommendations. Toxicology Letters. 2025 Jul 2. doi: 10.1016/j.toxlet.2025.06.010.
[32] Wu Z, He C, Han W, Song J, Li H, Zhang Y et al. Exposure pathways, levels, and toxicity of polybrominated diphenyl ethers in humans: a review. Environmental Research. 2020;187:109531.
[33] Montano L, Baldini GM, Piscopo M, Liguori G, Lombardi R, Ricciardi M et al. Polycyclic aromatic hydrocarbons (PAHs) in the environment: occupational exposure, health risks and fertility implications. Toxics. 2025;13(3):151.
[34] Choi S, Kim HJ, Kim S, Choi G, Kim S, Park J et al. Current status of organochlorine pesticides (OCPs) and polychlorinated biphenyls (PCBs) exposure among mothers and their babies of Korea-CHECK cohort study. Science of the Total Environment. 2018;618:674–681.
[35] Ge Y, Cui J, Zhang L, Zhang S, Baqar M, Cheng Z. Informal e-waste dismantling activities accelerated the releasing of liquid crystal monomers (LCMs) in Pakistan: occurrence, distribution, and exposure assessment. Science of the Total Environment. 2024;932:172987.
[36] Sridharan TB, Abilash D. Heavy metal exposure and its consequences on sperm quality: a cross-sectional perspective study. Indian Journal of Clinical Biochemistry. 2025 Sep 6. doi: 10.1007/s12291-025-01350-z.
[37] Viegas C, Gomes B, Dias M, Cervantes R, Pena P, Gonçalves S et al. Microbial occupational exposure in e-waste recycling: biological hazards, their sources, and potential toxic effects. International Journal of Environmental Research. 2025;19:39.
[38] Adu JT, Aneke FI. Evaluation of heavy metal contamination in landfills from e-waste disposal and its potential as a pollution source for surface water bodies. Results in Engineering. 2025;25:104431.
[39] Georgin J, Franco DSP, De oliveira JS, Dehmani Y, El Messaoudi N, Miyah Y et al. Decontamination of pollutants present in water, air, and soil through phytoremediation: a critical review. International Journal of Phytoremediation. 2025 Oct;27:1–27. doi: 10.1080/15226514.2025.2575792.
[40] Latiff NH, Kamoona S, Wan Sulaiman WS, Hatta FA, Ramya R, Othman R. Al and Fe heavy metal concentrations in the vegetative and root parts of Dicranopteris linearis, Nephrolepis bifurcata, Stenochlaena palustris and Acrostichum aureum grew in highly weathered soil. Environmental and Ecological Research. 2022;10(4):103–109.
[41] Singla L, Saini A, Kumar V, Singh G, Kumar A, Giri SK. Phytoremediation: treatment strategies for industrial and biomedical waste. In Bioremediation of Environmental Toxicants 2024 (pp. 250–271). CRC Press.
[42] Biswal T. Rhizofiltration: a sustainable green technology for remediation of heavy metals from aquatic systems. In Recent Advances in Bioremediation and Phytoremediation 2025 (pp. 1–21). Cham: Springer.
[43] Bai Y, Su J, Wang F, Cui H, Zhang H, Liu K. Harnessing synthetic biology for sustainable recovery of critical metal materials from electronic waste. Advanced Functional Materials. 2025;35(49):e09900.
[44] Qiao K, Wang Q, Liu X, Gong S, Wang J. Cadmium/lead tolerance of six Dianthus species and detoxification mechanism in Dianthus spiculifolius. Chemosphere. 2023;312:137258.
[45] Shao HQ, Zhang YW, Qu C, Li WH, Zhao YJ, Liu N et al. Analysis of heavy metal contamination in the soil and enrichment capabilities of terrestrial plants around a typical vanadium smelter area. Chinese Journal of Engineering. 2020;42(3):302–312.

[46] Mitton FM, Miglioranza KS, Gonzalez M, Shimabukuro VM, Monserrat JM. Assessment of tolerance and efficiency of crop species in the phytoremediation of DDT polluted soils. Ecological Engineering. 2014;71:501–508.

[47] Qiao K, Wang Q, Liu X, Gong S, Wang J. Cadmium/lead tolerance of six Dianthus species and detoxification mechanism in Dianthus spiculifolius. Chemosphere. 2023;312:137258.

[48] Zhu S, Zhao W, Sun S, Yang X, Mao H, Sheng L et al. Metagenomic analysis revealed N-metabolizing microbial response of Iris tectorum to Cr stress after colonization by arbuscular mycorrhizal fungi. Ecotoxicology and Environmental Safety. 2024;273:116157.

[49] Khan F, Ijaz I, Uzair M, Fatima N. Plant-assisted bioremediation of heavy metals. In Phytoremediation and Biofortification 2024 (pp. 147–158). Apple Academic Press.

[50] Skuza L, Szućko-Kociuba I, Filip E, Bożek I. Natural molecular mechanisms of plant hyperaccumulation and hypertolerance towards heavy metals. International Journal of Molecular Sciences. 2022;23(16):9335.

[51] Baughman OW, Kulpa SM, Sheley RL. Four paths toward realizing the full potential of using native plants during ecosystem restoration in the Intermountain West. Rangelands. 2022;44(3):218–226.

[52] Bian F, Zhong Z, Zhang X, Yang C, Gai X. Bamboo – an untapped plant resource for the phytoremediation of heavy metal contaminated soils. Chemosphere. 2020;246:125750.

[53] Sembada AA, Theda Y. Mass balance study of an integrated phytoremediation system using different types of constructed wetlands for chromium removal. Asia-Pacific Journal of Science and Technology. 2024;29(3):APST–29. doi: 10.14456/apst.2024.42.

[54] Song B, Xu P, Chen M, Tang W, Zeng G, Gong J et al. Using nanomaterials to facilitate the phytoremediation of contaminated soil. Critical Reviews in Environmental Science and Technology. 2019;49(9):791–824.

[55] Zhao S, Shang X, Duo L. Accumulation and spatial distribution of Cd, Cr, and Pb in mulberry from municipal solid waste compost following application of EDTA and (NH4)2SO4. Environmental Science and Pollution Research. 2013;20(2):967–975.

[56] Espada JJ, Rodríguez R, Gari V, Salcedo-Abraira P, Bautista LF. Coupling phytoremediation of Pb-contaminated soil and biomass energy production: a comparative life cycle assessment. Science of the Total Environment. 2022;840:156675.

[57] Zhang BY, Zheng JS, Sharp RG. Phytoremediation in engineered wetlands: mechanisms and applications. Procedia Environmental Science. 2010;2:1315–1327.

[58] Ferreira-Filipe DA, Oliveira L, Paço A, Fernandes AJ, Costa FM, Duarte AC et al. Biodegradation of e-waste microplastics by Penicillium brevicompactum. Science of the Total Environment. 2024;935:173334.

[59] Singhal P, Taduru AS, Fatima H, Kaushik G, Hussain CM, Chel AL. Electronic wastes, its impact on animal farming practices and health concerns. In Electronic Waste: Impact on Health, Animals, and the Environment 2025 (pp. 107–111). CRC Press.

[60] Shi L, Li J, Palansooriya KN, Chen Y, Hou D, Meers E et al. Modeling phytoremediation of heavy metal contaminated soils through machine learning. Journal of hazardous materials. 2023;441:129904.

[61] Magdziak Z, Gąsecka M, Goliński P, Mleczek M. Phytoremediation and environmental factors. In Phytoremediation: Management of Environmental Contaminants 2014 (Vol. 1, pp. 45–55). Cham: Springer.

Shah Tanvir Alam Rimon†, Md. Humayun Kabir Jim†,
Monjur Mourshed*

Transforming e-waste into value-added products: a pathway to achieve Sustainable Development Goals

Abstract: In order to further the Sustainable Development Goals (SDGs) of the United Nations, this chapter examines the possibility of turning electronic trash into goods with added value. It looks at modern recycling techniques, material recovery procedures, and creative reuse routes that have positive effects on the economy, society, and environment. Successful circular economy methods that turn used electronics into valuable resources are highlighted in case studies. The conversation emphasizes how efficient e-waste valuation lowers environmental pollution and resource depletion while promoting responsible production and consumption, climate action, and decent labor. Future directions, possibilities, and challenges for transforming e-waste sustainably are also discussed.

Keywords: E-waste, circular economy, value added product, sustainability, waste to energy

1 Introduction

Electronic waste (e-waste) refers to discarded electrical and electronic equipment, including all components, subassemblies, and accessories that are no longer functional, desirable, or economically repairable. This diverse category includes consumer electronics, information and communication technologies, household appliances, medical devices, and industrial machinery, each containing complex mixtures of metals, polymers, composites, ceramics, and hazardous chemicals [1]. With accelerating technological innovation, shortened electronic product lifespans, and increasing global digitalization, e-waste has become the world's fastest-growing waste stream. In 2024, global e-waste generation reached 62 million tons, rising five times faster than documented formal recycling activities [2].

***Corresponding author: Monjur Mourshed** Department of Mechanical Engineering, Rajshahi University of Engineering and Technology, Rajshahi 6204, Bangladesh,
e-mail: monjurmourshed@me.ruet.ac.bd
Shah Tanvir Alam Rimon, Md. Humayun Kabir Jim, Department of Mechanical Engineering, Rajshahi University of Engineering and Technology, Rajshahi 6204, Bangladesh

 | https://doi.org/10.1515/9783112225509-009

Despite this rapid increase, only about 22% of global e-waste enters regulated recycling systems, while the remainder is landfilled, stored within households, exported illegally, or processed through informal recycling operations particularly in developing countries. Informal recyclers often employ crude techniques such as open burning, manual dismantling, unregulated smelting, and acid leaching, which release hazardous contaminants, including lead, cadmium, mercury, brominated flame retardants, dioxins, furans, and microplastics into the environment. These pollutants lead to soil and water contamination, degradation of air quality, biodiversity loss, and severe health risks, including neurotoxicity, respiratory disorders, organ damage, and carcinogenic effects [3–5].

Globally, e-waste recycling relies on multiple processes, including mechanical pre-processing (shredding, sorting, grinding), pyrometallurgy (smelting, thermal treatment), hydrometallurgy (leaching, solvent extraction), electrochemical recovery, bioleaching, and emerging green technologies such as mechanochemical processing and supercritical fluid extraction [6, 7]. Countries with advanced recycling infrastructures, such as Japan, South Korea, Switzerland, and EU member states, operate sophisticated closed-loop systems supported by extended producer responsibility (EPR), digital product passports, take-back schemes, and high consumer awareness [8, 9].

Despite these environmental and health hazards, e-waste is increasingly recognized as a valuable secondary resource stream. Electronics contain high-value materials such as gold, silver, copper, cobalt, lithium, palladium, and rare earth elements, many of which are classified as critical raw materials due to supply vulnerabilities and geopolitical constraints [10]. The concentrations of precious metals in printed circuit boards (PCBs) and electronic components often exceed those of natural mineral ores, making urban mining an economically attractive and environmentally sustainable alternative to primary extraction [11, 12]. Economically, efficient e-waste management has significant implications for resource circularity, industrial innovation, and job creation. The annual economic value of materials contained in global e-waste exceeds USD 60–70 billion, with precious metals accounting for the majority of recoverable value [8, 9].

Effective recycling systems help mitigate raw material shortages, reduce environmental remediation expenses, and strengthen national resource security. Moreover, sustainable e-waste management contributes to multiple United Nations Sustainable Development Goals (SDGs), particularly SDG 3, SDG 6, SDG 9, SDG 12, and SDG 13 [12, 13]. However, developing countries face persistent challenges, including inadequate legislation, weak institutional enforcement, limited consumer awareness, inadequate recycling infrastructure, and the dominance of informal sectors. Addressing these barriers is essential as electronic consumption and waste generation continue to rise across South Asia, sub-Saharan Africa, and Latin America. Improving policy coordination, technology transfer, formalization of informal recyclers, and integration of circular economy principles is vital for long-term sustainability [4, 5, 7].

Although substantial research exists on e-waste generation, recycling technologies, and their environmental impacts, several critical gaps persist. Current studies often provide fragmented evaluations of global recycling technologies, focusing on isolated methods rather than integrating mechanical, chemical, biological, and advanced green processes into cohesive frameworks. Moreover, the majority of research is heavily skewed toward high-income regions, with limited attention to the socio-economic, infrastructural, and regulatory challenges faced by developing countries. Quantitative assessments of the economic benefits of e-waste management, such as urban mining, job creation, and resource security, remain insufficient, leaving policy-makers without robust economic evidence to guide decision-making. Similarly, the integration of circular economy principles, including eco-design, reparability, modularity, and EPR with recycling technologies is weak, and practical, region-specific implementation guidelines are scarce, limiting the applicability of research outcomes for policymakers, industries, and civil organizations. Finally, emerging environmental contaminants such as microplastics, flame retardants, and nano-materials derived from e-waste remain understudied, necessitating deeper long-term assessments of their environmental and health impacts [4, 13, 14]. These gaps collectively justify the need for a comprehensive chapter that synthesizes global literature, technological advancements, economic analyses, and policy insights to provide a more holistic understanding of e-waste management.

To address these identified gaps, this chapter aims to provide a comprehensive overview of global e-waste generation, composition, environmental impacts, and associated health risks. It reviews and compares major e-waste recycling technologies, encompassing mechanical, metallurgical, electrochemical, and biological processes, while analyzing the economic potential of e-waste through urban mining and its contribution to sustainable development. The chapter further evaluates global policy frameworks, regulatory models, and circular economy initiatives, proposing an integrated and sustainable e-waste management framework adaptable to diverse socio-economic contexts. Finally, it offers actionable recommendations for governments, industries, researchers, and civil organizations to enhance e-waste recycling efficiency and promote circularity.

2 Composition and potential of e-waste

E-waste usually comprises a wide range of products that are classified based on numerous factors such as lifespan, average weight, material composition, and functional features. In total, around 54 distinct product-centric categories fall into six distinct subcategories [1]. Heterogeneity is one of the key features of e-waste, which makes it difficult to list out its elemental compositions [5, 12]. However, e-waste consists of both hazardous and valuable components. The raw materials, such as metals, glass, and plastics, can be recovered through proper recycling. They are considered valuable components. About 15% of e-waste is made up of plastic, while about 60% is made up of recyclable metals [15]. Figure 1(a) shows a typical composition of materials present in e-waste [15]. In general, it can be anticipated that any e-waste comprises about 50% (w/w) of iron and steel and 10–30% (w/w) of plastics (either polycarbonates or polystyrene). Additionally, it contains 55 distinct metals, including Cu, Zn, Ni, Pb, As, Hg, Cr, Cd, Au, Ag, and Pd [5, 16, 17].

Resource recovery from e-waste is often referred to as urban mining. It is called so because the metals in e-waste have significant potential for economic value that is permanently lost from the market. Despite making up only 3–6% of all e-waste, PCBs are the most valuable component due to their high content of precious and base metals [17]. PCBs are composed of a mixture of metals (40%), plastics (30%), and ceramics (30%) [18]. Of the metals, most PCBs comprise copper (10–27%), nickel (0.3–2%), zinc (0.03–0.42%), tin (0.08–0.9%), aluminum (2–19%), and iron (8–38%) [4, 8–10]. In addition, high-grade PCBs can also contain metals of high value such as gold (250–2050 ppm), silver (110–4,500 ppm), and palladium (40–4,000 ppm) [17].

Recovering and reusing numerous metals and rare earth elements from secondary resources, such as e-waste, will be crucial as primary metal resources are depleted. It is anticipated that the demand for copper will rise by 275–300% by 2050, and existing virgin mineral mining is unlikely to be able to meet this need [5, 19]. Metals such as iron and aluminum are also in demand for the manufacture of electronic goods. Based on the waste generation data for 2019, it would be possible to recover metals with a combined value of US $57 billion [1].

Apart from its economic importance, recycling e-waste offers significant environmental benefits and contributes to energy conservation. By replacing virgin resources with recycled ones, energy consumption may be significantly decreased, and the environmental impact of mining and processing can be reduced. Energy savings from recycling can reach 95% for aluminum, 85% for copper, and 74% for lead and steel [19, 20]. It was also estimated that by recycling all e-waste generated in 2019, CO_2 emissions could be reduced by an additional 83 MT [1].

Despite its numerous economic and environmental prospects, however, if e-waste is not properly managed, it can threaten both human health and the environment. Moreover, it can significantly harm the quality of both soil and water. If the e-waste is treated improperly, such as by decomposing or being incinerated, it causes contamination of the natural resources [21, 22]. E-waste contains numerous hazardous and toxic chemicals (mercury, lead, nickel, cadmium, barium), and these elements can severely harm important organs, such as the brain, lungs, liver, heart, and kidneys [5, 23]. Table 1 represents the typical sources of e-waste, highlighting their potential economic value and potential human health risks.

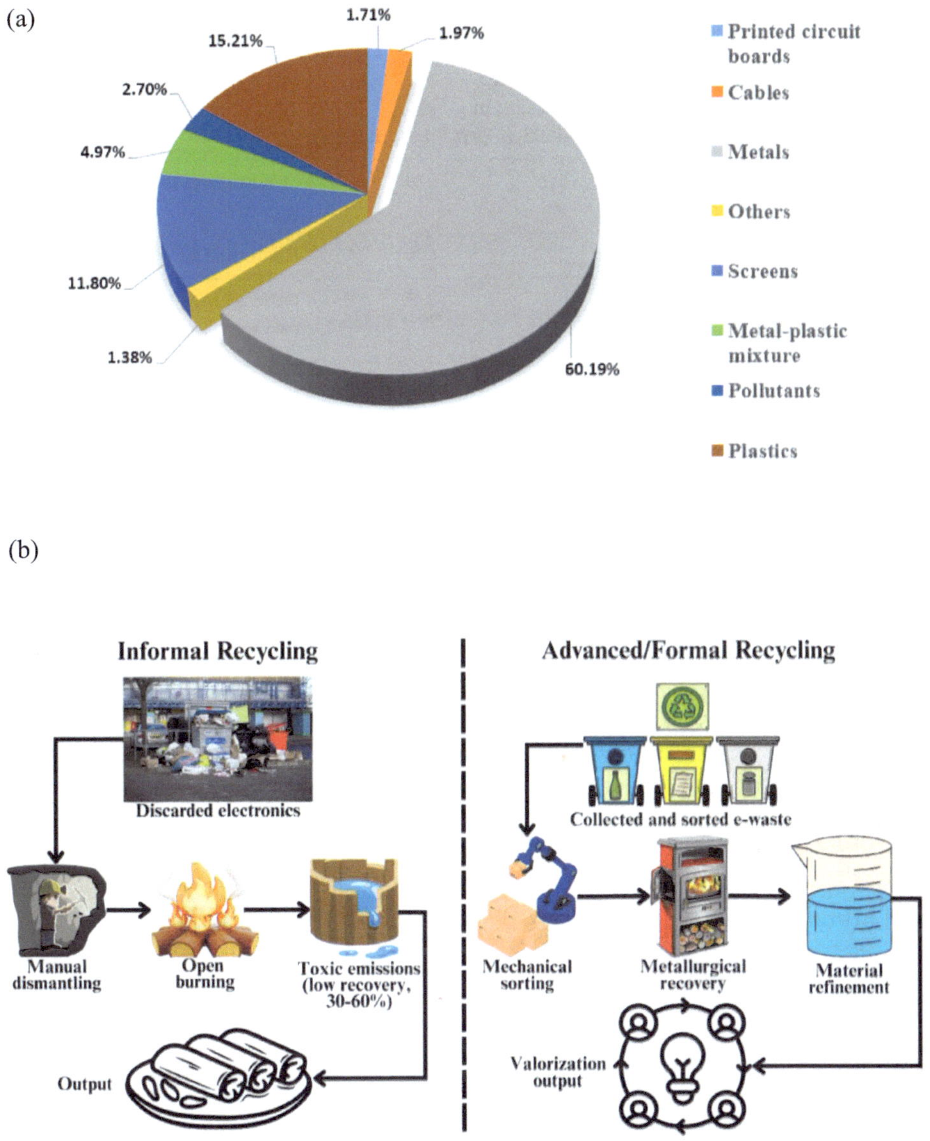

Figure 1: (a) A typical composition of materials present in e-waste [15]. (b) Comparative overview of informal and formal e-waste recycling pathways. Informal processes rely on manual dismantling and open burning, resulting in low recovery (30–60%) and toxic emissions, whereas advanced formal systems employ automated sorting and metallurgical processes to recover high-purity materials suitable for further valorization.

Table 1: Typical sources of e-waste and their corresponding economic potential and health risks.

Metal	Source of exposure	Amount present in e-waste (kT)	Potential value (USD million)	Potential human health risk
Ag	Circuit boards, switches, and batteries	1.2	579	Skin irritation and respiratory irritation
Al	Casings, heat sinks, and wiring	3,046	6061	Neurotoxicity and bone disorders
Au	Connectors, circuit boards, and bonding wires	0.2	9,481	Generally low toxicity, but refining fumes can cause respiratory issues
Bi	Solders and semiconductors	0.1	1.3	Kidney and liver effects at high exposure
Co	Rechargeable batteries (Li-ion) and magnetic components	13	1,036	Lung and heart damage and dermatitis
Cu	Wires, circuit boards, and cables	1,808	10,960	Liver and kidney damage from chronic exposure
Fe	Structural components, casings, and CRTs	20,466	24,645	Gastrointestinal irritation and liver toxicity at very high doses
Ge	Semiconductors and fiber optics	0.01	0.4	Kidney damage and respiratory irritation
In	LCDs, semiconductors, and photovoltaic cells	0.2	17	Lung fibrosis, emphysema, and respiratory system damage
Ir	Electrical contacts and electrodes	0.001	5	Skin and eye irritation
Os	Electrical contacts and catalysts	0.01	108	Severe eye, skin, and lung irritation
Pd	Printed circuit boards (PCBs), monitors, bulbs, CRTs, and cabling	0.1	3,532	Brain damage and nervous system issues
Pt	Hard drives, thermocouples, and catalysts	0.002	71	Allergic reactions and respiratory tract irritation
Rh	Electrical contacts and coatings	0.01	320	Eye, skin, and respiratory irritation; potential allergic sensitization
Ru	Electrical contacts and chip resistors	0.0003	3	Lung and skin irritation
Sb	CRT glass, plastic housing, solder alloys Spark	76	644	Stomach pain, vomiting, diarrhea, and stomach ulcers
Total			57,463.7	

3 Current recycling and recovery approaches

3.1 Manual dismantling practices

Manual dismantling refers to the process that involves disassembling various parts by using some basic tools. This is the most common starting approach to electronic waste processing. This is particularly evident in low- and middle-income countries, due to the availability of basic tools and affordable labor, which makes it a feasible option. Workers typically start the disassembly of devices manually. At first, the circuit boards are isolated, and then various connected wires are separated from each other. The motors, if present, are then removed. Finally, different components that may be reused or can be sold as spare parts are separated. This method has multiple advantages. Some of them include – low initial cost, easy entry into the process, having the capability of managing the diverse and chaotic nature of e-waste, retaining susceptible components of different machineries, possessing potential resale value, etc. [5, 7].

There are various disadvantages as well. Toxic materials like lead-based solder, mercury switches, and plastics with brominated flame retardants are often used by individuals who don't wear masks, gloves, or without ensuring there is enough air flow. Recovery rates can also be very different from one person to another. Processing rates are still low, sometimes only a few tons per day. The materials that are not usable again are often thrown away or burned, which adds to pollution and health problems [7, 16].

3.2 Mechanical recycling methods

Mechanical recycling is one of the most popular types of recycling for industrial purposes. This recycling process typically serves as an intermediate stage between manual sorting and metallurgical processing. This process is very efficient. It is possible to process significantly higher volumes up to 200 tons per day by using machinery such as shredders, crushers, screens, magnetic separators, eddy current units, and density-based systems. This method also improves workers' safety. This method makes more uniform batches of plastics, and ferrous and non-ferrous metals compared to the manual dismantling process [5, 12, 16].

This method has some disadvantages too. Many expensive parts are destroyed as soon as they reach a shredder, particularly PCBs and microchips. That means the item can never be used again. Trace metals and rare earth elements are also hard to get back mechanically [7, 16].

3.3 Metallurgical processing

The metallurgical processes of e-waste collection are the most sophisticated methods for the extraction of metals from electronic waste. There are normally two classifications: pyrometallurgical process and hydrometallurgical process. These processes are characterized by their complexity and high operational costs. The pyrometallurgical process is done by smelting the components by applying heat. The operating temperature during this process is typically between 1,000 and 1,500 °C [16].

The hydrometallurgical process is done by dissolving the materials in an appropriate solvent or liquid. This process uses electrochemical methods to selectively remove metals at low temperatures. The recovery rate is normally higher in a hydrometallurgical process. Almost 95% gold, platinum, and palladium can be obtained by using this process [9].

Recent studies show that by using bio-hydrometallurgical methods that use microorganisms to dissolve metals in waste streams, greener leaching agents, and ionic liquids [9, 12]. The proposed methods indicate an improvement in environmental performance. However, the metallurgical processing facilities required for large-scale operations are capital-intensive. These facilities need intensive environmental monitoring to handle emissions and residues in a safe manner [9, 13].

3.4 Formal and informal recycling systems

Electronic waste recycling is vastly classified as formal and informal recycling systems. There is a significant difference between these two recycling systems. Formal recycling operations usually follow standardized methods to ensure sustainability and safety. In a formal recycling system, 85–95% recovery efficiency can be achieved. High-purity secondary materials can be collected and reused from this recycling process. These materials are suitable for recycling into manufacturing cycles and are capable of maintaining high-quality standards without significant degradation. A formal recycling system is more popular in developed countries [1, 5].

Informal recycling is a common practice in developing countries. This process usually depends on small, unregistered workshops. The people who run these workshops use basic tools and manual methods. These systems are responsible for collecting a lot of electronic waste. They also provide jobs for many people. There are various disadvantages to this system. The amount of material recovered is very low, usually between 30% and 60%. This process uses burning cables and crude oil leaching, which cause environmental pollution [7]. The difference between a formal and informal recycling system is shown in Figure 1(b).

Some of the developed countries, including Japan, South Korea, and members of the European Union, have implemented EPR standards in their industries. They have already enhanced their collection rate as well as improved material quality by implementing these standards. On the other hand, the developing countries, including parts of the South Asian regions, Africa, and Latin America have not yet utilized these advanced procedures. They still heavily rely on informal recycling processes. This dependency leads to a decline in the environmental quality, increased risk to public health, and significantly lower recovery rate of materials. Developing countries frequently encounter challenges related to the importation of electronic waste. This e-waste may enter these nations through both legal and illegal channels. Such activities occur despite the existing restrictions imposed by the Basel Convention [4, 16].

Formal systems are stronger at recovery rates and environmental impact control, but informal systems offer vast collection networks and vital livelihood prospects. This is why hybrid vehicles are popular in many nations. Attero Recycling in India and the WEEE Centre in Ghana train and organize informal collectors to feed waste into formal processing plants instead of handling dangerous items [4, 16]. In fact, all the systems discussed above have their own advantages and disadvantages. The abovementioned recycling methods are listed in Table 2.

4 Pathways to value-added products

E-waste recycling processes help recover significant number of valuable materials. The valorization process of electronic waste is utilized to recover different components including metals, plastics, glass, and carbon. These are reused in functional materials and industrial products, thereby reducing the initial buying cost. This process is critical for resource recovery and sustainable material management. The implementation of these procedures leads to a reduction in environmental impact. It also decreases reliance on landfills. This, in turn, contributes to an improvement in the overall economy [6, 24]. There are different types of innovative pathways for e-waste valorization, such as recovered metals like alloys, electronics, casters, polymers, construction fillers or cement additives, electrodes for batteries and supercapacitors, etc. An overview of this system is shown in Figure 2.

4.1 Metals into new electronics and alloys

Metals represent the most significant economic factor in electronic waste. PCBs are particularly noteworthy, as they contain elevated concentrations of copper (Cu), gold (Au), palladium (Pd), silver (Ag), and nickel (Ni) [3, 9, 16]. Metallurgical processes produce high-purity metals that can be recycled into electronics manufacturing, alloy blending, and catalytic applications [25, 26]. Rare earth elements such as neodymium and dysprosium recovered from magnets and hard drives are essential for permanent magnets in electric vehicles and renewable energy systems [25]. Industrial examples

Table 2: Comparison of recycling approaches by key performance indicators.

Approach	Capital investment	Operating cost	Throughput capacity	Recovery rate	Environmental impact	Job creation potential	Technology requirement	References
Manual dismantling	Low (USD 10k–50k)	Low (USD 50–100/ton)	1–5 tons/day	60–70%	High (if uncontrolled)	Very high (50–100 jobs/ 1,000 tons)	Low	[3, 6, 9, 12, 13, 16, 24]
Mechanical processing	Medium (USD 0.5–2 million)	Medium (USD 100–200/ton)	50–200 tons/ day	70–85%	Low–medium	Medium (10–20 jobs/ 1,000 tons)	Medium	
Pyrometallurgy	Very high (USD 10–50 M)	High (USD 200–400/ton)	100–500 tons/ day	90–99%	Medium (with controls)	Low (5–10 jobs/ 1,000 tons)	Very high	
Hydrometallurgy	High (USD 2–10 M)	Medium–high (USD 150–300/ton)	20–100 tons/ day	95–99%	Low–medium	Low–medium (8–15 jobs/1,000 tons)	High	
Integrated approach	High (USD 5–20 M)	Medium (USD 120–250/ton)	50–300 tons/ day	85–95%	Low	Medium–high (15–30 jobs/1,000 tons)	High	

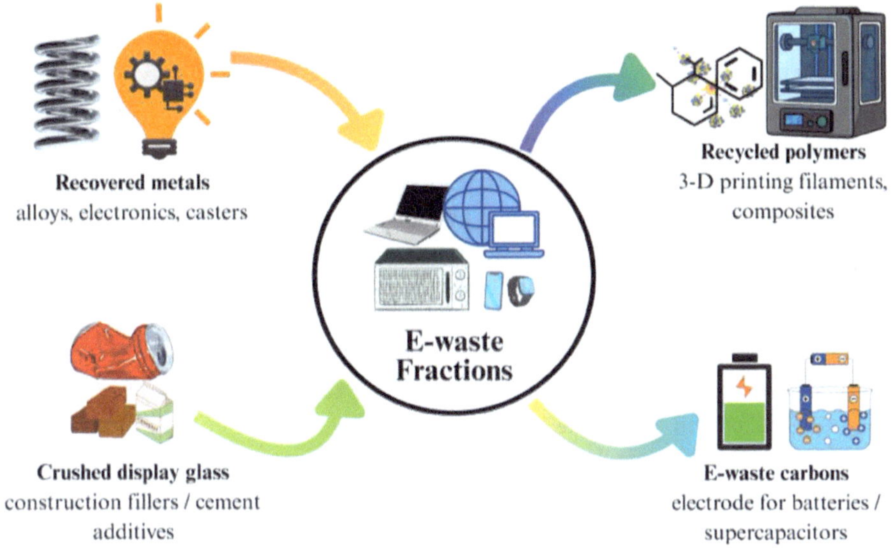

Figure 2: Pathways of e-waste valorization highlighting the transformation of major fractions into value-added materials through sustainable recycling technologies.

such as Umicore (Belgium) refine more than 20 metals that are used in electronics, catalysts, and materials for storing energy [13, 27]. Attero (India) has begun recovering Co, Ni, and Li from discarded batteries to support domestic electric mobility markets [12, 16]. Such pathways reduce reliance on mining and secure a supply of critical materials.

4.2 Plastics into 3D printing filaments and composites

Plastics contribute to 15–20% of e-waste and are often neglected because of their variability and contamination nature [28]. Plastics can be easily recycled into additive manufacturing cycles through advanced processing that breaks down and melts them into 3D printing filaments. The integration of recycled polylactic acid (PLA) with virgin PLA improves the tensile strength of plastic materials. This combination reduces the lifecycle environmental impacts [29]. At the same time, fibers can be mixed with plastic waste collected from electronic devices. This mixture forms composite materials which have longer durability than usual. Filabot in the USA specializes in the commercialization of desktop gadgets designed for the conversion of plastic waste into functional filaments. Experimental research indicates that plastic waste generated from 3D printing can be converted into electrodes, which are suitable for zinc-iodine batteries. This process effectively links polymer recycling with advancements in energy storage technology [29].

4.3 Glass into construction fillers

Glass materials can be collected from discarded electronic devices, which include outdated CRT monitors, LCD panels, and comparable display units. This is due to the potential presence of heavy metals and other contaminants, which complicate the processes associated with safe disposal [1, 6]. When the glass materials are being cleaned and processed, they produce valuable products within the construction sector. Ground electronic waste glass can function effectively as a filler material. It may serve as a partial additive in ceramic formulations within cement-based products. The incorporation of this material into concrete mixtures has improved compressive strength. It also improved durability and reduced clinker quantity [30].

Researchers conducted experiments to evaluate the effectiveness of crushed e-waste glass as a partial substitute for natural sand in the composition of mortar or concrete. The substitution of crushed electronic waste glass into natural sand in the composition of mortar or concrete provides significant advantages. It diverts glass from traditional disposal pathways, reduces the demand on virgin sand resources, etc. Several industrial projects across Europe and certain regions of Asia have already utilized the integration of glass derived from CRT into commercial construction materials. These applications indicate that, when subjected to suitable treatment processes, e-waste glass has the potential to transition from a disposal issue to a viable sustainable additive for infrastructure projects [13, 31].

4.4 Carbonaceous residues into electrodes for batteries and supercapacitors

Carbon-based residues are formed from the pyrolysis process applied to PCBs and plastic components. These residues can be transformed into activated carbons characterized by elevated surface area and conductivity [9, 26]. These carbon-based materials can be utilized in the fabrication of electrodes for batteries and supercapacitors. This application presents a viable approach to connecting electronic waste valorization with clean energy technologies. 3D-printed zinc-iodine batteries use carbon electrodes, which are derived from waste materials, have shown competitive performance metrics [29]. Many projects in Japan and Australia are engaged in the development of processes aimed at integrating carbon derived from e-waste into lithium-ion batteries and advanced capacitors [9, 29].

The high-temperature processing of PCBs and different plastic parts leaves behind carbon-rich residues, which are being used by electronic waste-recovering process such as valorization. Some of these materials can be broken down or heated up to make activated carbons, which have a lot of surface area and good electrical conductivity. These are important properties for the performance of electrochemical devices. The activated carbons have been utilized in experimental electrodes for slurry flow electrode-based energy storage, batteries, and supercapacitors. This application illustrates the potential for e-waste to contribute directly to the expanding clean-energy sector [9, 26].

Carbon materials obtained from electronic waste have also been utilized in the fabrication of 3D-printed zinc-iodine batteries with high-quality characteristics. In the Asia-Pacific region, there are ongoing developments involving companies in Japan and Australia. These companies are conducting experiments focused on the utilization of carbon residues within lithium-ion batteries and advanced capacitor technologies. Electronic waste has the potential to promote renewable energy systems. Moreover, it can contribute to the electrification of transportation networks [9, 29].

5 Case studies of successful models

There are a significant number of projects nowadays that are successfully using the electronic waste management process. Some of the examples of different established companies in different countries include the Attero project in India, the Umicore project in Belgium, Industrial Park in China, etc. These three international projects are discussed below, illustrating the influence of technology, business models, and governance structure on socioeconomic and environmental outcomes. Figure 3 presents the global diversity of these three projects and their geographical distribution.

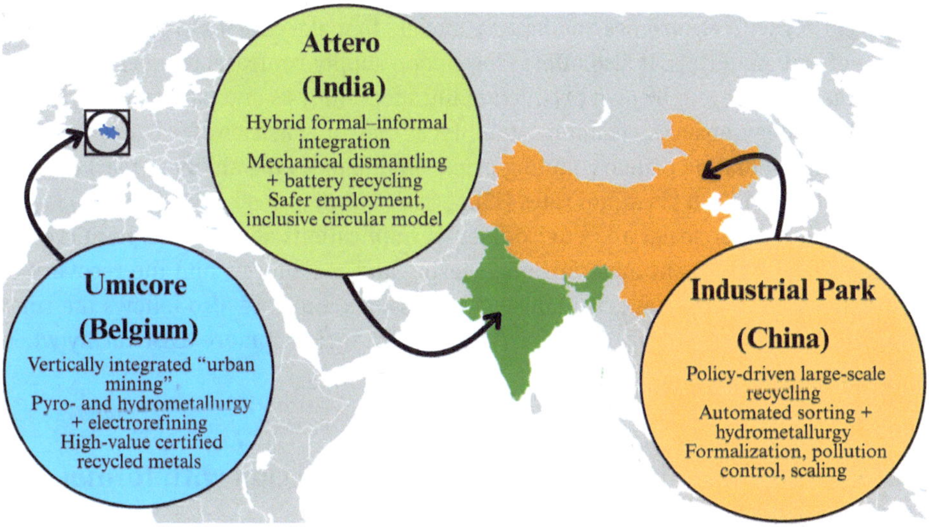

Figure 3: Global landscape of successful e-waste valorization models.

The map highlights the high-value integrated refining project Umicore in Belgium, which prioritizes technological advancement and closed-loop metal recovery. Then, a hybrid formal-informal-based project, Attero in India, which shows a collaborative approach, and the industrial recycling parks established in China, emphasize the effectiveness of regulatory measures and industrial expansion [3, 6, 7].

5.1 Umicore (Belgium): high-value refining and circular economy leadership

The Hoboken plant in Belgium, which is run by Umicore, is often recognized as one of the few places in the world where e-waste recovery has reached full industrial development. This operates on an urban mining business model. Urban mining model means it can manage difficult secondary materials, such as PCBs, used catalysts, industrial by-products, etc., in a similar process to how conventional ores are treated. It follows what the company describes as an urban mining model, that is, it tries to treat difficult secondary materials, PCBs, spent catalysts, industrial residues, and similar feedstock, in the same systematic way traditional mines handle ore. This process is a combination of pyrometallurgical smelting, hydrometallurgical refining, and electrochemical stages, which enable the recovery of metals [10, 11]. These include Au, Ag, Pd, Pt, Cu, and Ni, as well as several rare metals, which are essential for clean-technology manufacturing [1, 31].

The strength of this model is its extensive vertical integration. This company works closely with electronics and battery makers to make sure it always has a steady supply of raw materials. It also offers closed-loop supply contracts that send clean recycled metals back into its own production lines for catalysts and battery materials [9, 32]. This strategy ensures economic stability, access to consistent feedstock, reduces reliance on fluctuating primary mining prices, and enables the ability to market certified recycled metals at premium rates [13].

There are other social and environmental implications as well. The plant creates highly specialized jobs in materials engineering and metallurgy, and the recovery of metals from waste reduces greenhouse gas emissions. Umicore also follows the rules set by the European Union's WEEE Directive, which makes it more trustworthy when it comes to the environment [27].

5.2 Attero (India): integrating informal collection with formal processing

Attero recycling is an alternative example of an approach taken by a developing country to manage electronic waste. This was established in India in 2008. The majority of India's e-waste has historically been handled by informal collectors. Attero has progressively established a nationwide collecting network that directly involves the informal collectors rather than creating a system from beginning. The company trains and makes the workplace better for small-scale collectors, so they can work with the informal sector, which has a large market. This combination reduces the health risks that come with removing things without following the rules. It also maintains a high level of collection efficiency [7].

Attero employs a combination of mechanical dismantling, controlled component recovery, and collaborations with metallurgical processors for the ultimate extraction of metals. Recently, it has expanded into battery recycling to focus on the extraction of cobalt, nickel, and lithium, which are becoming increasingly vital for India's growing electronics and energy-storage sectors [12, 33].

The business collects, distributes, and sells high-value secondary metals by following India's EPR standards [16, 21]. These qualities help the company to operate effectively and reduce hazardous trash in informal landfills [13]. This project promoted small-scale enterprises, enhanced job safety, and improved domestic resource security. This hybrid model shows that accessible and sustainable recycling systems can exist in developing nations, where informal and formal systems can work in collaboration [13, 16].

5.3 China: from informal hubs to industrial-scale recycling parks

China produces the largest volumes of e-waste in the world. Its development in this area shows the opportunity for structured industrial transformation and the challenges of informal recycling [1]. Acid leaching, burning cables, and careless dumping were among the severely damaging behaviors that communities like Guiyu were known for a long time. These methods damaged the environment and public health. In the past decade, the Chinese government has relocated informal collectors to controlled industrial parks with mechanical dismantling units, automated sorting systems, and centralized metallurgical factories [5, 13, 20].

These parks have been upgraded with high-capacity crushing systems, density-based separation, magnetic and eddy-current separators, and enhanced hydrometallurgical processes that reduce emissions and achieve high recovery rates. These modifications enhanced material recovery and reduced the environmental impact on the local community. They also make it easier for China to obtain vital metals utilized in all of its manufacturing sectors [9, 34]. Significant governmental investment, national EPR programs, and private sector involvement are all included in this business model. The main objective is to regulate the industry, phase out criminal activity, and move previously unorganized workers into organized industrial zones where the surroundings are more closely monitored and less hazardous [27].

There are some mixed social and economic outcomes. Industrialization enhanced worker safety and produced more refined materials. But many unregistered workers lost their jobs as a result of the reorganization. Some people were able to move into formal employment as a result of training programs and cooperative activities, which reduced these effects [11, 13]. Table 3 presents the major characteristics of these three case studies. It highlights their technological selections, commercial frameworks, and socioeconomic consequences.

6 Contribution to Sustainable Development Goals (SDGs)

To promote sustainability, the United Nations proposed the 2030 Agenda, which, through the 17 SDGs, promotes and supports responsible practices in everything from production and consumption to the disposal of waste. These SDGs encourage recycling and the development of technological systems to minimize the detrimental effects of e-waste on the environment and society. E-waste management nearly correlates to most SDGs, including:

- SDG 3: To reduce illnesses and deaths from hazardous chemicals and pollution, with a focus on material recovery and safe recycling, reduce harmful pollutants and the health concerns.
- SDG 6: To improve the quality of the water by cutting down on dangerous releases and pollutants. Therefore, managing e-waste properly, heavy metals (Pb, Hg, and Cd) are prevented from leaking into soil and groundwater.
- SDG 8: To increase the efficiency of resources and separate environmental deterioration from economic growth. This subsequently helps the recycling, renovation, and materials recovery industries create green jobs.
- SDG 9: To renovate industries and infrastructure to make them more sustainable. Hence, it promotes creativity in smart recycling, sustainable technology, and circular manufacturing.
- SDG 11: By enhancing waste management, towns can lessen their negative environmental effects. Thus, an improvement over the sustainable urban waste systems through formal e-waste collection networks can be achieved.
- SDG 12: To create environmentally sound management of chemicals and waste. This ensures a sustainable production through resource recovery, reuse, and recycling.
- SDG 13: To include climate measures in planning and policy. Moreover, e-waste recycling lowers the need for energy-intensive mining and smelting, which lowers GHG emissions.
- SDG 14: To reduce and avoid land-based activities that pollute the sea; this keeps microplastics and hazardous leachates out of aquatic environments.

Table 3: Comparative analysis of successful e-waste recycling models.

Casestudy	Businessmodel	Keytechnologies	Scale	Socioeconomicoutcomes	Keylessons	References
Umicore (Belgium)	Vertically integrated with closed-loop supply contracts; multinational partnerships with electronics and battery industries	Pyrometallurgy, hydrometallurgy, electrochemical refining	>350,000 tons/year of complex waste streams	Highly skilled jobs, premium certified recycled metals, strong EU regulatory compliance	Vertical integration enhances profitability and supply security.	[1, 9–11, 26, 27]
Attero (India)	Hybrid model linking informal collectors with formal facilities under extended producer responsibility (EPR)	Mechanical dismantling, component recovery, metallurgical partnerships, lithium-ion battery recycling	~144,000 tons/year (mixed e-waste and batteries)	Safer employment, entrepreneurship opportunities, reduced informal dumping	Inclusive models balance efficiency with social equity.	[9, 12, 13, 16, 27, 33]
China (industrial parks)	State-led, large-scale industrial transformation integrating dismantling, sorting, and refining	Automated shredding, magnetic/eddy separation, hydrometallurgy, pollution control	Largest global capacity (>10 million tons/year across multiple parks)	Safer jobs, reduced pollution, displacement of informal workers mitigated through retraining	Regulation and scale can phase out informal hazardous hubs.	[5, 6, 9, 12, 13, 31, 34]

The main focus of SDG 3 is to promote and ensure healthy lives and foster well-being for everyone. Formal e-waste valorization prevents human exposure to harmful compounds and helps mitigate the release of harmful gases during open burning or informal recycling [13]. E-waste valorization not only reduces harmful emissions and water contamination but also helps improve environmental conditions and public health simultaneously [35]. Due to its impact on ecosystems, human health, and water resources, e-waste management is strongly connected to SDG 6. SDG 8 ensures decent work opportunities and economic growth worldwide, and these can be ensured through the economic opportunities within the recycling and materials recovery [36].

Sustainable Cities and Communities (SDG 11) is promoted through the integration of e-waste collection, segregation, and recycling systems into urban waste management frameworks. SDG 12 highlights the goal of ensuring sustainable production and consumption patterns. It promotes sustainable production and the use of eco-friendly gadgets. This practice contributes to a reduction in the environmental and social impacts involved in the production chain [25, 37]. Finally, Climate Action (SDG 13) is addressed through the climate mitigation potential of e-waste valorization. Significantly lowering greenhouse gas (GHG) emissions from primary mining, smelting, and manufacturing processes is achieved by recycling and reusing elements from electronic waste.

7 Challenges and barriers

Recovering valuable, rare earth- and usable elements while disposing of the harmful ones is a unique problem in managing e-waste. The limited number of environmentally friendly chemical liquids that are permitted for use in e-waste management, the absence of infrastructure facilities, the thermodynamic limits of separating complex mixtures of materials that result in cost-ineffective recovery challenges, the lack of financial and policy support, particularly for developing nations, and inconsistent laws that differ significantly from one nation to another are additional difficulties in managing e-waste [38, 39] Additionally, the informal recycling sector's involvement in addressing infrastructural deficits is a major challenge in managing e-waste.

E-waste is frequently handled by informal recyclers in developing nations, who lack the necessary tools and safety precautions. These workers extract important components using acid application or open burning, putting their health at risk and releasing hazardous pollutants into the environment. Due to a lack of official recycling mechanisms and economic need, the majority of e-waste recycling in many third-world nations is done by the unorganized sector. To reduce health hazards and environmental harm, investments in safe treatment systems and advanced recycling facilities are essential. Implementing EPR, a crucial tool for managing e-waste in developing nations, can be difficult due to the absence of formal recycling infrastructure. Investment in recycling plants can boost infrastructure development [5, 12].

Economic and cultural variations have a significant impact on the success of e-waste management techniques and related policy actions. Despite improved economic standing and higher per capita income, formal recycling and the implementation of e-waste policies have not increased significantly in developed nations. A recent case study in the USA revealed that, despite an average per-person e-waste generation of ~32.1 kg, the percentage of recyclables remains below 16% [3, 6]. It was observed that higher-income groups have easier access to recycling units and greater awareness of recycling policies, which has led to increased recycling of e-waste. Due to a lack of understanding and transportation, people from lower- or middle-class backgrounds often prefer to recycle, resell, or donate used electronics informally. These discrepancies highlight the importance of a balanced approach to e-waste recycling infrastructure and tailored awareness programs. Proper recycling is frequently too costly for certain businesses and governments, especially when it comes to institutional structures. Because of this, e-waste is either incorrectly disposed of or contracted out to the unorganized sector, where hazardous waste management methods result from cheap pay and insufficient safety precautions. Additionally, the hazardous yet valuable makeup of electronic items necessitates specific methods for effective material recovery and separation; however, these methods are expensive for recycling facilities to purchase and implement. To address these financial challenges and encourage safer, more efficient recycling, the governments and companies must think about incentives, subsidies, and stronger regulations.

The absence of e-waste legislation is a significant obstacle. E-waste legislation is still lacking in the majority of developing countries, aside from developed ones. E-waste legislation was in place in 61 nations, or 44% of the world's population, until 2014. About 71% of the world's population was covered by e-waste legislation in 78 nations in 2019. While China and India, the two most populous countries in Asia, now have e-waste legislation, other large nations with populations of more than 160 million, such as Bangladesh and Pakistan, do not currently have such laws. [38]. Green regulations, such as green product identification, should be mandatory, with producers being accountable for the chemical composition of their products and ensuring the retrieval of every ingredient once its life cycle ends. EPR enforcement is critical in encouraging stakeholders to design products responsibly, manufacture them, and process recovered e-waste. Such techniques can only be effective if the government provides the formal e-waste management system with financial aid, technical solutions, or expert assistance. A comprehensive SWOT analysis of global e-waste management and valorization systems reveals several interrelated strengths, weaknesses, opportunities, and threats that influence their development and sustainability.

Strengths

- E-waste contains a vast amount of valuable metals that have high potential to contribute to the global economy.
- Efficient recycling and recovery slow down the dependency on nonrenewable mineral resources.
- Creates employment opportunities and promotes sustainable practices.
- Promotes sustainability and contributes to lowering greenhouse gas emissions.
- Contribution to attain Sustainable Development Goals (SDGs).

Weaknesses

- The complex and heterogeneous composition of e-waste materials poses significant challenges to efficient recycling.
- Lack of adequate knowledge on disposal methods or the importance of recycling causes improper dumping in open sites.
- Lack of e-waste generation data makes the policy planning process difficult.
- Consumer awareness and education regarding the proper disposal of e-waste remain insufficient.
- Establishing modern recycling facilities requires a high initial investment.

Opportunities

- The growing global demand for critical and rare earth elements presents new avenues for resource recovery.
- Integration of digital technologies such as artificial intelligence (AI), the Internet of Things (IoT), and robotics can enhance smart waste management systems.
- Government incentives and EPR policies encourage formal recycling initiatives.
- Rising global awareness of sustainable consumption and production practices support market expansion.
- Urban mining is emerging as a promising new economic sector.
- Major electronics manufacturers are increasingly adopting circular supply chains, creating synergy with recycling industries.

Threats

- Informal recycling practices cause environmental pollution.
- The high cost of the cutting-edge technologies required for recycling discourages investment, particularly in developing nations.

- Informal sector workers remain concerned about health hazards.
- Effective waste management is threatened by poor policy implementation and a lack of accountability.
- Volatile market values of recovered metals and materials make recycling economically challenging.

8 Opportunities and future directions

If proper recycling practices are followed, e-waste could be transformed from a global challenge into an immense economic opportunity. A recent study showed the composition analysis of both smartphones and mobile phones. The composition analysis of smartphones revealed that they contained around 260 diverse nonmetallic and metallic components. On the other hand, the mobile phones had only around 150. A more detailed investigation showed that 1 kg of smartphones has, on average, approximately 1,600 mg of silver, 186.5 mg of gold, and 36.9 mg palladium (in an average), [40]. This highlights e-waste as a potential avenue for secondary mining and also opens the opportunities for the sustainable extraction of rare and precious metals. Moreover, metal recovery from e-waste, adopting the concept of urban mining, could be turned into a profitable business [41]. By applying the right business models, the recovery of these precious metals could significantly contribute to the economy.

Urban mining also facilitates the recirculation of metals and other resources recovered from e-waste into the supply chain of resources. Therefore, urban mining is a closed-loop supply chain technique that not only offers a sustainable use of mineral resources but also encourages circularity in the supply chain [41]. Moreover, in developing countries, urban e-waste mining also fosters job opportunities and potential for businesses to carry out different e-waste activities, such as sorting, dismantling, or recycling, in addition to the valuable supply chain.

E-waste management also has a significant impact on mitigating environmental pollution and offers human health benefits [40]. Nowadays, people are more likely to purchase new electronics as a result of the quick evolution of technology, leaving behind outdated ones that are thrown away or disposed of improperly. This behavior is often observed in high-income countries, such as the United States. In developing countries, however, e-waste management is given less priority and is never considered a serious issue. This poor practice and unaware act threatens the health and the environment around the dumping sites [3, 6]. However, if they take this seriously and adopt and implement e-waste management strategies, policies, and legislation, this could significantly change the overall scenario.

The circular economy is an exciting idea that offers a transformative solution to reduce waste and utilize resources sustainably. By emphasizing the constant reuse, refurbishment, and recycling of materials, this strategy reduces the need for raw materials while increasing the lifespan of gadgets. This can also dramatically reduce annual greenhouse gas emissions. To select the most effective circular economy methods for e-waste and equipment management, a multi-criteria decision-making (MCDM) approach is crucial [42]. By using MCDM, the selection and prioritization of circular economy strategies, such as the 9 Rs (refuse, rethink, reduce, reuse, repair, refurbish, remanufacture, repurpose, recycle, and recover), can be optimized [43]. Hence, the MCDM approach plays a crucial role in identifying the strategies that significantly reduce e-waste and enhance sustainability.

Blockchain is a cutting-edge technology, which could offer an innovative solution in managing e-waste. Some of the fundamental features of blockchain that have demonstrated its potential to transform traditional industries are decentralization, persistence, anonymity, and audibility [44]. Additionally, it has the potential to completely transform the lifespan of e-waste, from ensuring compliance with recycling standards to tracking and traceability [45].

Due to rapid technological advancements and innovative approaches, exploring efficient and innovative solutions for addressing e-waste has become a growing area of interest. AI is such a promising avenue that offers a wide range of potential applications for reducing e-waste. AI can enhance product designs by optimizing the ease of disassembly and identifying which components are most suitable for reuse or recycling. This strategy helps reduce waste generation during product design [46]. In recent years, machine learning has gained popularity as it offers the advantage of easy replacement by assessing specific products and identifying which components can be made modular, rather than discarding the entire device. This strategy helps in reducing e-waste by extending the lifespan of individual components. The device can forecast the remaining operational life of its parts with the help of sensors and intelligent algorithms [47]. Modern devices, such as the iPhone, monitor battery health and alert users when it is time to replace the battery. This not only benefits the consumers' finances by preserving the device but also reduces unnecessary waste [48].

9 Conclusion

The transformation of electronic waste into value-added resources represents a critical lever for advancing global sustainability, strengthening circular economy systems, and mitigating the environmental burdens associated with rapid technological growth. As this chapter has demonstrated, electronic waste is not merely a waste stream but a complex reservoir of metals, polymers, and functional materials whose recovery can substantially reduce dependence on primary resource extraction. Current research and industrial practices illustrate that, with appropriate technological, regulatory, and economic frameworks, discarded electronics can be reintroduced into productive value chains through recycling, upcycling, and innovative material conversion processes.

The chapter's analysis of material composition, recovery techniques, and emerging nanomaterial applications underscores the broad technological potential embedded within electronic waste streams. Examples from diverse national and industrial contexts further reveal that successful electronic waste valorization requires coordinated action: robust policy instruments, formalized collection and recycling infrastructure, industry and academia collaboration, and public awareness initiatives. When integrated effectively, these elements support responsible consumption and production practices, enhance resource security, and create opportunities for green employment, aligning closely with the United Nations Sustainable Development Goals.

Despite these promising developments, significant challenges persist. Global disparities in electronic waste governance, informal and hazardous recycling practices, limited enforcement of extended producer responsibility, and technological barriers in complex material separation continue to hinder progress. Addressing these gaps will require targeted investments in advanced recycling technologies, harmonized regulatory frameworks, and stronger incentives for producers to design electronics with end-of-life recovery in mind.

Overall, this chapter contributes to the growing body of literature emphasizing that sustainable electronic waste management is not solely a waste mitigation strategy but an essential component of a resilient, resource-efficient, and circular global economy. By reframing electronic waste as a strategic material stock and by advancing scalable valorization pathways, societies can reduce environmental pollution, stimulate innovation, and move closer to achieving long-term sustainability objectives.

References

[1] Forti V, Balde CP, Kuehr R, Bel G. The Global E-waste Monitor 2020: Quantities, flows and the circular economy potential. In United Nations University (UNU), International Telecommunication Union (ITU) & International Solid Waste Association (ISWA) 2020 (Vol. 120). Bonn/Geneva/Rotterdam.

[2] The global E-waste Monitor 2024–Electronic Waste Rising Five Times Faster than Documented E-waste Recycling: UN The global E-waste Monitor 2024-Electronic Waste Rising Five Times Faster than Documented E-waste. Recycling: UN; 2024.

[3] Masud MH, Akram W, Ahmed A, Ananno AA, Mourshed M, Hasan M, Joardder MUH. Towards the effective E-waste management in Bangladesh: A review. Environ Sci Pollut Res. 2019;26:1250–1276.

[4] Masud MH, Mourshed M, Mahjabeen M, Ananno AA, Dabnichki P. Global electronic waste management: current status and way forward. Paradigm Shift in E-waste Management. 2022;9–47.

[5] Mourshed M, Khatun S, Islam K, Masuk NI, Masud MH. Generation, composition, collection, and treatment of E-Waste Management of Electronic Waste: Resource Recovery. Technology and Regulation. 2024;39–72.

[6] Masud MH, Mourshed M, Hossain MS, Ahmed NU, Dabnichki P. Generation of waste: problem to possible solution in developing and underdeveloped nations. In Waste management and resource recycling in the developing world 2023 (pp. 21–59). Elsevier.
[7] Hsu J, Wang J, Stern M. E-waste: A global problem, its impacts, and solutions. J Glob Inf Manag. 2024;32:1–28.
[8] Mourshed M, Masud MH, Rashid F, Joardder MUH. Towards the effective plastic waste management in Bangladesh: a review. Environ Sci Pollut Res. 2017;24:27021–27046.
[9] Kwon G, Yoon K, Kwon E, Park J, Lee H, Song H. Technical advancement in valorization of electronic waste and its contribution to establishing economic value-chain. Chem Eng J. 2024;494:153154.
[10] Hagelüken BC. Recycling the platinum group metals: A European perspective. Platinum Meta Rev. 2012;56:29–35.
[11] Nowakowski P. Urban Mining of e-Waste: Conversion of Waste to Wealth. In Management of Electronic Waste Priya A (ed). 2024 (Vol. 2024, pp. 152–172). Resource Recovery, Technology and Regulation, Wiley.
[12] Quinto S, Law N, Fletcher C, Le J, Antony Jose S, Menezes PL. Exploring the E-Waste Crisis: Strategies for Sustainable Recycling and Circular Economy Integration. Recycling. 2025;10:72.
[13] Srivastava AN, Sikarwar VS, Bisen D, Fathi J, Maslani A, Lopez Nino BN et al. E-Waste Unplugged: Reviewing Impacts, Valorization Strategies and Regulatory Frontiers for Efficient E-Waste Management. Processes. 2025;13. 2014.
[14] Horvath Z. Transforming our world-new agenda and goals for sustainable development Hungarian YB. Int'l L & Eur L. 2016;167.
[15] Ilankoon I, Ghorbani Y, Chong MN, Herath G, Moyo T, Petersen J. E-waste in the international context–A review of trade flows, regulations, hazards, waste management strategies and technologies for value recovery. Waste Manag. 2018;82:258–275.
[16] Islam MK, Khatun MS, Mourshed M. An in-depth analysis and review of management strategies for E-waste in the south Asian region: A way forward towards waste to energy conversion and sustainability. Heliyon. 2024;10.
[17] Kaya M. Recovery of metals and nonmetals from electronic waste by physical and chemical recycling processes. Waste Manag. 2016;57:64–90.
[18] Sum EY. The recovery of metals from electronic scrap. Jom. 1991;43:53–61.
[19] Cui J, Forssberg E. Mechanical recycling of waste electric and electronic equipment: a review. J Hazard Mater. 2003;99:243–263.
[20] Widmer R, Oswald-Krapf H, Sinha-Khetriwal D, Schnellmann M, Böni H. Global perspectives on e-waste. Environ Impact Assess Rev. 2005;25:436–458.
[21] Khatun MS, Islam MK, Sultana A, Mourshed M. Application of plastics for production of fuel. Environmental Hazards of Plastic Wastes. Elsevier. 2025;79–96.
[22] Sarmah D, Nirala A, Hazarika H. Heavy metal-induced neurotoxicity and endocrine disruption from e-waste: emerging health challenges and intervention strategies. J Toxicol Environ Health Sci. 2025;1–18.
[23] Islam MR, Mostafa M, Rubaida NJ, Mourshed M. Impact of plastic pollution on agricultural ecosystems: Challenges and sustainable solutions. In Microplastics in Agriculture and Food Science 2025 (pp. 3–20). Elsevier.
[24] Kibria MG, Masuk NI, Safayet R, Nguyen HQ, Mourshed M. Plastic waste: challenges and opportunities to mitigate pollution and effective management. Int J Environ Res. 2023;17:20.
[25] Hamzat AK, Murad MS, Subeshan B, Asmatulu R, Asmatulu E. Rare earth element recycling: a review on sustainable solutions and impacts on semiconductor and chip industries. J Mater Cycles Waste Manag. 2025;1–24.
[26] Phogat P, Kumar S, Wan M. A scientometrics study of advancing sustainable metal recovery from e-waste: processes, challenges, and future directions. RSC Sust. 2025;3:2434–2454.

[27] Serpe A, Purchase D, Bisschop L, Chatterjee D, De Gioannis G, Garelick H et al. 2002–2022 20 years of e-waste regulation in the European Union and the worldwide trends in legislation and innovation technologies for a circular economy. RSC Sustain. 2025;3:1039–1083.
[28] Bhoi NK. Advancements in E-waste recycling technologies: A comprehensive overview of strategies and mechatronics integration for future development. Sust Mater Technol. 2024;42:e01182.
[29] Sonigara KK, Vaghasiya JV, Mayorga-Martinez CC, Pumera M. Point-of-use upcycling of 3D printing waste for developing 3D-printed Zn–I 2 batteries. J Mater Chem A. 2025;13:11804–11816.
[30] Malaiškienė J, Bekerė K. Impact of Electronic Waste Glass on the Properties of Cementitious Materials. Buildings. 2024;14:1218.
[31] Goyal S, Gupta S. A comprehensive review of current techniques, issues, and technological advancements in sustainable E-waste management e-Prime-Advances in Electrical Engineering. Electronics Energy. 2024;9:100702.
[32] Abedin T, Pasupuleti J, Paw JKS, Tak YC, Islam MR, Basher MK, Nur-E-Alam M. From waste to worth: advances in energy recovery technologies for solid waste management. Clean Technol Envir Pol. 2025;1–27.
[33] Firmansyah ML, Rizki IN, Ullah N. Recent advances in urban mining technology: A focus on electronic waste recycling potential in Indonesia. Cleaner Waste Syst. 2025;10:100239.
[34] Zholdasbay EY, Dosmukhamedov N, Argyn A, Kurmanseitov M, Tazhiev YB, Daruesh G, Koishina G. Technological and environmental aspects of electronic waste recycling. Metallurgist. 2024;68:274–281.
[35] Rautela R, Arya S, Vishwakarma S, Lee J, Kim K-H, Kumar S. E-waste management and its effects on the environment and human health. Sci Total Environ. 2021;773:145623.
[36] Yang B-C, Lee C-H, Suryawan IWK. Financial support in E-waste management within a circular economy with structural equation modeling analysis from Taiwan. Circ Econ Sustain. 2025;5:949–973.
[37] Abd Aziz MA, Hairunnaja MA, Arifin MA, Isa KM. Advancements in Metal Recovery from Industrial Wastes: A Comprehensive Overview. Controlling Environmental Pollution: Practical Solutions. Elsevier. 2025;225–245.
[38] Baldé CP, Forti V, Gray V, Kuehr R, Stegmann P. The global e-waste monitor 2017: Quantities, flows and resources. In United Nations University (UNU), International Telecommunication Union (ITU) & International Solid Waste Association (ISWA) (ed) 2017 (pp. 1–109).
[39] Sahle-Demessie E, Glaser J, Richardson T (eds). Electronics waste management challenges and opportunities. In ACS, National Meeting 2018 Boston, MA.
[40] Singh N, Duan H, Yin F, Song Q, Li J. Characterizing the materials composition and recovery potential from waste mobile phones: A comparative evaluation of cellular and smart phones. ACS Sustain Chem Eng. 2018;6:13016–13024.
[41] Xavier LH, Giese EC, Ribeiro-Duthie AC, Lins FAF. Sustainability and the circular economy: A theoretical approach focused on e-waste urban mining. Resour Policy. 2021;74:101467.
[42] Fetanat A, Tayebi M, Shafipour G. Management of waste electrical and electronic equipment based on circular economy strategies: navigating a sustainability transition toward waste management sector. Clean Technol Envir Pol. 2021;23:343–369.
[43] Hunger T, Arnold M, Ulber M. Circular value chain blind spot–A scoping review of the 9R framework in consumption. J Clean Prod. 2024;440:140853.
[44] Samiha I, Forkan MIU, Sarker AS, Das A, Sultana N (eds). Blockchain-Based Solutions for E-Waste Management: A Review of Opportunities and Challenges. In 2025 IEEE International Conference on Interdisciplinary Approaches in Technology and Management for Social Innovation (IATMSI) 2025 IEEE.
[45] Palagan CA, Joe SSA, Mary SJ, Jijo EE. Predictive analysis-based sustainable waste management in smart cities using IoT edge computing and blockchain technology. Comput Ind. 2025;166:104234.

[46] Lakhouit A. Revolutionizing urban solid waste management with AI and IoT: a review of smart solutions for waste collection, sorting, and recycling. RINENG. 2025;104018.
[47] Prentice C, Dominique Lopes S, Wang X. Emotional intelligence or artificial intelligence–an employee perspective. J Hosp Mark Manag. 2020;29:377–403.
[48] Agrawal A, Gans JS, Goldfarb A. Artificial intelligence: the ambiguous labor market impact of automating prediction. J Econ Pers. 2019;33:31–50.

Fahad Rahul*, Md. Shamsur Rahman, Suneha Safayet Audrika, Sumaia Akter Lamisa

Harnessing biological systems for sustainable e-waste valorization: pathways to a circular economy

Abstract: E-waste is the fastest growing global waste stream, and it poses a significant threat to human health and the environment because of toxic metals and risks involving persistent organic pollutants. Biological methods such as bioleaching, phytoremediation and microbial biotransformation are relatively inexpensive and environmentally friendly methods of metal recovery and detoxification as compared to conventional processes. This chapter focuses on the recent progress in the field of biological methods of e-waste treatment, as well as discussing the possibility of its application to circular economies. Case studies show how things can be applied and policy views show how things can be expanded into more sustainable solutions. The discussion provides biotechnology as a game-changing technology transforming electronic waste materials into resources within closed-loop model.

Keywords: E-waste, bioleaching, phytoremediation, microbial biotransformation, circular economy, resource recovery, sustainability

1 Introduction

Electronic waste (e-waste) is currently experiencing the most rapid growth among global waste streams worldwide, and the volume of e-waste is projected to grow to over 62 million tons annually in 2022, as compared to 33 million tons of e-waste 10 years ago [2]. The growth is exponential due to technological innovation and the relatively short life cycles of its products and the need to possess the new product by the consumers. Unfortunately, approximately 20% of this wastage is systematically gathered and recycled and the rest 80% is sent to landfills, incineration, or informal disposal with severe environmental and health consequences (Julee Bane-

***Corresponding author: Fahad Rahul**, Department of Environmental Science and Disaster Management, Gopalganj Science and Technology University, Gopalganj 8100, Bangladesh, e-mail: fahadrahul18@gmail.com
Md. Shamsur Rahman, Department of Environmental Science and Disaster Management, Gopalganj Science and Technology University, Gopalganj 8100, Bangladesh
Suneha Safayet Audrika, BRAC University, Dhaka, Bangladesh
Sumaia Akter Lamisa, Institute of Social Welfare and Research, University of Dhaka, Bangladesh

 | https://doi.org/10.1515/9783112225509-010

rji et al., 2026). Lead, cadmium, mercury, persistent organic pollutants (POPs), and others make soil and water poisonous; uncontrolled recycling often leads to the release of carcinogenic chemicals, including dioxins, into the air. The phenomenon is especially puzzling in underdeveloped countries, where the level of e-waste is growing at approximately two times the pace of industrialized countries[3]. Conventional recycling methods of e-waste, including mechanical shredding, and chemical leaching, are energy-intensive, generate secondary pollutants, and are relatively expensive. In this respect, the biological systems are more sustainable and ecologically viable. Bioleaching, phytoremediation, and microbial biotransformation provide effective approaches to recover high-value metals, including copper, gold, and rare earth elements (REEs), and have a low environmental impact. Under optimum conditions, taking into consideration economic and environmental factors, a recovery rate of above 90% has been achieved with acidophilic bacteria with bioleaching in small-scale work [4]. Together with the recovery of metals, biological therapeutics are also applied to clean up the toxic organics through biodegradative conversion of the harmful organics to less harmful substances. The two functions of recovery and purification of raw materials are closely connected with the concept of a circular economy (CE). Biotechnology has the potential of converting e-waste management systems into circular processes that use waste to create new and sustainable applications instead of causing degradation to the environment [5]. This chapter examines the challenges associated with e-waste management, the potential of biological valorization methods, and their integration within a closed-loop circular economy framework. Drawing on recent technological advances, case studies, and policy perspectives, it highlights the role of biotechnology in transforming e-waste into valuable resources for sustainable management systems.

2 Literature review

The growing amount of e-waste being produced around the world has led to an impressive literature on its management, valorization pathways, and sustainable recovery. The latter was more focused on the mechanical and chemical processing (including, but not limited to, shredding, smelting, and acid leaching) that enabled the extraction of colossal quantities of metals and generated colossal volumes of secondary pollution at the same time [6]. With more segments of the population becoming aware of the environmental effects of e-waste, more academic attention started to be drawn to environmentally friendly solutions, including the use of biological systems in e-waste management. The biological methods are attractive due to their potential cheapness and sustainability. One well-researched method of accessing the copper, gold, zinc, and rare earths within discarded circuit boards is bioleaching, or making use of microorganisms to recycle the metals. Some types of bacteria, such as *Acidithio-*

bacillus ferrooxidans, can recover metals at comparable rates as the established processes in the right environment and can do so with a significantly smaller environmental footprint [7]. Phytoremediation research has demonstrated the use of phytoabsorption and phyto-sequestration of e-waste metals within polluted soil dumpsites by plants, as a natural process of decontamination and rehabilitation of degraded ecosystems. There is an enormous amount of literature devoted to microbial biotransformation where the microorganisms are used to transform a toxic substance into a harmless one. This is an extremely effective way of mitigating risks of POPs within plastics, flame retardants, and other electronic components. Biotechnological applications are always mentioned related to microorganisms in the context of bioremediation and resource recovery [8].

The literature shows that such biological approaches could be implemented in the models of the CE. Researchers also point out that biovalorization can not only reduce waste but also create economic opportunities; the recovery of valuable materials during the process benefits industries utilizing cobalt and palladium. Moreover, the practice of microbial leaching is presented with reference to the pilot studies performed in various continents, such as Asia, Africa, and Europe, with the emphasis on the pilot-scale case studies and phytoremediation policy-related initiatives [9]. Finally, existing literature shows that biotechnology is a promising bridging technology to transform working practices in the e-waste industry and connect environmental conservation with sustainable industrialization issues and risks of e-waste to the environment (Table 1).

The global increase in e-waste has led to an increasing number of literature explaining management approaches, valorization mechanisms, and sustainable recovery techniques. The focus of historical studies on metal recovery has been mainly on mechanical and chemical methods (shredding, smelting, and acid digestion), often with significant secondary pollution, and often with high costs of energy use [10]. As interest in green solutions, such as bio-based systems, to discard e-waste increased, the realization of degradation increased. The importance of biological techniques as a potential method has been based on its low cost and its ability to be environmentally friendly. The solubilization of the metals by the microbial activity is called bioleaching and has been widely investigated in the recovery of copper, gold, zinc, and REEs in waste printed circuit boards [11].

Application of *Acidithiobacillus ferrooxidans* in bacterial water treatment has shown a similarity in metal recovery rates compared to traditional methods, except that it has low environmental impact. Given that extensive studies have been conducted on the absorption and sequestration of contaminated soil at informal e-waste dumpsites, it is evident that plants through natural detoxification and rehabilitation of degraded ecosystems could offer a viable eco-friendly alternative to metal phytoremediation [16]. There is a large literature on microbial biotransformation where microorganisms are used to mediate the chemical treatment of toxic substances into less toxic molecules. It will help reduce the environmental risk of the remaining organic pollutant impregnated in plastics, flame retardant, and other fragments of electronic

Table 1: E-waste challenges and environmental risks.

Challenge	Description	Environmental risk	Human/health impact	References
Rapid growth of e-waste	Annual global generation exceeds 60 million tons, overwhelming recycling capacity.	Accumulation in landfills, leading to soil and water contamination.	Communities near disposal sites face higher exposure to toxins and waste hazards.	[12]
Toxic heavy metals	Lead, cadmium, and mercury leach from discarded electronics.	Long-term contamination of soil, groundwater, and aquatic ecosystems.	Bioaccumulation leads to neurological, kidney, and developmental disorders.	[13]
Informal recycling practices	Open burning, acid leaching, and unsafe dismantling dominate in many regions.	Release of carcinogenic dioxins, greenhouse gases, and persistent pollutants.	Workers suffer from respiratory diseases, cancers, and chronic health conditions.	[14]
Loss of valuable resources	Gold, silver, palladium, and rare earth metals wasted due to poor recovery.	Increased mining demand and intensifying ecological degradation.	Economic loss for communities and industries face supply shortages of critical raw materials.	[15]

devices. It has been argued as an advantage of biotechnological solutions that microorganisms act as a resource reservoir and pollution-reducing agent. The integration of biology into the model of a CE can be considered one of the most popular topics of modern research. The researchers stress that waste minimization relies on biological valorization and the reuse of rare elements like cobalt and palladium, which can be used in certain industries, has economic potential [17]. On a personal scale, examples of application can be offered in Asia, Africa, and Europe, where microbial leaching has been used on a pilot scale, up to development of policies on phytoremediation. In all these research articles, it has been noted that bio technology is what has transformed the process of e-waste management, as well as the interface between the environment and welfare of the industrial development [18].

The use of biological systems can be considered as a partial solution to the e-waste management problem is considered [19]. Compared to more conventional recycling techniques, which require energy or the application of harmful chemicals to complete the process, biological solutions harness the potential of plant and bacteria to clean environments and produce useful products simultaneously with limited environmental impact [20].

Bioleaching is a well-studied technique to extract metals in e-waste making use of microorganisms, including bacteria and fungi. Through metabolic activity, these microorganisms synthesize organic acids and other metabolites that dissolve metals such as copper, gold, zinc and cobalt. Optimized conditions can give bioleaching a recovery rate similar to traditional methods and no hazardous by-products are produced [21]. The capacity of microorganisms to survive under harsh conditions has allowed degrading shredded electronic trash, leading to a scalable urban mining of needed metals. In addition, bioleaching systems can be designed in bioreactors, and processes are controlled and adjusted to meet industrial requirements.

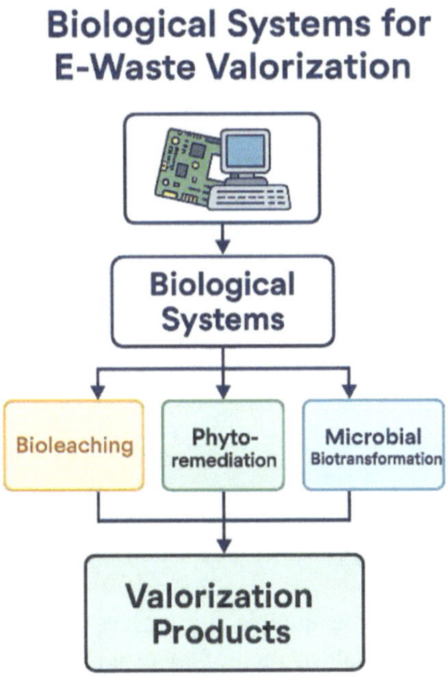

Figure 1: Schematic representation of biological approaches for sustainable e-waste valorization.

This image explains the major biochemical routes to e-waste valorization, specifically focusing on bioleaching, phytoremediation, and microbial biotransformation as the most important interventions. Bioleaching utilizes microbes to extract metals from e-waste instead of using chemical leaching processes. Heavy metal-polluted soils can be phytoremediated by plants that extract these elements thereby helping to regenerate the ecosystem. Bioremediation is the process of converting toxic chemical compounds into less toxic or non-toxic forms by degradation [22]. All these machines transform the dangerous e-waste materials into commodities and reduce environmental harm. The figure shows the sustainability and the CE of wastewater in terms of resource recovery and the reduction of ecological impact [23].The other possible approach is the use of phytoremediation where plants are used to trap and store metals in soils that are polluted by ill-disposed e-waste. Some hyperaccumulator plants can accumulate metals in their tissues up to dozens or even hundreds of thousands of times higher than in the typical plant, allowing them to clean up polluted soil and transform poisonous metals into valuable products. Indian mustard and sunflower are effective in the removal of heavy metals. The biomass can then be burnt or otherwise refined to harvest concentrated metal values [24]. Combined activities of remediation and the recovery of metallic elements make phytoremediation a cost-efficient approach to treating polluted soils in developing countries where the number of problems associated with soil pollution is growing (Figure 1). Biotransformation of microbes is suggested as a mechanism of cleansing of harmful organic pollutants in plastics, flame retardants, and other components of e-waste [25]. Microbial metabolism can biologically convert it into a form that is less harmful to the environment and to humans. Other than detoxifying, some types of bacteria also help to solubilize metals, making them more accessible to be recovered . These organisms have the potential to solve e-waste management challenges and they have many benefits. They operate effectively at ambient temperature, have a low-energy consumption, and produce a very low secondary emission level. Biological valorization will help shift to the CE and turn waste streams into value-added products, reduce dependence on virgin mining, and lessen the impact of the electronic industries on the environment. Biological systems create a new frontier as the demand of metals increases and pollution management becomes more popular. Subsequent hybrid systems with inheritance of not only biological but also mechanical processes will allow greater valorization of e-waste at large scale and sustainably in future [26].

3 New developments and innovations

The last several years have seen the emergence of new sustainable alternatives to traditional recycling processes, in particular the biological valorization of e-waste (Table 2). The programs aimed to improve efficiency, scalability, and flexibility of bio-based technologies in expanding their utilization in industry. More sophisticated biotechnologies have been combined with traditional microbiological techniques to increase the yield of valuable metals and to effectively remediate hazardous substances [27].

Engineering microorganisms to undergo bioleaching is also an impressive achievement. The change in certain metabolic pathways to enhance the solubilization of metal has affected microbial strains that generate high levels of organic acids and oxidizing agents. Genetically modified microbes have shown their ability to effectively recover such metals in complex mixtures of metals such as mine tailings. This enhancement solves the main problem that has been claimed against conventional bioleaching, which is that it is far too slow as compared to chemical reactions.

There have also been parallel developments in phytoremediation. More recent studies have found special hyperaccumulator plants that are able to purify REEs in soils, which is one of the biggest challenges in biological clean-up processes. Plant biotechnology has helped scientists to increase the efficiency of metal uptake by plants through selective breeding and genetic modification, leading to plants that are able to endure high levels of harmful chemicals and grow very fast [32]. Plant-microbe hybrid

Table 2: Recent advances and innovations in biological e-waste valorization.

Innovation	Description	Impact on e-waste management	References
Genetically engineered microbes	Modified microorganisms produce more acids and oxidizing agents to accelerate bioleaching.	Faster recovery of metals, such as copper and gold, reducing processing time.	[28]
Enhanced phytoremediation	Discovery and engineering of hyperaccumulator plants for rare earth and heavy metal uptake.	Efficient recovery of scarce resources and detoxification of contaminated soils.	[29]
Microbial consortia applications	Use of multiple microbial species for simultaneous metal recovery and pollutant breakdown.	Holistic treatment of e-waste combining detoxification and resource extraction.	[30]
Digital and bioreactor integration	AI and automated bioreactors optimize growth conditions and process efficiency.	Scalable and predictable operations suitable for industrial adoption.	[31]

systems where plants and beneficial rhizospheric microbial communities are combined have, in this regard, been found to be more efficient in the accumulation of metals and in the remediation of soil. Considerable progress has been made on microbial biotransformation, especially using microbial consortia as opposed to separate strains. These consortia might offer a more diverse range of ways in which metal and PAH can be degraded and, therefore, enable a more refined treatment approach toward e-waste [33]. Recent developments in metagenomic studies have allowed researchers to discover previously unknown microbial consortia that live in harsh conditions, including acid mine drainage, which have the innate ability to effectively decompose e-waste constituents.

The key tendency is the combination of bio-based approaches and digital technology. Optimized bioleaching conditions: With the use of tools like machine learning and artificial intelligence (AI), the parameters of bioleaching, including pH, temperature, and microbial growth conditions can be optimized, leading to improved reliability and effectiveness in the recovery process [34]. One of the most important steps toward achieving large-scale commercial processing under controlled conditions is the creation of bioreactors with electronic control monitoring systems that practically eliminate the previously described bottleneck in scale-up. With the emerging new technologies, a paradigm shift in e-waste as it is known will be experienced. They optimize the efficiency, speed, and scaled-up industrial recycling, and thereby enable the incorporation of biological processes. The latter is directly responding to the growing e-waste issue in an environmentally friendly manner, thus broadening the base of a healing economy where outdated devices would be turned into new ones [35].

3.1 Analyses and implementations of cases

Biological systems for e-waste valorization have received interest, with numerous pilot projects and case studies demonstrating their viability at both laboratory and field scales. The given case studies in the real world explain how the newly developed biological technologies can be successfully applied to different geographical and socio-economic settings and help to develop the waste management system that can be called sustainable [36].

Of particular interest in this respect is the research on bioleaching of printed circuit boards which form one of the most valuable and difficult parts of e-waste. Laboratory-scale tests have demonstrated that bacteria, including *Acidithiobacillus ferrooxidans* and *Leptospirillum ferrooxidans*, can effectively leach copper and zinc from shredded boards, achieving yields of up to 90% [37]. These experiments have progressed to pilot-level studies, where regulated bioreactors can handle increased e-waste flow, on a regular basis. These examples show to what extent bioleaching can be implemented as a substitute or a complement to traditional smelting in urban mining [38].

Phytoremediation initiatives have been set up in less developed countries in order to reclaim land that has lost its value due to informal e-waste processing. As an example, hyperaccumulator plants have addressed the problem of close-to-source open burning sites; To date, hyperaccumulator plants, such as sunflower (*Helianthus annuus*) and Indian mustard (*Brassica juncea*), have been widely utilized to extract heavy metals in soil as biomass. Besides improving the soil quality, these plants can also be used to extract metals during the biomass processing. The usefulness and low cost of phytoremediation as a tool to support economically disadvantaged regions affected by the negative effects of e-waste disposal cannot be underestimated. Bioremediation with the help of microorganisms: The viability of microbial bioremediation has been proven in soil contaminated with POPs of plastics, flame retardant, etc. Pilot studies have also tried to break down some of the most dangerous flame retardants by mixed microbial cultures to yield less toxic products, and thus reduce any health risk to surrounding residents [39]. These studies show that biological systems can be used to simultaneously extract metals and detoxify toxic organic substances, a bi-value chemistry that is almost impossible to accomplish with chemicals. Bio-based methods have been tested at a technical level through integration with mechanical pretreatment in several countries. The pre-sorted shredded waste is treated in a bioreactor where microbial leaching takes place and at the same time allows the recovery of plastic and glass. Hybrids are cheap and also healthy [40].

All these case examples together prove the beauty and effectiveness of biological systems in use. These applications reflect ecological and economic returns, and thus offer compelling arguments that biological mineral recovery could soon be a possibility, with lab-on-a-chip through to large-scale e-waste recovery at industrial levels. They highlight that context-specific solutions are required where all three elements (technology, policy, and citizen mobilization) have a synergistic relationship that improves outcomes [41].

3.2 Institutional and policy frameworks guiding e-waste regulation.

Laws and policies play a fundamental role in dictating the way the community tackles the growing problem of e-waste. Although the most promising technology, biological valorization cannot move beyond research and pilot stages to the level of industrial implementation without a solid legal framework. E-waste laws have been developed, mostly with a focus on extended producer responsibility (EPR), recycling, and promoting the principles of the CE, in many countries over the last 10 years, but much remains to be done regarding the scope and implementation of these laws. Over the past few years, there has been a substantial movement toward the concept of EPR whereby the responsibility of collecting and recycling e-waste is transferred over to the producers and importers of electronic goods [42]. Producer financial and legal obligations are imposed under the EPR schemes, which in turn encourages eco-design, material recovery, and use of sustainable recycling methods. (Table 3) Formal recycling rates have risen dramatically in areas that have already implemented EPR; however, there are still problems in ensuring compliance all the way along the global supply chain [43].

Table 3: Policy and regulatory perspectives in e-waste management.

Policy/regulatory focus	Description	Impact on e-waste management	References
Extended producer responsibility (EPR)	Assigns responsibility for collection and recycling to producers and importers.	Encourages eco-design, increases formal recycling rates, and reduces burden on governments.	[44]
Recycling standards and guidelines	Establishes rules for safe recycling practices, including biological methods.	Ensures environmental safety, builds public trust, and enables large-scale adoption of bio-based solutions.	[45]
International cooperation	Global and regional agreements restrict unsafe e-waste exports and promote collaboration.	Reduces illegal dumping in developing countries and supports fair global resource recovery.	[46]
Support for research and innovation	Policies that fund pilot projects, incentivize private investment, and link academia with industry.	Accelerates commercialization of biotechnology, fosters circular economy integration, and drives sustainable innovation.	[47]

Another focus of regulatory regimes is the creation of transnational standards of recycling services. The strict environmental regulations on pollutant emissions apply to traditional mechanical and chemical purification processes, and similar regulations are now being implemented on biological processes. Bioleaching and phyto-removal are considered as environmentally friendly technologies, but have to be regulated to be safe when applied in large scale, especially involving genetically modified microorganisms or foreign plant species. Risk assessment and continuous monitoring standards must be transparent and rigorous in an attempt to create a certain level of public trust and ensure that biological valorization is not counterproductive to environmental protection goals [48].

Policy needs to be an international joint venture. Selling e-waste to other countries is a major problem in international trade because the processing plants in the developed countries sell e-waste to developing countries that lack well-developed systems to control the waste .Such abuses are prohibited by international laws and regional obligations; however, because of poor enforcement, they continue to exist. Biological policies integrated into international policies can offer affordable and sustainable ways of handling imported and locally generated e-waste to underdeveloped countries [48].

Also, a good research and innovation policy is needed to facilitate the use of biotechnology in integrating waste management with biotechnology [49]. The scale of the biological solutions is promoted through pilot works, encouragement of public-corporate investment, and academic-corporate partnerships. They do this by incorporating biotechnology into national circular (green) economy efforts, allowing governments to address the waste management concern, and improve their credibility as champions of sustainable innovation [50]. Another purpose of the regulation is to make the implementation of the law feasible; the aim is to have the right amount of regulation that is acceptable within the political context and can be adopted. This includes integrating the collection, recycling, valorization processes with the social integration of persons in the informal recycling sector. By harmonizing rules with the natural biological processes in the living ecosystem, governments would accelerate the transition to the CE by transforming e-waste into an asset to ensure sustainable growth [51].

4 Routes to a circular economy

CE is the redesign of the linear system of take-make-dispose to reuse, recycle, and regenerate resources. E-waste conversion is especially important because of the high turnover rates of the technology and the high concentrations of valuable elements in discarded electronic products [52]. A new approach to waste disposal, resource preservation, and reduction of environmental impact is the integration of biological systems with the principles of a CE. Biological valorization (resource recovery) is one of them. Metals that are especially relevant to bioleaching and metal extraction include copper, gold, and REEs used in electronics in the modern age. Compared to traditional mining or the chemical leaching method, the microorganism- and plant-based approach to recovery is less toxic and requires less energy [53]. The metals that have been recovered can be reused in the production process and hence reduce primary extraction and make the supply chain more circular and sustainable. This would be in line with the precepts of the CE whereby the use of items is encouraged over time [54].

A second activity is the purification of garbage, which is a type of environmental cleanup and a form of material salvaging. Plastics and flame retardants contain some dangerous synthetic chemicals that microbial biotransformation can help neutralize; soils contaminated with informal e-waste recycling are being rehabilitated using phytoremediation. Such biological processes convert dangerous wastes to reduce toxicity in order to completely revert treated land and water systems to productive conditions. The mutual recovery of resources and restoration of ecology is essential to the health of natural systems, which explains the relevance of regeneration in the CE [55].

Figure 2: Conceptual model illustrating sustainable resource flows within a circular economy system.

The graphic is a metaphorical illustration of the shift to an electronic trash CE. The idea is simple. The model is not an open loop; the arrows represent a cyclic interaction of raw materials and ideas that do not allow garbage and waste to turn into new finished products. The first strategy, biological valorization, is based on bioleaching and bio-systems to extract useful metals in order to recycle. The second food chain, which includes purification of garbage and restoring the environment, demonstrates the applicability of microbial biotransformation and phytoremediation to remove toxicity and reconstruct the environment [56]. The third strategy, circular design, emphasizes that products must be modular, durable, and recyclable to ensure compatibility with biological recovery systems. The fourth strategy, policy and governance, identifies the use of frameworks, incentives, and global cross-border efforts to encourage sustainable waste management. All of these routes form an active loop in which technology, biology, and policy are connected in a single entity. With these strategies in place, e-waste can move beyond a linear pattern of production and consumption to a regenerative process of resource efficiency, resource ingenuity, and sustainability, which will benefit not only the industry but also the environment. The Figure 2 above illustrates the routes toward the CE [57].

It is becoming clear that electronic equipment will have to be more readily recyclable and more susceptible to bio-recovery processes. The increased recyclability of bio-based materials will be made possible by modular design, less hazardous chemicals, and eco-friendly materials. Production of dismantled and recyclable electric-powered tool increases the functioning of biological systems and allows recovering a higher portion and less pollution [58].

There is also the policy and governance channel. Policies can create legislative contexts that require producers to have greater accountability and provide incentives to businesses that use bio-based recycling methods and promote innovation using tax credits and research grants. International cooperation requires e-waste to be treated fairly and the export of hazardous waste should be banned to countries that do not have proper infrastructure. Including biological systems in national policy on CE can allow policymakers to create the best environment to innovate and invest in the CE; community and stakeholder involvement is necessary to realize the potential of these avenues [59]. The informal sector processes a large proportion of e-waste in the poor countries. The informal labor can be incorporated into the formal recycling business through training, safety requirements, and monetary rewards to increase resource recovery. Due to the reduced risk, other highly prioritized biological strategies can be used at the community level and stakeholders engaged to address the impact on human health and the environment. At the same time, there are new opportunities created by technological convergence and innovation. It is considered more efficient to use mechanical pretreatment coupled with a biological recovery system (hybrid system) when e-waste is sorted and fragmented before microbial or phytoremediation. With the help of AI, especially machine learning, such biological operations can be improved and made more predictable and scalable to industry use [60]. The combination of technology and biology is perfectly consistent with the concept of sustainable innovation that drives the CE. These directions all provide a guide on how to achieve the CE in biological systems. Recovery of resources, waste cleaning, circular design, policy enhancement, community investment, and integration of current technology will make the growing global menace of e-waste an opportunity to sustainable growth. In this respect, the threats to the environment could be minimized, and the prospects of economic and social progress could be created, in such a way, that the advantages of circularity could be distributed among trading companies, nature, and people [61].

5 The prospective developments and integration of technology

Modern technology-based bioprocessing of e-waste has reached a level of efficiency, scale, and resiliency never seen before. Three of the most established recycling methods, namely bioleaching, phytoremediation, and microbial biotransformation have been recognized as environmentally sustainable; however, they are viewed as slow and not durable enough to use in an industrial setting [62]. These kinds of processes are becoming stronger and more profitable solutions as digital technologies, automation, and biotechnology continue to grow. The combined development of biology, engineering, and data science has brought e-waste valorization to the sustainable and profitable stage. One of the most important integration domains is the use of AI and machine learning to improve biological processes. These experimental and operational data can then be analyzed with algorithms to build models that predict the optimal bioleaching conditions, such as temperature, pH, nutrient content, and parameters of microbial growth [63]. AI models can evaluate the effectiveness of metal recovery, which can be continuously improved. Predictive analytics could predict the result of recovery in various situations, thereby helping the industry reduce the number of experiments and errors and increase the efficiency of the procedure [64].

It would be interesting to develop an automated bioreactor system. Modern bioreactors are also sensor-mounted and give real-time feedback on oxygen, nutrients, and waste in contrast to open systems or semi-controlled biological systems. These solutions do not modify the activity of microbes that exist and thus they minimize variability of recovery. Automation is a desirable implementation, since it improves reproducibility, and enables scaling of laboratory tests to large-scale industrial applications, which is required to become commercially viable. SynBio is coming up as the biological future of e-waste. Genetic engineering of microbes is expanding microbial biotechnology to endow microbes with tolerance to increased levels of dangerous metals, produce effective leaching agents at a high rate, and degradation of complex organic contaminants [65]. Such bioengineered microorganisms can be used to carry out specific tasks to improve the efficiency and precision of metal recovery. More productive hyperaccumulator plants are also being produced by plant biotechnology, again in theory with far higher concentrations of REEs, thus generating more phytoremediation opportunities. Biotechnology is gaining relevance and is being complemented by hybrid systems that combine mechanical, chemical, and biological processing. The mechanical shredding and sorting of e-waste improves surface exposure to microbial activity by exposing more surface area and removing contamination. A first chemical process sometimes follows bioleaching to remove the surface coating, so that it becomes easier to extract metals in an environmentally safe way [66]. These hybrid systems present a balance between recovery rates and environmental impact, and, consequently, are the most advantageous. The use of blockchain and digital traceability technologies will continue to gain importance in the future. Using the chain in the recycling and collection of e-waste can offer a clear and traceable system to ensure that the reclaimed materials are well reused and returned to manufacturing. It discourages dumping and promotes a responsible attitude toward the environment by ensuring that business enterprises adhere to the environmental standards. These system-integrating methods show a high propensity of worldwide application [67]. As renewable energy technologies, electric vehicles, and consumer electronics continue to drive the need to consume more of the most critical raw resources, including REEs, sustainable recycling will become an economic necessity. Pens will be a leading driver of a green sector, based on the concept of a CE, with bio-based and hybrid recycling technologies supported by a research and legislative framework. Besides this, the technical methods can also be improved according to the programs that can mobilize the communities within the system to transform the inclusive and sustainable development to enable the informal jobs in the stable recycling systems to formal and safe jobs [68].

Biovalorization combined with modern digital technologies, automation, and synthetic biology is one of the key ways to realize its potential. Hopefully, the result of such systems will be to provide efficiency and scalability, and recycling as a response to global requirements of security of supplies and environmental sustainability. Economic growth: Biological technologies will enable the circular economies of the future when they focus on being more sophisticated, and e-waste is the first step on the path to innovation resilience [69].

6 Sociological-economic implications of biological valorization

The use of bio-systems to produce e-waste has important socioeconomic implications on the environment. Bioleaching, phytoremediation, and microbial biotransformation of waste can have an important economic, livelihood security, and inclusive development impact. The biological processes justify the co-determination of sustainability and economic empowerment, particularly in regions with a high informal recycling rate, compared with traditional recycling that, in addition to consuming energy, is toxic in certain situations [70].

The creation of job opportunities is probably the most important thing. Millions of people across the world are informally recycling e-waste, usually in unsafe environments, including open burning or acid leaching. Biological valorization is a less threatening option that can be implemented easily on community or industrial level. Training local employees in the use of bioleaching equipment, how to grow hyperaccumulator plants, or how to carry out a microbial detoxification can be done. Professionalization of these trades has not only protected workers against unhealthy exposure, but also helped stabilize incomes and improve working conditions. Social inclusion: Improved and formalized recycling services have the potential to include marginalized groups, such as women and children, in the recycling process [71].

Besides the health consequences, there is a huge financial cost. Extraction of precious metals, including gold, silver, copper, and rare earth magnetic elements, improves the economy and minimizes logistical and environmental effects of mining activities. The reuse of resources promotes both local and global chains of value, thus reducing the need to import raw materials and increasing the security of resources [72]. This is an advantage to poor countries because bad waste technology can be turned into a commodity. Then, valorization helps to offset the primary economic cost of environmental degradation, including soil remediation costs or local communities spending in response to such environmental damage, usually imposed on the shoulders of the public authorities [73].

A significant social and economic gain is in terms of startups and inventineering. Compared to industrial smelting or leaching of chemicals, biological systems are much more economical and available to small to medium enterprises. Bio-tech has an opportunity to help in the creation of local recycling locations, managed by start-ups and local programs. These businesses are not only profitable but also drive green innovation, which helps to promote new and sustainable industries. The growing need to commercialize more processes and products worldwide in a more sustainable way leads to more commercialization opportunities in biological valorization industries [74].

The improvement in the health of people is vital to the society. Informal recycling subjects employees and immediate communities to dangerous emissions, polluted soils, and water that cause respiratory illnesses, cancer, and birth defects. The burden of illness caused by these harmful technologies can be relieved and it is possible to reduce healthcare costs by substituting them with biological alternatives which will contribute to a better lifestyle of the whole society. In this way, a positive cycle is formed. The growth of production and community empowerment result in social welfare and protection of nature [75].

There are also policy and governance implications in the biological valorization design, i.e., national waste management program. Governments, which incorporate such systems in their national waste management programs, are given the opportunity to reduce the socioeconomic gap between the developed and developing countries. The latter can be prevented by transferring technology to the industrialized countries to the less developed countries to prevent the perilous path of recycling to the less harmful and more sustainable recycling methods. This will not only improve equity globally, but also increase cooperation between countries in resolving the international aspect of the e-waste disaster [76]. With its socioeconomic implications, biological valorization offers future relevance that goes beyond its technical potential as a fix. It is a path to sustainable employment, equitable development, and healthy economies. E-waste biological systems: Bio-adoptive e-waste is capable of changing the world issue to the origin of green growth (Vazques & Barbosa, 2021). Each of these strategies can change the social and economic equations around recycling, with the support of policy, investment, and community action, to make the wealth created by a CE more fairly distributed between people and the planet [77].

7 Equitable and sustainable approaches to electronic waste management

E-waste is gradually being viewed as an environmental as well as a justice problem. The harm done by discarded devices is not equally distributed the world over. Despite the fact that the developed world produces a considerable amount of e-waste, they are usually shipped to developing countries that do not have proper recycling facilities [78]. Therefore, poor people who are exposed to work with dangerous substances, destruction of the environment, and health hazards end up bearing the negative consequences. The environmental justice framework also lends rhetorical power to technically viable, yet equitable, inclusive, and socially responsible strategies [79].

Inequality of risk allocation, is, in fact, one of the main factors of environmental justice in e-waste management. There are informal recycling plants in vulnerable communities of some other parts of Asia, Africa, and Latin America, where their citizens do not have an alternative means of making ends meet, but instead, they resort to hazardous health recycling activities. Women, men, and children breathe in toxic vapors and heavy metal and toxic-contaminated soil and water [80]. Such societies experience increased cases of respiratory disease, neurological disease, shorter life, etc., where we in the first world have cheap electronics sugus, and none of the above. Such differences also need to be fixed and removed when it comes to e-waste disposition solutions .Routes of valorization based on biology help to find a solution that is more equitable. Alternative systems such as the biological systems are also very attractive to communities that still rely on the toxic recycling processes, because they can be operated in environmentally friendly conditions, consume less toxicants, and can be designed to be locally used. Both bioleaching and phytoremediation are systems that may be used at a relatively small scale and low tech by the local population in order to exploit metals in a manner not in direct contact with hazardous chemicals. This reduces health risks among weak women and gives them a chance to make money in better terms [81].

Equity of e-waste governance includes distribution of benefits. Recycled precious metals in e-waste end up going back to the world supply chains, where they mainly benefit industries in the developed countries with little benefit to the community that handles the waste. The benefits of recycling and green recycling, in particular, should therefore be distributed equally to the affected local communities. This is possible by way of cooperatives or community business ventures, and policy structures that ensure that profits are recycled and used locally to develop. Biological recycling is in this respect a blue and green solution [82].

It is important to be involved in the decision-making. The communities that are affected have not been properly involved in drafting legislation or technology relating to e-waste. Environmental justice models also insist on community participation in the development of policies and programs regarding recycling. Marginalized communities can play a role in adapting interventions to alternative cultural, social, and economic contexts, and consequently improving the viability and acceptability of interventions within the population [83]. Moreover, the equality across borders must result in stricter laws regarding transportation of cross-border e-waste. The rich countries should assume the responsibility of disposing their own garbage, not funneling that to poorer countries like the transfer of technology, funding of local infrastructures, and constructions to assist the developing countries to dispose e-waste in the safe and profitable manner. This is not unfair to the planet, and this creates a planetary solidarity in solving a common problem [84]. Finally, cleaning up the e-waste environment allows us to reform the systems that brought human suffering while protecting our ecosystems. Biological valorization approaches offer cost-efficient ways to de-risk and share value and have an element of participation. Once equity becomes the spine of e-waste work and is an end in itself, which in this instance is sustainable technologies at work to energize communities and a step toward global justice improvement, then, and only then, we shall finally be able to begin imagining that the burden of waste is not only resting squarely on the shoulders of the most vulnerable [85].

8 Education and capacity development dimensions

To achieve successful incorporation of biotechnologies in e-waste valorization, technological developments and favorable policies cannot suffice; education and capacity building are also required. The transfer of knowledge and skills is what can and will make even the most promising solutions achieve the desired impact. The necessary skills in managing sustainable e-waste and the proper use of biological technologies should be taught and developed in people, communities, and business [86].

Such a program is mainly concerned with social education and environmental training. Unfortunately, very few people understand the risks of not properly disposing e-waste or recognize the importance of electronics in the recycling stream. The population must be educated about the health, economic, and environmental impacts of e-waste through awareness campaigns, curricula, and community workshops. Moreover, increasing awareness of safe disposal procedures, benefits of recycling, and the benefits of biological systems in reducing harm could lead to behavioral changes at both the household and community levels. Finally, participants will learn that their old devices are not wastes but they are resources and they can contribute so much to the environment [87].

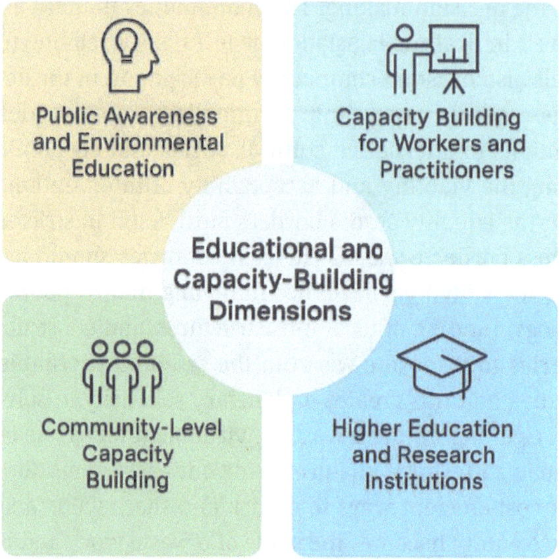

Figure 3: Education and institutional capacity development for effective e-waste governance. [88].

The main dimensions of education and capacity building needed to develop bio-mediated e-waste valorization are depicted in Figure 3. The underlying idea is to share knowledge and experience with people and organizations, so that they can adopt sustainable solutions. Training and environmental education is one of those initiatives, the purpose of which is to create awareness of the importance of recycling and the harmful consequences of improper disposal of garbage to the environment. By accomadating a large number of workers and practitioners, the second building highlights the training of the waste pickers, mechanical engineers, entrepreneurs, and other professionals in safe and technologically appropriate bio-based practices. The third element is the role of universities and research centers in the development of scientific knowledge and innovation, as well as the education of future specialists [89]. Finally, Community Economic Development (CED) enables civil society organizations, cooperatives, nongovernmental organizations (NGOs) and communities to manage and control the waste stream as an important economic resource. The result is diversity, security, and long-term environmental sustainability [90]. Biological valorization would move beyond being considered a secondary technology to being a secondary practice by merging and intertwining knowledge, technology, and academic research with grassroot empowerment. The image shows that education and training are needed to empower society, industry, and government to adopt the CE, such that the social co-benefits of sustainable recycling are realized across society [91].

In conjunction with increased knowledge among the population, there is a pressing need to enhance the capacity of practitioners and employees. Informal recycling is common in most low-income countries; people use risky methods to recover valuable minerals like open burning or using acids or other chemicals to leach precious minerals. The training courses can also provide them with information about safer and more cost-effective biological valorization techniques [92]. Workshops and vocational training facilities can be organized to provide practical experience with operational bioleaching plants, phytoremediation, or small-scale bioreactors. Their benefits go beyond making the workplace safe, as they increase the economic opportunities of workers by institutionalizing their roles within sustainable recycling systems.

In this respect, many universities and research facilities play a major role. E-waste management and biological recycling should also be part of the curriculum of colleges and technical institutes to develop a new generation of environmental scientists, engineers, and policymakers. Besides projects by students that can support local solutions, research can drive microbial engineering, phytoremediation, and hybrid systems. Academia, industry, and government collaborations can work to enhance the pipeline of knowledge transfer that will help translate innovative research into practice [93].

The other equally important point is that of training in governance and policy literacy. In order to develop socially and commercially sustainable bio-systems, the decision-makers should have an understanding of the business of science and the scientific principles of business [94]. Capacity-building measures are needed to equip government officials, regulators, and policymakers with greater expertise in creating and implementing beneficial regulations, promoting pilot projects, and promoting cross-sectoral partnerships. An informed policy framework is needed to align educational efforts with long-term national objectives related to the CE and the accomplishment of sustainable development [95].

Community-level capabilities increase the ability to create a more inclusive and robust society. Intermediaries between advanced technologies and grassroots lifestyles may be local institutions, cooperatives, and NGOs. Training to empower the residents to set up small biological recycling facilities to locally manage their waste and monitor the environment. Capacity-enabling activities should also be integrated into community organization to ensure that e-waste solutions become more feasible as well as relevant and to encourage a sense of ownership and responsibility toward change by the people participating in the local community [96].

Capacity-building works on a global scale. Information flow, training of support staff, and technological transfer between countries can help developing countries become part of the sustainability process. Cross-national learning opportunities could be achieved through workshops, trade fairs, and global networks of scholars and practitioners abroad. When a bioleaching model has been well developed in one country, it can be adopted and modified in another country with similar socioeconomic features, thus helping to spread best practices across countries [97]. With the introduction of e-learning and digital technology, the capacity-building possibilities have been extended. To communicate information to distant areas and provide lifelong learning to professionals, students, and policymakers online courses, virtual laboratories and mobile applications can be used [98]. These tools can give resource constrained locations access to high-level bio-e-waste valorization information. Finally, biological treatments should not be used to make e-waste disposal effective without education and capacity building. Society can create the right framework, in which sustainable recycling facilities can emerge, through improving public awareness, employee education, reinforcing educational programs, empowering local organizations, and supporting legislators [99]. This program together with international cooperation and improved digital learning tools will make the biological valorization technologically viable, socially inclusive, and broadly available. This element will capitalize on the potential of biotechnology, make it more inclusive, and empower citizens at every level with the competencies and knowledge required to participate in a real-life CE actively [100].

9 Conclusion

Electronic trash (e-waste) management is a major challenge both in environmental and socioeconomic perspectives, and it requires greater effort than traditional reuse and recycling strategies. Bioleaching, phytoremediation, and microbial biotransformation are inexpensive, environmentally friendly, and sustainable biological technologies to recover useful metals simultaneously with detoxification of toxic substances, as a part of a CE. Examples based on application, biotechnology innovations, enabling policies, and alignment with digital technology are being developed to facilitate the translation of research questions into scalable solutions. The socioeconomic impacts of these processes can be observed since, on the one hand, valorized biology can offer a chance to earn secure livelihoods, become entrepreneurs, and develop, and, on the other hand, solve the problems of justice and equity related to the communities affected by dangerous recycling. Training and education can provide an individual and organizations with the information and skills required to participate in a sustainable waste management system. With this in mind, this issue of viewing biological systems as not only technical solutions, but also transformative processes of converting e-waste into an innovation, resilience and sustainable development platform on a global scale.

Abbreviations

E-waste	Electronic waste
EPR	Extended producer responsibility
AI	Artificial intelligence
ML	Machine learning
IoT	Internet of Things
REEs	Rare earth elements
Au	Gold (chemical symbol, often used in recovery studies)
Cu	Copper (chemical symbol)
Zn	Zinc (chemical symbol)
Pb	Lead (chemical symbol)
Cd	Cadmium (chemical symbol)
Hg	Mercury (chemical symbol)
POPs	Persistent organic pollutants
NGO	Nongovernmental organization

References

[1] Yusuf AAR et al. Prototipe Video Motion Graphic untuk Materi Tipografi Kelas XI. Jurnal Dinamika Vokasional Teknik Mesin. 2025;1(1).

[2] Azzahra A. Implementasi Program Kartu Identitas Anak (KIA) di Dinas Kependudukan dan Pencatatan Sipil Kabupaten Bandung Barat. Jurnal Pendidikan Tambusai. 2024;8.

[3] Agyekum PB, Oyortey D, Lutterodt HE, Brako F, Ofosu IW, Olayide OE.Circular Economy for Sustainable Development in Africa: Advances and Applications in Food Processing. In Proceedings of the 5th Annual Circulariity Africa Conference 2024 2024.

[4] Nisa RZ, Kurniawati D, Faruq UA. Pengaruh Rasio Keuangan Terhadap Harga Saham Lq-45 Di Bursa Efek Indonesia. EKOMA : Jurnal Rkonomi, Manajemen, Akuntansi. 2024;4(1).

[5] Anindita SP et al. sahan Balance Sheet As of December 2014. Journal of Islamic Accounting and Business Research. 2024;00(1).

[6] Turna T, Yıldız Y. Treatment of vegetable oil industry wastewaters with coagulation-flocculation methods TT – Bitkisel Yağ Sanayii Atıksularının Koagülasyon – Flokülasyon Yöntemleriyle Arıtılması. Dicle Üniversitesi Mühendislik Fakültesi Mühendislik Dergisi. 2024;15(2).

[7] Tursunbayeva A et al. A spotlight on environmental sustainability in view of the european green deal. Sustainability (Switzerland). 2024;16(1).

[8] Fitriana R et al. Dampak E-commerce bagi Perekonomian Indonesia. Kementrian Keuangan Direktorat Jenderal Kekayaan Negara.

[9] Bhattacharjee S, Dwivedi A, Tiwari SP. Development of Biodegradable Substrates and Synaptic Transistors for Next-Generation Transient Electronics. Advanced Materials Technologies. 2025;10(7). doi: 10.1002/admt.202401494.

[10] Santolin J, Vlaeminck SE, Appiah-Twum H, Van Winckel T, Spiller M. Consequential LCA of NPK fertilizers from microbial, animal, plant, and mineral origin highlights resource constraints and environmental impacts. Journal of Cleaner Production. 2024;457. doi: 10.1016/j.jclepro.2024.142312.

[11] Andrei V, Wang Q, Uekert T, Bhattacharjee S, Reisner E. 2022, 55, 3376–3386 Downloaded via Rensselaer polytechnic inst on May 22. Accounts of Chemical Research. 2024;17(19).

[12] Malik M, Yunus N, Niazi HR, Ayub U. Article ID 36 1 of 8 Pakistan? 2025.

[13] Bai Y, Su J, Wang F, Cui H, Zhang H, Liu K. Harnessing Synthetic Biology for Sustainable Recovery of Critical Metal Materials from Electronic Waste. 2025. doi: 10.1002/adfm.202509900.

[14] Muñoz-Ruiz V et al. Ecotoxicity assessment of sustainable magnesium phosphate cements (Sust-MPCs) using luminescent bacteria and sea urchin embryo-larval development tests. Journal of Environmental Chemical Engineering. 2024;12(5). doi: 10.1016/j.jece.2024.113995.

[15] Hassan OM. Integrated Digital, Biological, and Human Capital Innovations For Circular And Sustainable Waste Management: A Critical Review. 2025. doi: 10.1007/s42452-025-07904-3.

[16] Srivastava P, Srivastava S. Comprehensive approaches to waste management for sustainable development. International Journal of Advanced Research in Science, Communication and Technology. 2025. doi: 10.48175/ijarsct-26675.

[17] Elizabeth OA, Olalekan BM, Adeola AA. Waste-to-Value Technologies in the Global South. Sustainable Bioeconomy Development in the Global South: Volume II Bioeconomy Techniques. 2025. doi: 10.1007/978-981-96-0305-3_8.

[18] Roy SK, Rao AS, Kumar Singh S, Somasundaram R, Dash S. Harnessing bio-based products for Sustainable E-waste management in biomanufacturing: Green solutions. Environmental Applications of Carbon-Based Materials. 2024. doi: 10.4018/979-8-3693-3625-0.ch013.

[19] Albay M, Ozbayram EG, Botana LM. Advancing aquatic biotechnology in the circular bioeconomy era: Innovations, challenges, and opportunities. Environmental Management. 2025;75(9). doi: 10.1007/s00267-025-02205-2.

[20] Ammayen I et al. From marine and agricultural waste to soil health: optimized co-composting of sardine by-products and tomato plant biomass for sustainable agriculture and environmental protection. Compost Science & Utilization. 2025. doi: 10.1080/1065657X.2025.2532374.

[21] Sayago UFC. The design of a sustainable industrial wastewater treatment system and the generation of biohydrogen from E. crassipes. Polymers (Basel). 2024;16(7). doi: 10.3390/polym16070893.

[22] Lin CH, Chen Y, Chang KI, Chi PY. The evaluation of economic impact of the Basel Convention on the trade of waste electrical and electronic equipment. Journal of Asian Economics. 2025;100. doi: 10.1016/j.asieco.2025.101996.

[23] Tewary A, Upadhyay P. Proposing a 6R framework promoting circular strategies for platform organizations. Digital Policy, Regulation and Governance. 2025;27(2). doi: 10.1108/DPRG-11-2023-0169.

[24] Deepika P, Malvi NB, Sahana B, Gowda L, Harshitha BJ. Advancing e-waste management through circular economy principles: A pathway to sustainable innovation. Integrated Approaches for Sustainable E-Waste Management. 2025. doi: 10.4018/979-8-3693-7383-5.ch003.

[25] Garcia R. Localized E-waste recycling framework for general santos city: a circular economy approach. Journal of Interdisciplinary Perspectives. 2025;3(2). doi: 10.69569/jip.2024.0691.

[26] Dagwar PP, Iqbal SS, Dutta D. Sustainable recovery of rare Earth elements from industrial waste: A path to circular economy and environmental health. 2025. doi: 10.1016/j.wmb.2025.02.004.

[27] Phogat P, Kumar S, Wan M. A Scientometrics Study of Advancing Sustainable Metal Recovery from e-Waste: Processes, Challenges, And Future Directions. 2025. doi: 10.1039/d5su00049a.

[28] Singh Y, Chauhan MS, Pappu A. Novel hybrid E-plastic composite panels and its materials performance- A sustainable and environmentally friendly approach for circular economy. Ceramics International. 2025;51(26). doi: 10.1016/j.ceramint.2025.08.225.

[29] Mpuya MR, Joyce M, Abdulbatswa H, Fahamu KS, Respikius M, Camilius S. Practices of E-Waste management among people living in African Cities: The Case of Dar Es Salaam, Tanzania. East African Journal of Environment and Natural Resources. 2025;8(2). doi: 10.37284/eajenr.8.2.3082.

[30] Moita Neto JM, Leal RC, dos NL, Araújo S, Da silva EA. Challenges and Opportunities for the Development of Urban Mining in Brazil. 2025. doi: 10.3390/min15060593.

[31] Mohamed D, Fayad A, Mohamed A-MO, Nahyan MTA. The role of E-waste in sustainable mineral resource management. Waste. 2025;3(3). doi: 10.3390/waste3030027.

[32] Hu X, Wang J, Feng C, Yuan J, Wei Q, Wang H. A comprehensive review of laser-induced-graphene for sensor applications: fabrication, properties, and performance evaluation. 2025. doi: 10.1039/d4tc03547j.

[33] Ahossouhe MS, Andrianisa HA, Damba DD, Kouassi D, Narra S, Sanou A. Scenario-Based Extended Cost–Benefit Analysis for E-Waste Metal Recovery in Low-Income Countries: Evidence from an Integrated Model in Burkina Faso. Sustainability (Switzerland). 2025;17(18). doi: 10.3390/su17188351.

[34] Theerthaana P, Raj SNV. An investigation on adoption of AI-enabled e-waste recycling kiosks in the post-pandemic era. Electronic Government. 2025;21(1). doi: 10.1504/EG.2025.143120.

[35] Yu X, Sun Z, Sun D, He R. Sustainable development assessment of household e-waste reverse supply chains from an environmental ethic perspective. Humanities and Social Sciences Communications. 2025;12(1). doi: 10.1057/s41599-025-05012-6.

[36] Copara M, Pilamunga A, Ibarra F, Oyaque-Mora SM, Morales-Urrutla D, Córdova P. A Systematic Review and Bibliometric Analysis for the Design of a Traceable and Sustainable Model for WEEE Information Management in Ecuador Based on the Circular Economy. 2025. doi: 10.3390/su17146402.

[37] Fivaz E, Makohliso S, Schönenberger K. Circular and net-zero opportunities in the management of electronic waste in low- and middle-income countries: the example of Ghana. 2025. doi: 10.1080/19376812.2024.2353642.

[38] Neves LF, De oliveira MJG, Carvalho CS, Castanon JAB. Logística Reversa. Periódico Técnico e Científico Cidades Verdes. 2025;13(41). doi: 10.17271/23178604134120255765.

[39] Srivastava SK, Shrivastava R. Modeling for copper recovery from E-waste by using machine learning technique: An approach for the circular economy. Current Analytical Chemistry. 2025. doi: 10.2174/0115734110343714241130170537.

[40] Cheng S et al. Sustainable electromagnetic interference shielding materials from cellulose-grafted n-type polymers. Nature Communications. 2025;16(1). doi: 10.1038/s41467-025-63665-y.

[41] Jiang K, Yan J, Ouyang R, Mao Y, Huang Z, Long C. Bisphenol analogues in soils from four circular economy E-waste dismantling parks: emission characteristics and health risks. Journal of Environmental Exposure Assessment. 2025;4(4). doi: 10.20517/jeea.2025.46.

[42] Heinrich V, Bansal H, Fröhling M. Electronic waste management in Northern India: a regional case study of Chandigarh. J Mater Cycles Waste Manag. 2025;27(3). doi: 10.1007/s10163-025-02194-8.

[43] Baffoe G, Green H. Circular economy and social equity: Addressing the complexities of E-waste management in Accra, Ghana. 2025. doi: 10.1016/j.envdev.2025.101380.

[44] Ferreira-Filipe DA, Duarte AC, Hursthouse AS, Rocha-Santos T, Silva ALP. Biobased Strategies for E-Waste Metal Recovery: A Critical Overview of Recent Advances. 2025. doi: 10.3390/environments12010026.

[45] Garcia L, Morell J, Lao C, Solé-Sardans M, Dorado AD. Metal recovery from discarded lithium-ion batteries by bioleaching coupled with minimal mechanical pre-treatment. Minerals. 2025;15(6). doi: 10.3390/min15060566.

[46] Tran AV, Khoa BT. Global Research Trends in Circular Economy: A Bibliometric Analysis in E-Commerce. 2025. doi: 10.1155/hbe2/8645845.

[47] Golpeo JM, Andres GL, Faronal IN, Casuat CD. Sustainable E-Waste Management Through a Circular Economy Framework: Integrating Smart Technologies for Resource Recovery in Urban Environments. In E3S Web of Conferences, 2025. doi: 10.1051/e3sconf/202564301005.

[48] Santhuja P, Suresh A. CNN-Based Circular Economy in E-Waste Management with Blockchain and IoT Integration. In Proceedings of International Conference on Visual Analytics and Data Visualization, ICVADV 2025, 2025. doi: 10.1109/ICVADV63329.2025.10961209.

[49] Sandwal SK, Jakhar R, Styszko K. E-Waste Challenges in India: Environmental and Human Health Impacts. 2025. doi: 10.3390/app15084350.

[50] Lubeck MA. Assessment of ARPA-E Energy Storage Program: Capability and Capacity to Solve Battery Waste Issues. 2025. doi: 10.1007/s43615-025-00590-8.

[51] Porto WS, Souza ICG, Freitas MAL. DO Conflito À Circularidade: Repensando os E-Resíduos Sob Múltiplas Lógicas De Sustentabilidade E Valor. Boletim de Conjuntura – BOCA. 2025;22(65).

[52] Kang D, Choi M. AI-assisted smartphone trading and the digital divide among older adults. Journal of Aging and Environment. 2025. doi: 10.1080/26892618.2025.2506054.

[53] Choudhary D, Kumar A, Gong Y, Papadopoulos T. Examination of risks in circular supply chains using transition management lens: towards a circular economy in emerging markets. Journal of the Operational Research Society. 2025. doi: 10.1080/01605682.2025.2459252.

[54] Dominguez-Villanueva S, Rojas-García MA, Sánchez-Ruiz FJ, Hernández-Reyes H, Cuautle-Gutierrez L. Life cycle analysis of waste management of a brake pad manufacturing company. Cleaner Waste Systems. 2025;11. doi: 10.1016/j.clwas.2025.100314.

[55] Bairaktari M. Deep Eutectic Solvents (DESs) in Waste Reduction, By-Product Valorization, and Advancing the Circular Economy. Sustainable Chemical Engineering. 2025. doi: 10.37256/sce.6220256332.

[56] Cunha L, Monteiro J, Futuro A, Regufe MJ, Soeiro J, Sousa R. Recycling PCBs for nanoparticles production with potential applications in cosmetics, cement manufacturing, and CO2 capture. Waste Management. 2025;191. doi: 10.1016/j.wasman.2024.11.018.

[57] Yin X, Liu R, Yue Y, Li J, Yang Y. Ultra-fast and selective recycling of gold from electronic waste based on triiodide ionic liquids. AIChE Journal. 2025;71(6). doi: 10.1002/aic.18773.

[58] Kumar V, Shukla OJ. Unravelling sustainable impediments in E-waste supply chain management: A Fuzzy-WINGS decision approach. Journal of Environmental Management. 2025;389. doi: 10.1016/j.jenvman.2025.126208.

[59] Jadoun S, Chinnam S, Jabin S, Upadhyay Y, Jangid NK, Zia J. Recovery of Metals from E-waste: Facts, Methods, Challenges, Case Studies, and Sustainable Solutions. 2025. doi: 10.1021/acs.estlett.4c00696.

[60] Shams H, Ab Rahman MN, Hishamuddin H, Rafique MZ. A Circular Consumption Behavior Model for Addressing and Reducing Product Demand and Disposal. Resources. 2025;14(9). doi: 10.3390/resources14090148.

[61] Shamsuddin MS, Mokhtar NF, Mokhlis S, Hasan ZRA, Aziz NA, Yusof MN. E-commerce waste management: a systematic review. International Journal of Advances in Applied Sciences. 2025;14(3). doi: 10.11591/ijaas.v14.i3.pp817-827.

[62] Jaiswal SK, Mukti SK. Sustainable Remanufacturing of E-waste: An Integrated FDM and AHP Approach in Indian Context. Circular Economy and Sustainability. 2025;5(3). doi: 10.1007/s43615-024-00486-z.

[63] Jungi H, Virani AA, Podder S, Girase H, Mitra J. Sustainable Combination of Waste with Waste: Utilization of Biomass to Recover Critical Metals from Spent Lithium ion Batteries. 2025. doi: 10.1002/batt.202400518.

[64] De carvalho MCN, de Souza APS, de Oliveira Neto JF, Mendonça Silva M, Florencio L, Machado Santos S. Management and recovery of critical and strategic raw materials from E-Waste: A case study in Brazil with a focus on printed circuit boards. Journal of Hazardous Materials Advances. 2025;17. doi: 10.1016/j.hazadv.2024.100544.

[65] Faradillah P, Sultan AAM, Samat KF, Kamarulzaman T. Strengthening e-waste governance: A decision framework for sustainable transboundary movements under the Swiss-Ghana Amendments. Chinese Journal of Population Resources and Environment. 2025;23(3). doi: 10.1016/j.cjpre.2025.07.003.

[66] Bhattacharjee J, De A, Kamila B, Mandal A. Recent scenario of e-waste recycling: Chemical engineering. 2025. doi: 10.1515/ijcre-2024-0058.

[67] Asif AH, Li C, Lim H, Sharma R, Sun H. Advancing Australia's lithium and rare earth value chain: recent trends and innovations in functionalised adsorbents and their perspective applications for circular economy. 2025. doi: 10.1016/j.mineng.2025.109325.

[68] Singh S, Yadav V, Routroy S, Dasgupta MS. Assessment of risk propagation in an e-waste collection system using Bayesian networks. J Mater Cycles Waste Manag. 2025;27(3). doi: 10.1007/s10163-025-02185-9.
[69] Ben Abdelhamid M, Casadella V, Tahi S. Optimizing Urban Mining Investments: Leveraging Sequential Compound Real Option Analysis for Circular Economy Transition. Environmental Modeling and Assessment. 2025;30(4). doi: 10.1007/s10666-025-1003`0.
[70] Grima-Carmena L, Giner-Sanz JJ, Pocoví-Martínez S, Bosch-Mossi F, Pérez-Herranz V. Study of the effect of the pretreatment and the concentrations of iron (II) and chloride, on the electrochemical leaching of waste printed circuit boards using ANOVA analysis. Journal of Industrial and Engineering Chemistry. 2025;152. doi: 10.1016/j.jiec.2025.04.054.
[71] Nguyen KH, Tran MD, Duong TTT. Global Knowledge Networks on Circular Economy and Sustainable Consumption: Sociological Insights into Emerging Environmental Discourses. Circular Economy and Sustainability. 2025. doi: 10.1007/s43615-025-00658-5.
[72] Singh Y, Chauhan MS, Pappu A. Rising trends and real-world potential: Bibliometric analysis and quantitative benchmarking of decentralized micro-recycling systems for e-waste plastics. 2025. doi: 10.1016/j.scowo.2025.100122.
[73] Kumar S et al. Enhanced selective gold recovery from e-waste via synergistic hetero-atom controlled quasi-planar benzoxazine-based covalent organic frameworks. Materials Horizons. 2025. doi: 10.1039/d5mh00698h.
[74] Karthick A, Priya A. Review on utilization of electronic waste in construction industry as binder, mortar and concrete. 2025. doi: 10.1007/s42107-025-01373-0.
[75] Sayyed SF, Sailaja N. Smart Care Solution for Smart Home Devices. In Conference on Human Factors in Computing Systems – Proceedings, 2025. doi: 10.1145/3706599.3719956.
[76] Hasan MA, Das S, Tiwari V. Advancing Sustainable Procurement in E-Waste Recycling Through Grey Integrated Decision Analysis. Sustainable Development. 2025. doi: 10.1002/sd.70343.
[77] Bukhari S, Ahmad I, Mustafa G, Nabi AA, Zaman K. Digitalization, finance, and resource efficiency: advancing Pakistan's circular economy. SN Business and Economics. 2025;5(12). doi: 10.1007/s43546-025-00994-6.
[78] Ferreira N et al. A recycling-oriented approach to rare earth element recovery using low-cost agricultural waste. Metals (Basel). 2025;15(8). doi: 10.3390/met15080842.
[79] Barnabas SG. Towards an efficient E-waste management regime in Nigeria. Journal of Environmental Law and Policy. 2025;5(1). doi: 10.33002/jelp050105.
[80] Mascarenhas J, Freitas DR, Fernandes SDC. Exploring sustainable consumer behavior: the Circular Behavior Integration Framework (CBIF). In Proceedings of the Design Society, 2025. doi: 10.1017/pds.2025.10183.
[81] Awasthi M, Vaibhav K, Choudhary AK, Gautam AK, Chandra A. E-Waste Management: An Essential Deed to Safeguard Future. Global Waste Management. 2025. doi: 10.1002/9781394318414.ch5.
[82] Crozier MN, Ntsala PG, Schenck C, Crozier mccleland J, Van dyk EE, Petersen J. Drivers, barriers and enablers of a South African circular solar PV module supply chain: A focus on reuse. Waste Management and Research. 2025. doi: 10.1177/0734242X251340306.
[83] Vives Pons J, Comerma A, Escobet T, Dorado AD, Tarrés-Puertas MI. Optimizing Bioleaching for Printed Circuit Board Copper Recovery: An AI-Driven RGB-Based Approach. Applied Sciences (Switzerland). 2025;15(1). doi: 10.3390/app15010129.
[84] Singh Y, Chauhan MS, Pappu A. Enhanced performance of epoxy-modified E-waste plastic–fly ash hybrid composites: Mechanical, spectroscopic, and thermal characterization. Construction and Building Materials. 2025;501. doi: 10.1016/j.conbuildmat.2025.144135.
[85] Kurian S, Bauri R, Subramaniam P. Dissolution and Recovery of Metallic Copper from e-Waste Through a Sustainable Route Using Thymol-Capric Acid Deep Eutectic Solvent. Journal of Sustainable Metallurgy. 2025;11(3). doi: 10.1007/s40831-025-01204-1.

[86] Jones ME, Barragán-Paladines MJ, Hunt CA. What Goes in the Galapagos Does Not Always Come out: A Political Industrial Ecology Case Study of E-Waste in Island Settings. Sustainability (Switzerland). 2025;17(19). doi: 10.3390/su17198704.

[87] Bertelli B, Kocollari U, Merzi L, Torricelli C. The role of circular economy in reducing corporate default probability: insights from recent research. 2025. doi: 10.1016/j.cosust.2025.101555.

[88] Chatterjee B, Das S. Sustainable Management of Discarded Solar Photovoltaic Material: Enhancing Electronic Waste Recycling in India. Lecture Notes in Civil Engineering. 2025. doi: 10.1007/978-981-97-6024-4_4.

[89] Tavala AN, Tabaroei A. Sustainable clay soil stabilization using constructions waste: Mechanical behavior, engineering parameters and microstructural analysis. Results in Engineering. 2025;27. doi: 10.1016/j.rineng.2025.106247.

[90] Chetry AB, Thapa S, Bhudathoki K, Raj Regmi S, Ohto K. Revolutionizing Urban Mining: Solvent Extraction for Sustainable Metal Recovery from Electronic Waste. Journal of Material Sciences & Manufacturing Research. 2025. doi: 10.47363/jmsmr/2025(6)204.

[91] Maqsood S et al. Valorization of plant-based agro-industrial waste and by-products for the production of polysaccharides: Towards a more circular economy. 2025. doi: 10.1016/j.afres.2025.100954.

[92] De souza RS, Guarnieri P, De oliveira vieira B. THE Challenges Of The Transition To The Circular Economy In The Electrical And Electronic Waste Sector In Brazil. RAE Revista de Administracao de Empresas. 2025;65(4). doi: 10.1590/S0034-759020250401x.

[93] Paternina Martínez L, Perdomo Useche YL. Convirtiendo las basuras en oportunidades: gestión de residuos solidos en institución educativa rural. Gestión y Ambiente. 2025;27(2). doi: 10.15446/ga.v27n2.118992.

[94] ElObeid T et al. Red beet process waste: A sustainable glucose syrup alternative for gummy confectionery. Journal of Food Science. 2025;90(5). doi: 10.1111/1750-3841.70262.

[95] Kumar R et al. Pioneering zero-waste technologies utilization and its framework on sustainable management: international, national and state level. 2025. doi: 10.1007/s42452-025-06693-z.

[96] Licastro A, D'Alessandro C, Szopik-Depczyńska K, Arbolino R, Ioppolo G. From data scarcity to strategic action: a managerial framework for circular economy implementation in mediterranean small towns. Sustainability (Switzerland). 2025;17(14). doi: 10.3390/su17146474.

[97] Cheng YZ et al. Light-promoted extraction of precious metals using a porphyrin-integrated one-dimensional covalent organic framework. Angewandte Chemie – International Edition. 2025;64(2). doi: 10.1002/anie.202414943.

[98] Dada SR, Suleiman A, Adu EO, Adebayo A, Seriki AL, Oluleke CO. The future of E-waste recycling: emerging technologies and practices evaluating the social and economic benefits of advanced E-waste recycling technologies. International Journal of Current Science Research and Review. 2025;08(02). doi: 10.47191/ijcsrr/v8-i2-32.

[99] Shahrabifarahani S, Torabi SA, Rahiminia M. Circular sustainable supply chain network design for electronic devices. Environment, Development and Sustainability. 2025. doi: 10.1007/s10668-025-06068-9.

[100] Del vargas rosas CC. Manejo y gestión de los residuos sólidos urbanos en la Ciudad de México. European Public & Social Innovation Review. 2025;10. doi: 10.31637/epsir-2025-1712.

[101] Julee Banerji, V. G., Sunil Kumar Parisa, & Pawan Whig. (2026). Waste management, informal recycling, environmental pollution, and public health. In *Sustainable Solutions for Environmental Pollution* (pp. 345–374). Elsevier. https://doi.org/10.1016/B978-0-443-33145-9.00012-2

Tijo Cherian*, Sijo Asokan, G Balasubramani, N Sharmila Devi, Smitha Vijayan, R Mohanraju, Shehin Sadaka MC, Shibin Eranhottu, Fahmeeda Parveen PS, Beena Cherian, Sini Kurien

E-waste alchemy: turning old tech into sustainable solutions

Abstract: The intricate connection between the management of electronic trash, or "E-waste," and the concept of the circular economy is of paramount importance since the electronics business has expanded rapidly in recent decades. Numerous consumer electronics produce E-waste, which has a negative impact on the environment, human health, and animal welfare. All of the negative impacts of E-waste are covered in this chapter, with a primary emphasis on the environmental pollution caused by improper disposal techniques. This chapter's main focus is on how conventional wisdom about managing E-waste is being altered by the circular economy. Electronics are examined in relation to key circular economy tenets, including design for recyclability, repairability, and longevity. This chapter highlights effective approaches that have adopted circular practices and provide concrete solutions to reduce the adverse environmental effects associated with E-waste. This chapter also discusses the obstacles to the electronics industry's broad adoption of circular methods, looking at both financial and ecological variables. It emphasizes how important it is for producers, consumers, and legislators to work together to overcome these obstacles and promote a circularity paradigm change. The study's conclusion synthesizes the main findings and emphasizes how crucial it is to move toward a circular economy with the goal to ensure sustainable e-trash management. The chapter concludes with a strong call for participation, urging people, organizations, and policymakers to take

***Corresponding author: Tijo Cherian**, Post Graduate Department of Food and Industrial Microbiology, Bishop Kurialacherry College for Women, Amalagiri 686561, Kottayam, Kerala, India, e-mail: tvarghese891@gmail.com

Sijo Asokan, School of Biosciences, Mar Athanasios College for Advanced Studies Tiruvalla (MACFAST – Autonomous), Pathanamthitta, Kerala, India

G Balasubramani, Department of Biotechnology, Saveetha Institute of Medical and Technical Sciences (SIMATS), SIMATS engineering, Thandalam, Chennai 602105, Tamil Nadu, India

N Sharmila Devi, Post Graduate Department of Food and Industrial Microbiology, Bishop Kurialacherry College for Women, Amalagiri 686561, Kottayam, Kerala, India, e-mail: tvarghese891@gmail.com

Smitha Vijayan, Beena Cherian, Sini Kurien, Department of Food Technology & Quality Assurance, Mar Athanasios College for Advanced Studies Tiruvalla (MACFAST – Autonomous), Pathanamthitta, Kerala, India

R Mohanraju, Shehin Sadaka MC, Shibin Eranhottu, Fahmeeda Parveen PS, Department of Ocean Studies and Marine Biology, Pondicherry University, Port Blair, Andaman and Nicobar Islands, India

 | https://doi.org/10.1515/9783112225509-011

proactive steps to promote a healthier and more sustainable future by implementing circular practices in the electronics sector. This chapter provides a comprehensive analysis of e-trash and a roadmap for stakeholders to successfully navigate the complex facets of this problem. This chapter provides insightful viewpoints on implementing a more resilient and sustainable approach to managing E-waste.

Keywords: Electronic waste, plastics, recycling, sustainable management, value-added products

1 Introduction

The demand for electrical and electronic (EE) equipment has grown in recent years due to advancements in information and communication technologies [1, 2]. Gaming consoles, home appliances, surgical instruments, electric equipment, cell phones, and computing are just a few examples of the many EE gadgets that fall under the broad category of "E-waste" (electronic waste). Asia is the largest supplier of E-waste among the various nations, followed by America, Europe, Africa, and Oceania (Figure 1) [3]. One of the biggest obstacles to recycling is the difficult process of sorting the many polymers found in WEEE (waste electrical and electronic equipment) and then converting them into goods with added value. The elimination of color compounds (cadmium) and brominated flame retardants (BRFs) from WEEE presents additional difficulties.

Only 17% of the 54 million metric tons of E-waste produced yearly in 2019 were collected and recycled (The Global E-waste Monitor 2020). The remainder, believed to be worth US$64 billion, is either illegitimately shipped to Asian and African nations or disposed of in landfills, where it is handled and reprocessed by unpaid laborers. The majority of these workers are women and children, and they run the risk of being exposed to dangerous substances like polycyclic aromatic hydrocarbons, BRFs, nickel, mercury, and lead, which could have a negative impact on their health [4]. Another important concern is the poisoning of ecosystems and soil in E-waste regions by organic compounds and heavy metals. E-waste generation is expected to reach 74.7 million metric tons by 2030 due to the growing usage of cell phones, computers, and other electronics, as well as their quick depreciation [3]. A circular economy for electronics manufacturing is one of the ideas that will eventually help to mitigate climate change. According to a new analysis by the Global E-waste Statistics Partnership, 17.4% of E-waste was recycled in 2019, preventing the emission of about 15 million tons of carbon dioxide analogs into the environment. According to the World Health Organization (WHO) report on "Children and Digital Dumpsites," recycling and proper E-waste collection systems are essential for preserving the environment and lowering emissions of greenhouse gases [5]. E-waste is one of the most rapidly expanding residential waste streams; consequently, environmentalists and governments, particularly in developing nations, are becoming increasingly concerned about the increasing volumes of E-waste.

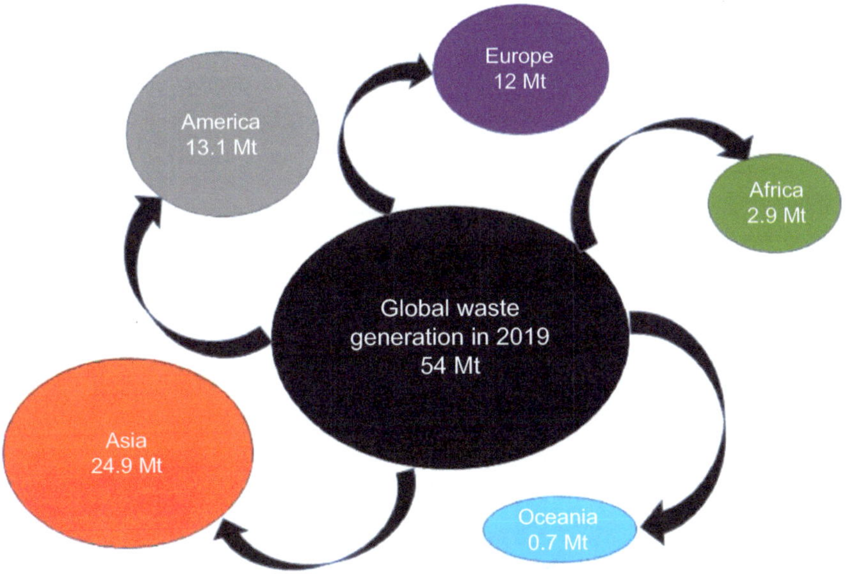

Figure 1: Global E-waste quantities per continents (reproduced from [1]).

Inadequate infrastructure for managing E-waste, incorrect disposition, and unregulated reprocessing of E-waste are further issues that pose a major environmental risk [1]. Additionally, the unofficial E-waste chain of value involves a number of entities [6] such as:

1. Distributors: Those who purchase new or used equipment and sell it straight to customers
2. Customers
3. Collectors, sometimes referred to as garbage collectors, handle E-waste and typically operate in unsanitary environments
4. Refurbishers and repairs
5. Dismantlers, sometimes known as segregators, are those who disassemble machinery into parts that can be sold
6. Recyclers turn E-waste into secondary raw materials and products that are supplied to industrial providers
7. Downstream suppliers purchase the various parts that recyclers sorted out, disassembled, and retrieved

E-waste is made up of a range of materials, including glass, heavy metals, and plastics, of which plastics make up roughly 15% (Figure 2). Additionally, there are persistent organic contaminants like BRFs that must be isolated from the fractions of plastic since they are carcinogenic in nature. Chemical additives may also be incorporated, including thermal stabilizers, antistatic agents, light and heat stabilizers, pigments, antioxidants, and plasticizers [7]. As plastics must first be isolated from metals and then sorted into various groups, the recuperation and recycle procedure becomes extremely complicated and prohibitively exorbitant. In order to recover plastics from E-waste and turn them into a supply of raw materials for a range of uses, copious investigations have been accomplished recently [8].

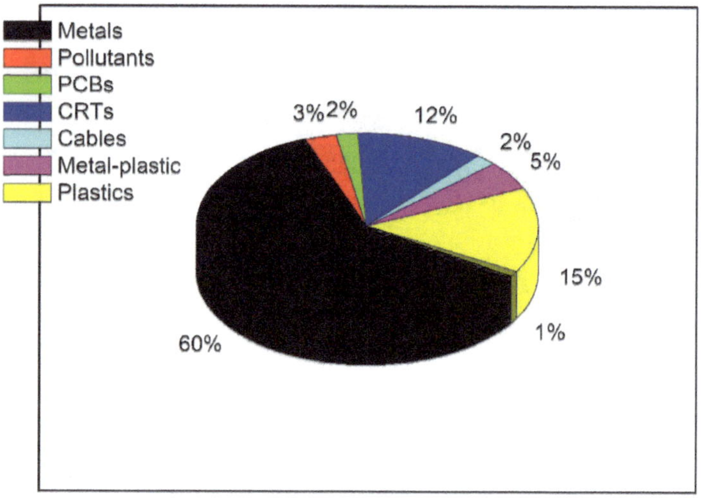

Figure 2: Components of E-waste (reproduced from [1]).

There are now three ways to recycle plastic from E-waste: thermal, chemical, and mechanical. However, over 80% E-waste plastics ends up in landfill due to inadequate rates of recycling. In order to supplement the current technologies, novel non-destructive methods for recovering and recycling plastics from E-waste are needed. The research that deals with turning E-waste polymers into lucrative composites and components for use in the building industry is the main topic of this chapter. The necessary laws, rules, and policies that make it possible to implement the E-waste circular economy have also been discussed. The disparities between current recycling techniques and new advancements will also be covered in this chapter.

1.1 Novelty of the study

This chapter is unique in that it takes an integrative and value-added approach to managing electronic trash (E-waste), pushing beyond traditional topics of disposal and recy-

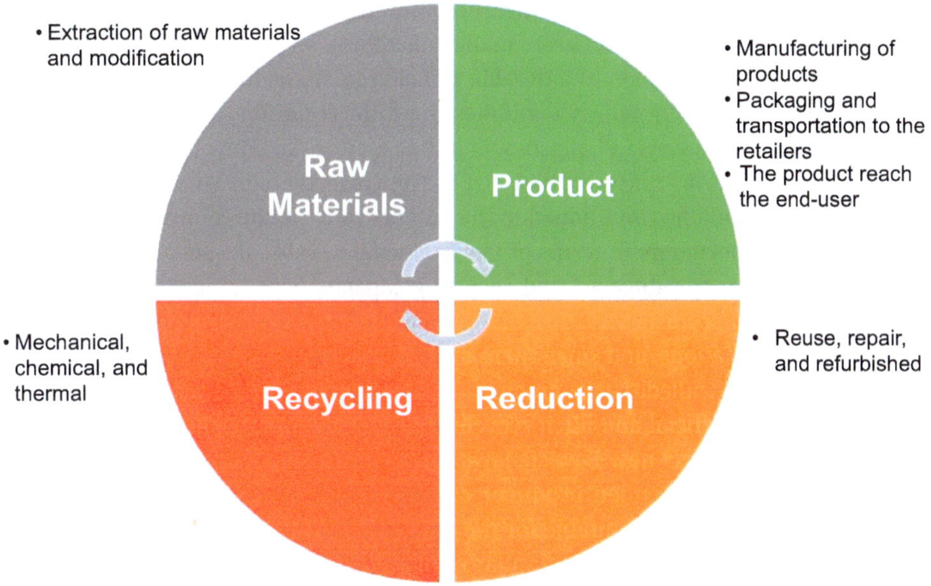

Figure 3: Circular model for E-products (reproduced from [1]).

cling. Unlike many previous efforts, which focus largely on metal recovery or hazards to the environment, this chapter positions E-waste plastics at the heart of circular economy initiatives, stressing their transformation into value-added commodities rather than down-cycled materials. A crucial unique feature is the systematic connection between realistic technology pathways like mechanical, chemical, and thermal recycling and circular economy concepts (design for recyclability, expanded producer responsibility, and material loop closure). This chapter focuses on new applications of E-waste plastics in microfactory-based valorization, additive manufacturing, polymer composites, and construction materials, establishing E-waste as a secondary resource for sustainable material development. Furthermore, this chapter combines policy frameworks, socioeconomic elements, and technical advancements into a single cohesive narrative. The particular focus on developing-country contexts, informal recycling chains, and legislative inadequacies adds to its uniqueness and relevance, providing a comprehensive roadmap rather than a solely technical analysis.

2 Policies, laws, regulations, and E-waste management

Every nation must address the serious issue of E-waste contamination and its detrimental effects on humanity and the biosphere. Each nation's regulatory agencies and legislators are responsible for creating and enforcing laws, regulations, and policies pertaining to the appropriate E-waste management as well as addressing these issues in a formal and legal manner. The notable global regulations include the EU directive on WEEE governance [9], which mandates the differentiation and division of plastic with BFRs from other WEEE plastics, and the recent Basil Convention, which guarantees that worldwide transportation of E-waste (regardless of whether it is potentially harmful) should be conditioned to the authorization of the country of origin and any state of transportation. In terms of E-waste creation rates, the stakeholder expectations, existence of "producers," and recycling capacity, it is critical to adjust these methods to local circumstances. The management of E-waste is still in its infancy in many developing nations, and since there are no laws specifically addressing E-waste, it is inadequately handled. Large amounts of unaccounted-for E-waste have resulted in the growth of unofficial and illicit recycling facilities [10]. First and foremost, developing nations must enact new E-waste laws that prioritize the following:

1. Encouraging the extended producer responsibility (EPR) principles, which will make the producer accountable for the product after its useful life is over
2. To promote a cradle-to-cradle circular economy in order to reduce post-consumer waste products and eliminate the constant usage of raw materials (Figure 3)
3. To close material loops by promoting the 10Rs (Refuse, Rethink, Reduce, Reuse, Repair, Refurbish, Remanufacture, Repurpose, Recycle, and Recover)
4. Establish buy-back centers, drop-off locations, and collecting banks to formalize the E-waste management system. This encourages businesses and families to eliminate of their undesirable E-waste and barter for profits
5. Analyzing and incentivizing the various phases of E-waste management in order to standardize the system
6. To guarantee that the laws pertaining to E-waste are upheld and that the newest technologies are enforced

3 Plastic recycling from electronic waste

Electronic device collection, physical cataloguing, disassembly, and shredding are the initial steps in the E-waste plastic recycling. Non-metallic components (plastics, ceramics, and glass) and metals are mechanically separated from the shredded E-waste. Polymers such as polyethylene, polypropylene, polyamides, high-impact polystyrene (HIPS), acrylonitrile-butadiene-styrene (ABS), and polycarbonate (PC) are among the various plastics that can be recovered from E-waste; these must be sorted into grades before being recycled into products or converted to energy (Figure 4). Despite the fact that E-waste contains premium industrial plastics like ABS, recycling E-waste plastics remains a challenge because of the ignorance of potential compatibilities of various plastics during melt extrusion process and wide range of chemical additives like plasticizers BRFs, polymers. Large amounts of graded plastics and appropriate recycle infrastructure are also necessary for a sturdy recycle operation [11].

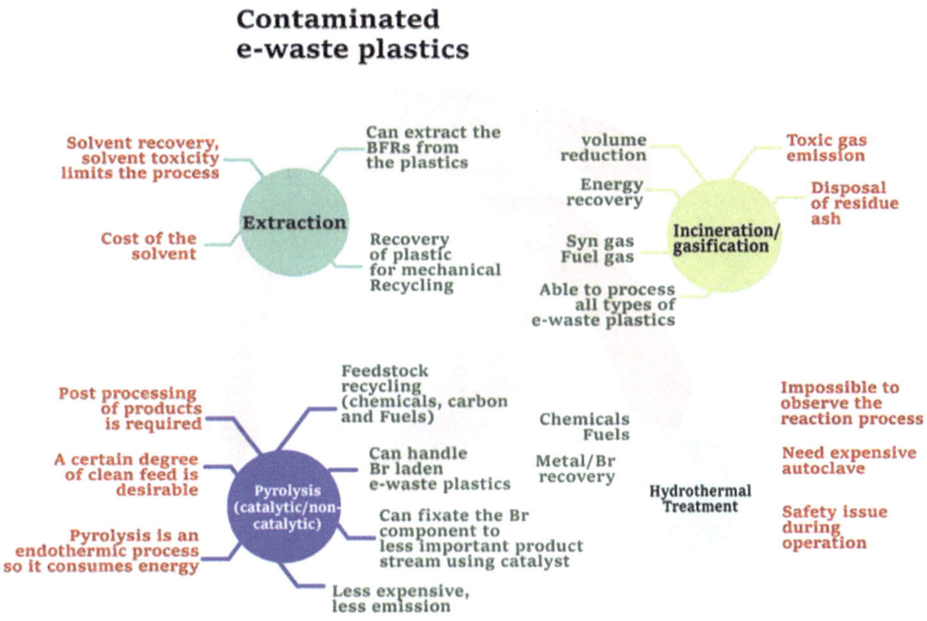

Figure 4: E-waste plastics to finished products (reproduced from [11]).

3.1 Primary and secondary recycling (mechanical recycling)

The most popular method for recuperating E-waste plastics is mechanical recycling, which involves melting the materials and shaping them into new items. In primary recycling, recovered plastic is utilized to create goods that are comparable to those manufactured with virgin plastics in terms of performance. When recovered plastic is utilized to create new goods with lower performance criteria than the original materials, this is known as secondary recycling. Following the separation of the plastic fractions from the E-waste, the plastics are granulated into pellets, cleaned, and sorted (either through optical sorting or manually using magnetic separators and eddy-current separators, depending on the desired end product). Following sorting, the polymers are put through melt processing methods like injection molding, hot pressing, and melt extrusion [11]. E-waste plastics are increasingly being transformed into green goods through the use of environmentally conscious technologies (Figure 5) like additive manufacturing. The possibility of converting E-waste plastics into sustainable 3D-printing filaments has been demonstrated by a number of investigations [12].

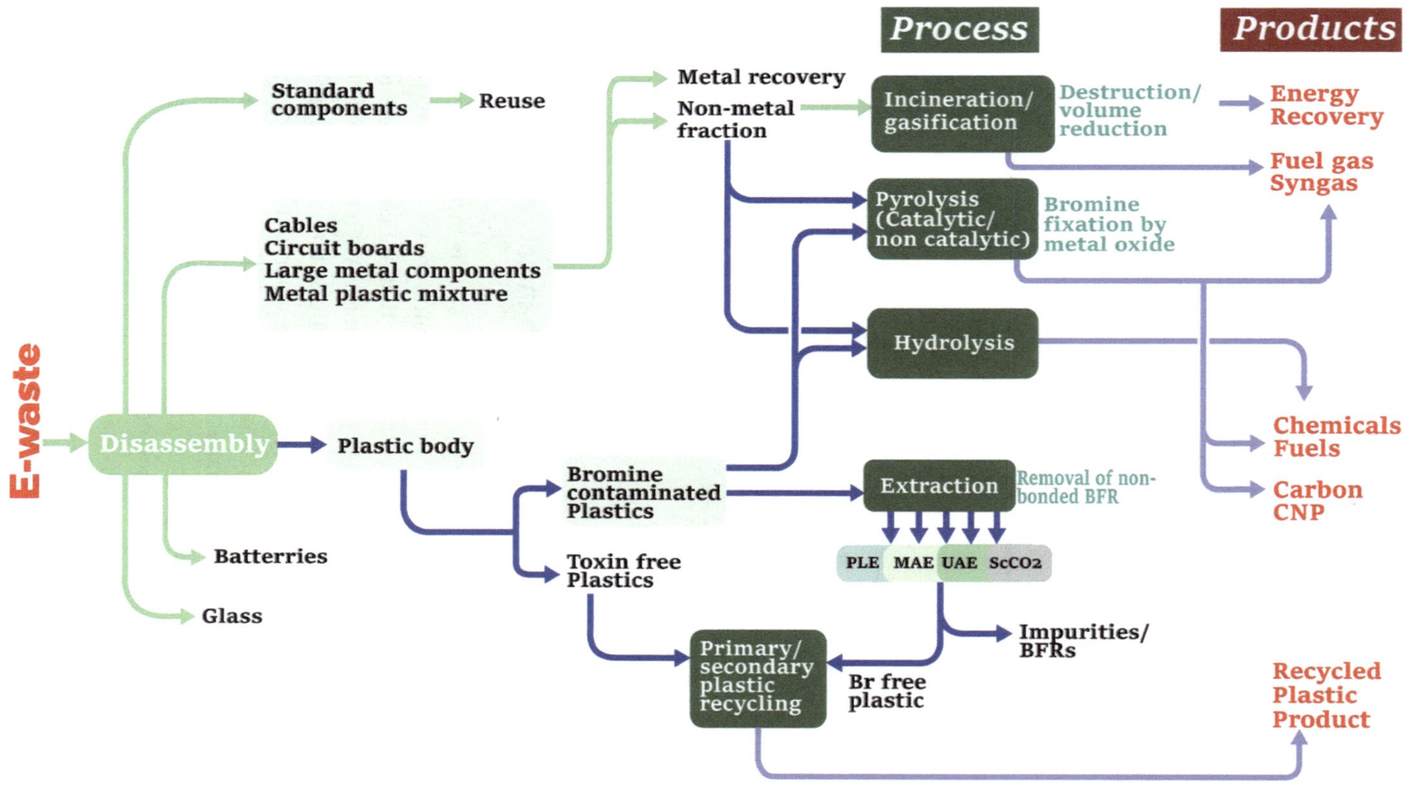

Figure 5: Pros (green text) and cons (red text) of extraction process in the recycle of contaminated E-waste plastics (reproduced from [11]).

3.2 Tertiary recycling (chemical and thermal recycling)

In order to recover fuels and chemicals via depolymerization processes, plastics from E-waste undergo thermal and chemical treatments [13]. Chemical recycling makes it possible to dispose of polluted polymers from E-waste sans the need for complex prior-treatment procedures. After that, the plastic portions are repurposed to create useful things. Catalytic cracking, pyrolysis (Figure 6), gasification, hydrogenation, and dissolution, are common methods. In order to produce products like char, combustible gases, and oil, the sample is burned to high temperatures (400–800 °C) under inert conditions. Pyrolysis produces fewer contaminants than conventional thermal treatments. Optimized operating factors, including heating rate, time, feedstock composition, and temperature, determine the products of pyrolysis [14]. High-value products can be produced using catalytic pyrolysis, which reduces residence time and temperature [15]. Dissolving methods using different solvents, such as carbon dioxide, acetone, water and SCFs (supercritical fluids) [14, 16, 17], have been used in studies to investigate the production of monomers, all of which were conducted on a laboratory scale. The CreaSolv technique, which uses solvents to extract BRFs from E-waste and precipitates polymers that may be recycled using extrusion technologies, is a noteworthy innovation with the possibility of upscaling [18].

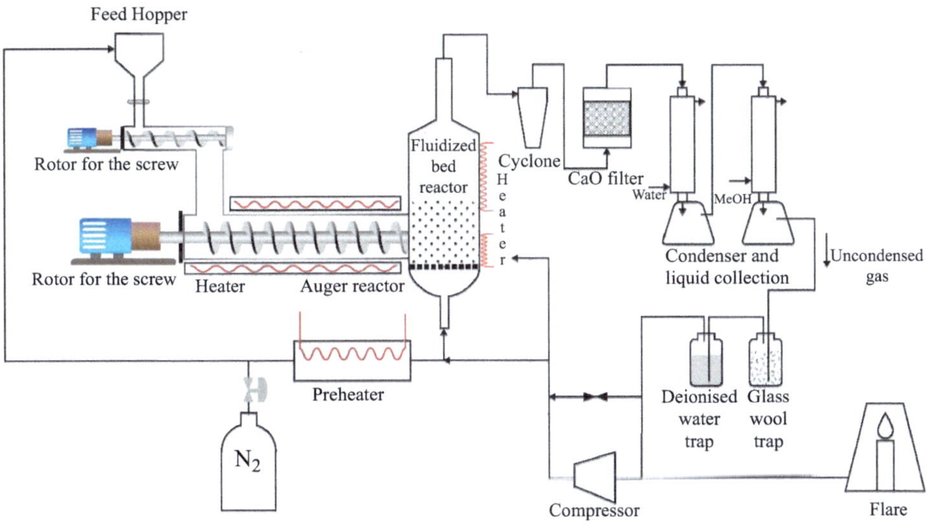

Figure 6: Diagrammatic representation of two-stage continuous pyrolysis process (reproduced from [11]).

4 New advances in the transformation of plastics from E-waste

4.1 Utilizing plastics from E-waste in construction sector

For usage in construction, E-waste polymers are now thought to be a good alternative to concrete with coarse and fine particles [19]. Given that E-waste plastics are thought to prevent the degradation of natural resources and a solution to E-waste contamination, they are being used to replace concrete, either completely or partially [20]. Using E-waste plastics instead of commercially available concrete has two benefits: it lowers the end product's cost and produces lightweight concrete, which uses a smaller amount of fuel during transit and conserve energy [21]. Tensile flexural and compressive strengths increase when concrete is partially replaced with E-waste plastics by adding loadings of up to 16% by weight [21]. Higher loadings of E-waste polymers have also been shown to cause a negative trend in the mentioned characteristics. By adding nanofillers such as graphite platelets, the mechanical and hardness characteristics of cementitious composites invigorated with E-waste plastics can be enhanced. These nanofillers are important for strengthening the microstructure and decreasing porosity [22]. When compared to conventional concrete, the resulting composite materials strengthened with E-waste polymers had far better workability and durability [20, 23]. The new "microfactories," where E-waste polymers can be converted into a brand-new, superior-quality product instead of being recycled back into recyclable plastics, are another type of recycling. One example is the Sydney microfactory that turns E-waste plastic into "Grenew carbon," a unique premium commodity utilized in metallurgical sectors as an alternate top-charging substitute for carbon [24].

4.2 Utilizing E-waste as reinforcements in composite

E-fillers, or E-waste plastics, are increasingly being used as reinforcement in polymers for composites usage. When E-filler is added to polymeric materials, the resulting composites' impact, flexural, and tensile properties increase [25]. In a recent study, Sun et al. [26] created composite boards using nonmetal particles from waste printed circuit boards and ABS waste plastic. Compatibilizers like maleic anhydride-grafted ABS have the potential to significantly increase the interfacial adhesion between the nonmetal particles and the ABS plastic. The creation of wood plastic composites using wood fibers and NWPCBs (nonmetals from waste printed circuit boards) was the subject of another study. The epoxy and phenolic resins make up the majority of NWPCB, and the resulting composites showed strong tensile strength with filler content and storage modulus. Reusing NPCB in high-strength wood composites has been proposed to be a sustainable way to produce value-added products and address environmental degradation [27].

5 Strengths and limitations

One of the chapter's main features is its extensive and diverse coverage. It successfully integrates sustainability studies, policy analysis, materials engineering, and environmental science, making it accessible to a wide range of professional and academic audiences. The chapter shows good conceptual clarity, especially when explaining:

1. Size and seriousness of the worldwide E-waste problem
2. Material and chemical complexity of E-waste plastics
3. Comparative advantages of thermal, chemical, and mechanical recycling routes

Another major strength is the wide use of current literature, which has been endorsed by contemporary peer-reviewed researches and authoritative reports (e.g., Global E-waste Monitor, WHO). This increases scientific legitimacy and places the chapter into the current global discourse. The focus on value-addition product development, such as E-waste-based construction materials and composites, has a significant impact. This chapter advances from problem diagnosis to solution-oriented science by exhibiting real applications with better mechanical, durability, and sustainability. Furthermore, the inclusion of policy tools (WEEE Directives, Basel Convention, and EPR) and their practical ramifications increases the chapter's relevance to policymakers and regulatory authorities, rather than only academicians.

Notwithstanding its positive attributes, the chapter has several limits that should be acknowledged. One drawback is that many of the advanced valorization and recycling methods presented are still in the laboratory or on a pilot scale. While this chapter acknowledges this, a more in-depth technoeconomic examination of scalability, electrical consumption, and lifetime implications will help to reinforce the discussion. Another restriction is the relatively limited quantitative performance comparison of E-waste-derived materials to traditional commercial alternatives. Although developments in durability and mechanical attributes are examined, more standardized comparison datasets would improve industrial applicability. This chapter focuses on plastic fractions of E-waste, with rare earth elements, metals, and hybrid material recovery receiving comparatively little emphasis. Because E-waste management is intrinsically multi-material, a more integrated approach to cross-material recovery chains could give greater completeness. Finally, while legislative frameworks are considered, implementation issues at the regional and local levels, particularly in informal recycling economies, might be investigated with more empirical detail.

6 Future perspectives and research directions

The foreseeable future of E-waste management, notably the valorization of E-waste plastics, is at the crossroads of policy-driven change, systems-level thinking, and technology innovation. While enormous strides have been made at the laboratory and pilot scales, the next phase of development must prioritize sustainability, scalability, and socioeconomic integration to ensure that scientific achievements have actual-world impact.

6.1 Scaling up and industrializing recycling technologies

One of the most pressing future issues is moving from laboratory-validated techniques to financially feasible industrial-sized systems. Many advanced recycling processes, including as composite reprocessing, chemical depolymerization, and selective dissolution, have proven technically feasible but are limited by process complexity, energy intensity, and cost.

Future studies should prioritize:

a. Process intensification to decrease energy usage and material losses.
b. Modular and dispersed recycling facilities, including microfactories, can operate near waste generating sources.
c. Automation and AI-driven sorting systems can enhance feedstock purity and process efficiency.

Decentralized recycling options are especially interesting in underdeveloped nations, where centralized infrastructure is generally scarce. These methods can dramatically cut transportation costs, increase local value generation, and incorporate informal recycling industries into institutional circular economy models.

6.2 Revolutionary materials engineering for E-waste-derived polymers

E-waste plastics are intrinsically heterogeneous, including mixtures of PC, HIPS, ABS, and flame-retardant chemicals. Overcoming compatibility difficulties remains a key research challenge. Future materials research should center on:

a. Advanced compatibilization technologies include grafted polymers, block copolymers, and reactive compatibilizers
b. Nanofiller integration (e.g., graphene, nanoclays, and bio-based fibers) to improve flame retardancy, thermal stability, and mechanical strength
c. Structure-property relationship studies are used to optimize formulations for specific applications

These methods can enable the creation of high-performance recycled polymer composites suited for additive manufacturing, electrical housings, automotive, and construction transforming recovered E-waste plastics from low-value reuse to practical engineering materials.

6.3 Green chemistry and next-generation recycle approaches

Sustainability requires a transition toward ecologically friendly recycling technology. Conventional thermal and solvent-based procedures frequently use dangerous chemicals or require significant energy inputs. Future directions include:

a. Green solvent systems, such as deep eutectic solvents and ionic liquids, for selective polymer dissolution
b. Use SCF technologies, such as CO_2, for polymer purification and additive extraction
c. Enzymatic and catalytic depolymerization provides good selectivity under mild circumstances

These technologies are consistent with the principles of green chemistry, reducing hazardous byproducts and facilitating closed-loop recycling.

6.4 LCA (life cycle assessment) and TEA (technoeconomic analysis)

Quantitative sustainability assessment is critical for gaining industrial and policy approval for E-waste recycling solutions. Future studies should include technoeconomic analysis (TEA) and life cycle assessment (LCA) as main components rather than supplemental assessments. The key research priorities are:

a. Comparative LCAs of mechanical and chemical recycling routes
b. Energy payback and carbon footprint analysis
c. Cost-benefit analysis of decentralized versus centralized recycling systems

These data-driven assessments will help inform investment decisions, promote evidence-based policymaking, and guide technology selection.

6.5 Integration into regulation, policy, and circular economy models

Technological innovation is insufficient without enabling financial and regulatory foundations. Future advancement is dependent on the alignment of policy, science, and economic systems. Important policy-oriented directions include:

a. Enhancing and enforcing EPR programs
b. Creating standards and certification programs for recovered E-waste materials
c. Promoting eco-design and design-for-circularity within the manufacturing process

Including scientific developments in regulatory structures ensures long-term system stability and industrial acceptance.

6.6 Socioeconomic and ecological impacts

Moving ahead, E-waste must be re-casted as a strategic urban resource rather than an environmental burden. E-waste plastics can be effectively valorized as follows:
a. Reduce the reliance on virgin petrochemical resources
b. Reduce greenhouse gas emissions
c. Create green jobs and promote inclusive economic growth

Future study should include socioeconomic effect assessments that consider community-level benefits, informal sector integration, and labor conditions.

6.7 Transdisciplinary and interdisciplinary collaboration

The intricacy of E-waste concerns needs cross-disciplinary and sector collaboration. Social scientists, policymakers, economists, environmental scientists, material scientists, and engineers, must collaborate to create solutions that are socially fair, commercially viable, and technically sound. Collaborative platforms that connect government, industry, and academia will be critical for converting research findings into scalable, real-world applications.

In a nutshell, the prospects of E-waste recycling and valorization reside in systemic transformation rather than incremental advancement. E-waste has the potential to become a pillar of the circular economy by combining policy innovation, sustainability assessment, green chemistry, and modern material science. The view presented in this chapter sees E-waste not just as a waste stream to be controlled, but also as a key resource for material security, climate resilience, and long-term development.

7 Conclusion

E-waste production is expected to increase to 74.7 metric tons by 2030. It is obvious that a coordinated strategy including all parties involved in the value chain is required to address this issue. The following is a summary of the actions required for E-waste management:
1. Investigations on turning E-waste plastics into environmentally friendly and sustainable products with added value are expanding on rapid pace and scale. Examples include the use of E-waste polymers as fillers in composite materials and as a partial substitute for concrete in construction applications.
2. This chapter has covered chemical, thermal, and mechanical recycling of plastics from E-waste. The majority of this, meanwhile, is still done in laboratories and

needs to be expanded. Investigations must concentrate on creative conversion techniques as E-waste recovery rates are quite low.
3. To lessen the amount of E-waste disposed in landfill in developing nations, laws and regulations devoted to creating appropriate E-waste management methods are necessary.

References

[1] Mtibe A, Mokhena TC, John MJ. Sustainable valorization and conversion of e-waste plastics into value-added products. Current Opinion in Green and Sustainable Chemistry. 2023;40:100762. https://doi.org/10.1016/j.cogsc.2023.100762.

[2] Sengupta D, Ilankoon IMSK, Kang KD, Chong MN. Circular economy and household e-waste management in India: integration of formal and informal sectors. Minerals Engineering. 2022;184:107661. https://doi.org/10.1016/j.mineng.2022.107661.

[3] Forti V, Baldé CP, Kuehr R, Bel G: The global e-waste monitor. Quantities, flows, and the circular economy potential. 2020. https://www.itu.int/en/ITU-D/Environment/Documents/Toolbox/GEM_2020_def.pdf.

[4] Cao P, Fujimori T, Juhasz A, Takaoka M, Oshita K. Bio accessibility and human health risk assessment of metal(loid)s in soil from an e-waste open burning site in Agbogbloshie, Accra, Ghana. Chemosphere. 2020;240:124909. https://doi.org/10.1016/j.chemosphere.2019.124909.

[5] https://www.who.int/news/item/15-06-2021-soaring-e-waste-affects-the-health-of-millions-of-children-who-warns

[6] Avis W: Responsible E-waste value chains in Africa. 2022. https://opendocs.ids.ac.uk/opendocs/handle/20.500.12413/17152

[7] Kike O, Sindiku O, Osibanjo O, Weber R. Polybrominated diphenyl ethers (PBDEs) concentrations in soil, sediment and water samples around electronic wastes dumpsites in Lagos, Nigeria. Emerging Contaminants. 2022;8:206–215. https://doi.org/10.1016/j.emcon.2022.03.003.

[8] Kaliyavaradhan SK, Prem PR, Ambily PS, Mo KH. Effective utilization of e-waste plastics and glasses in construction products – a review and future research directions. Resources, Conservation and Recycling. 2022;176:105936. https://doi.org/10.1016/j.resconrec.2021.105936.

[9] https://environment.ec.europa.eu/topics/waste-and-recycling/waste-electrical-and-electronic-equipment-weee_en

[10] Uhunamure SE, Nethengwe NS, Shale K, Mudau V, Mokgoebo M. Appraisal of households' knowledge and perception towards e-waste management in Limpopo province, South Africa. Recycling. 2021;6:39. https://doi.org/10.3390/recycling6020039.

[11] Das P, Gabriel JC, Tay CY, Lee JM. Value-added products from thermochemical treatments of contaminated e-waste plastics. Chemosphere. 2021;269:129409. https://doi.org/10.1016/j.chemosphere.2020.129409.

[12] Mohammed M, Wilson D, Gomez-Kervin E, Petsiuk A, Dick R, Pearce JM. Sustainability and feasibility assessment of distributed E-waste recycling using additive manufacturing in a Bi-continental context. Additive Manufacturing. 2022;50:102548. https://doi.org/10.1016/j.addma.2021.102548.

[13] Charitopoulou MA, Kalogiannis KG, Lappas AA. Achilias DS: Novel trends in the thermo-chemical recycling of plastics from WEEE containing brominated flame retardants. Environmental Science and Pollution Research. 2021;28:59190–59213. https://doi.org/10.1007/s11356-020-09932-5.

[14] Jadhao PR, Ahmad E, Pant KK, Nigam KDP. Advancements in the field of electronic waste Recycling: critical assessment of chemical route for generation of energy and valuable products coupled with

metal recovery. Separation and Purification Technology. 2022;289:120773. https://doi.org/10.1016/j.seppur.2022.120773.
[15] Jung S, Lee S, Song H, Tsang YF, Kwon EE. Sustainable valorization of e-waste plastic through catalytic pyrolysis using CO_2. ACS Sustainable Chemistry & Engineering. 2022;10:8443–8451. https://doi.org/10.1021/acssuschemeng.2c01469.
[16] Preetam A, Jadhao PR, Naik SN, Pant KK, Kumar V. Super critical fluid technology – an eco-friendly approach for resource recovery from e-waste and plastic waste: a review. Separation and Purification Technology. 2023;304:122314. https://doi.org/10.1016/j.seppur.2022.122314.
[17] Anderson L, Yu E, Chen WT. Chemical recycling of mixed plastics in electronic waste using solvent-based processing. Processes. 2022;10:66. https://doi.org/10.3390/pr10010066.
[18] https://www.ivv.fraunhofer.de/en/recycling-environment/recycling-plastics-creasolv.html.
[19] Mane KM, Nadgouda PA, Joshi AM. An experimental study on properties of concrete produced with M-sand and E-sand. Materials Today Proceedings. 2021;38:2590–2595. https://doi.org/10.1016/j.matpr.2020.08.086.
[20] Ullah Z, Qureshi MI, Ahmad A, Khan SU, Javaid MF. An experimental study on the mechanical and durability properties assessment of E-waste concrete. Journal of Building Engineering. 2021;38:102177. https://doi.org/10.1016/j.jobe.2021.102177.
[21] Shinu NMMT, Needhidasan S. An experimental study of replacing conventional coarse aggregate with E-waste plastic for M40 grade concrete using river sand. Materials Today Proceedings. 2020;22:633–638. https://doi.org/10.1016/j.matpr.2019.09.033.
[22] Ahmad F, Qureshi MI, Ahmad Z. Influence of nano graphite platelets on the behavior of concrete with E-waste plastic coarse aggregates. Construction and Building Materials. 2022;316:125980. https://doi.org/10.1016/j.conbuildmat.2021.125980.
[23] Ahmad F, Jamal A, Mazher KM, Umer W, Iqbal M. Performance evaluation of plastic concrete modified with e-waste plastic as a partial replacement of coarse aggregate. Materials. 2022;15:175. https://doi.org/10.3390/ma15010175.
[24] https://www.smart.unsw.edu.au/technologies-products/microfactorie-technologies/e-waste
[25] Jayamani E, Ki COS, Rahman MR, Bakri MKB. Recycled polymer and plastic waste and its biocomposites. In Recycled plastic biocomposites. Woodhead Publishing Series in Composites Science and Engineering. 2022;81–96. https://doi.org/10.1016/B978-0-323-88653-6.00009-2.
[26] Sun Z, Shen Z, Zhang X, Ma S. Co-recycling of acrylonitrile-butadiene-styrene waste plastic and nonmetal particles from waste printed circuit boards to manufacture reproduction composites. Environmental Technology. 2015;36:160–168. https://doi.org/10.1080/09593330.2014.940399.
[27] Yang S, Jiang J, Duan W, Bai S, Wang Q. Production of sustainable wood-plastic composites from the nonmetals in waste printed circuit boards: Excellent physical performance achieved by solid-state shear milling. Composites Science and Technology. 2020;200:108411. https://doi.org/10.1016/j.compscitech.2020.108411.

Neelesh Kumar Maurya*

Phytoremediation: a green environmental technology for management of e-waste-contaminated sites

Abstract: The rampant growth of electronic waste (e-waste) has caused an alarming environmental disaster as global e-waste generation increased 62 million tons annually in 2022 and is expected to reach 82 million tons in 2030 (Global E-waste Monitor, 2024). E-waste is still not managed correctly, with hotspots like Guiyu, China, and Agbogbloshie, Ghana, specializing in the recycling of useful e-waste and dumping the toxic e-waste contaminants (e.g., heavy metals (lead, cadmium, mercury) and polybrominated diphenyl ethers, polychlorinated biphenyls, and rare earth elements). Such practices pose extreme threats to human and environmental health, as conventional methods of excavating/removing subsoils with the remaining toxic contamination through leaching (chemicals) are overpriced and environmentally unsustainable ($100,000–500,000/ha). Thus, alternatives are sorely needed. Solar-driven, rhizosphere plant-based technologies utilize $5,000–15,000/ha, integrate multiple contaminants to harmoniously extract, stabilize, and degrade contaminants. Between 2023 and 2025, this chapter aims to evaluate waste, contamination chemistry, hyperaccumulators, and the microbial-assisting ways of degradation (filled with chelating agents, biochar, CRISPR, and nanomaterials) and, finally, to apply them in the real field. Annual metal depletion of 20–50% and degradation of 40–70% of organic contaminants were confirmed in the cited studies. Other innovations, such as phytomining, offer precious metal restoration valued at $10–150k/ha. Limitations such as variable biomass production, regulatory gaps, and insufficient stakeholder involvement necessitate integration of AI-driven monitoring, extended producer responsibility (EPR) policies, and circular economy strategies for scalable deployment. Phytoremediation can help avert the e-waste crisis on a larger scale sustainably, thus necessitating the need for interdisciplinary research and policy integration according to the review's findings.

Keywords: E-waste, phytoremediation, hyperaccumulators, heavy metals, persistent organic pollutants, phytomining

***Corresponding author: Neelesh Kumar Maurya**, Department of Nutrition and Dietetics, School of Allied Health Sciences, Sharda University, Greater Noida, Uttar Pradesh, India, e-mail: neeleshkumar.maurya@gmail.com

 | https://doi.org/10.1515/9783112225509-012

1 Introduction

1.1 The global e-waste crisis: scale, trajectory, and transboundary implications

The twenty-first century digital revolution has precipitated an unprecedented environmental challenge: the exponential accumulation of electronic waste (e-waste) worldwide [1]. The Global E-waste Monitor 2024 estimated that 62 million metric tons (Mt) of e-waste were generated in 2022, equivalent to about 7.8 kg per capita globally, with projections reaching approximately 82 Mt by 2030 – a 32% increase driven by accelerated device obsolescence, rising consumer demand, and the proliferation of Internet of Things devices [1]. Asia dominates e-waste generation, contributing nearly 25 Mt annually, while parts of Africa and South Asia function as primary receivers of transboundary e-waste flows, processing an estimated 100,000–215,000 tons per year in informal recycling hubs such as Moradabad (India), Guiyu (China), and Agbogbloshie (Ghana) [1, 2]. These largely unregulated practices mobilize a high proportion of toxic constituents – often approximated at 90% – into surrounding environmental matrices, generating legacy contamination that can persist for decades in soils, sediments, and groundwater [3].

The transboundary movement of e-waste from high income countries (e.g., European Union member states and the United States) to low- and middle-income countries (LMICs) exacerbates global environmental injustice, as roughly 80% of exported waste undergoes crude recycling via open burning, acid leaching, and manual dismantling, releasing persistent toxicants into soil, water, and air [2, 3]. This "toxic trade" runs counter to the intent of the Basel Convention on the Control of Transboundary Movements of Hazardous Wastes and Their Disposal, yet enforcement remains weak and traceability limited, with only about 22.3% of e-waste managed through formal recycling channels globally [1, 2]. The resulting contamination hotspots commonly exhibit soil lead (Pb) concentrations 23–41 times local background levels, groundwater chromium (Cr(VI)) concentrations exceeding World Health Organization (WHO) guideline values by up to an order of magnitude, and atmospheric dispersion of $PM_{2.5}$-bound metals over regional to transboundary distances [2–4]. The economic burden associated with these impacts is substantial, with health-related costs estimated at approximately USD 15 billion per year and an additional USD ~62 billion per year in unrecovered precious metals and critical raw materials lost through informal disposal and inefficient recovery practices [1, 5].

1.2 Chemical composition and ecotoxicology of e-waste

E-waste streams are typically composed of 40–50% metals by weight, with metal concentrations often 50–100 times higher than those encountered in natural ores, which

Figure 1: Negative effects of e-waste on the environment (reprinted from Sandwal et al. [1]).

positions discarded electronics as a potential "urban mine" but also as a significant toxicological liability when mismanaged [5]. Major metallic toxicants include Pb (approximately 5–15 kg/ton from solder and cathode ray tube glass), cadmium (Cd; 0.1–1 kg/ton from nickel-cadmium batteries), mercury (Hg; 0.01–0.1 kg/ton from switches and lamps), copper (Cu; up to ~128 kg/ton from wiring and printed circuit boards), and hexavalent chromium (Cr(VI)); 0.5–2 kg/ton from coatings and surface treatments) [3, 5]. Rare earth elements (REEs) such as neodymium (Nd; 50–200 g/ton from permanent magnets), dysprosium (Dy), and indium (In) contribute substantial economic value but have been associated with teratogenic effects, oxidative stress, and hepatic accumulation in experimental biota [5]. Precious metals – gold (Au; ~0.25 kg/ton), silver (Ag; ~3.6 kg/ton), and palladium (Pd) – underscore the magnitude of resource loss when e-waste is processed through primitive techniques that fail to recover these high-value constituents [1, 5].

Organic fractions account for roughly 20–30% of total e-waste mass and are dominated by flame retardants and dielectric fluids such as polybrominated diphenyl ethers (PBDEs) and polychlorinated biphenyls (PCBs) used in plastics and capacitors [3, 6]. PBDE concentrations in certain e-waste plastics can range from 10 to 120 kg/ton, and these compounds function as persistent endocrine disruptors, with epidemiological and toxicological studies reporting thyroid hormone reductions of 30–50% and associated neurodevelopmental impacts [6]. PBDEs exhibit long environmental half-lives (approximately 10–20 years) and high hydrophobicity (log K_ow values of 6–8), yielding bioconcentration factors (BCF) above 10^5 and facilitating bioaccumulation and biomagnification along aquatic food webs [6]. During uncontrolled combustion or low temperature incineration of e-waste, polychlorinated dibenzo-*p*-dioxins and dibenzofurans are formed; congeners such as 2,3,7,8-tetrachlorodibenzo-*p*-dioxin possess toxicity equivalence factors (TEFs) of 1, indicating extremely high toxic potency [2]. Additional contaminants such as phthalates and polycyclic aromatic hydrocarbons (PAHs) are released from plastic casings and insulation during thermal or photooxidative degradation, contributing to cumulative carcinogenic and mutagenic risks [3, 6]. The persistence, mobility, and mixture toxicity of these chemicals generate multigenerational contamination, with documented groundwater plumes extending kilometers downgradient from informal recycling clusters and sediment burial processes sequestering a large fraction of contaminant loads over time as reflected in Figure -1 [2–4].

1.3 Health and ecological risks: a planetary health perspective

The human health consequences of e-waste exposure are profound and disproportionately affect vulnerable populations in LMICs, including children, pregnant women, and informal sector workers [2–4]. Chronic Pb exposure from contaminated soil, dust, and food pathways is associated with neurodevelopmental deficits; pooled analyses suggest that each 10 µg/dL increase in blood Pb concentration is linked to IQ reductions of roughly 4–7 points in children, as well as increased risk of renal dysfunction and hypertension in adults [2, 3]. In communities adjacent to major informal recycling hubs such as Agbogbloshie, mean pediatric blood Pb levels exceeding 70 µg/dL – approximately 10 times above widely referenced "safe" thresholds – have been documented, while adult workers frequently report high rates of respiratory disease, reproductive disorders (including two- to threefold higher miscarriage rates), and dermatological problems [2]. PBDE exposures, primarily via inhalation of contaminated dust and ingestion of contaminated food, have been associated with thyroid hormone dysregulation and neurocognitive impairment, with some cohorts exhibiting IQ reductions of 10–15 points in relation to high body burdens [3, 6]. Cr(VI) is well established as a human carcinogen with a substantial oral slope factor, and Hg released from broken components can be converted to methylmercury in aquatic systems, leading to Minamata-like neurological syndromes in fish-dependent populations [2–4].

Ecologically, e-waste-derived contaminants exert severe pressure on biodiversity and ecosystem functioning. PBDE and PCB concentrations in fish, invertebrates, and sediments collected near e-waste recycling sites routinely exceed European Union environmental quality standards by up to two orders of magnitude, and are associated with approximately 50% reductions in invertebrate diversity and endocrine disruption phenomena such as intersex conditions in fish [6]. REEs and heavy metals can alter soil and sediment microbial communities, with some studies reporting up to 70% reductions in microbial diversity and significant impairment of critical biogeochemical processes, including nutrient cycling, carbon turnover, and denitrification [3, 5]. The combined economic costs of these health and ecological impacts – including medical care, productivity losses, and degradation of ecosystem services such as fisheries and agriculture – are estimated at tens of billions of dollars annually [1–3, 5]. In this planetary health context, there is growing recognition of the need for integrated, nature-based remediation strategies, among which phytoremediation is increasingly highlighted as a cost-effective stabilizer-extractor capable of addressing multicontaminant legacies at e-waste sites [3, 6–8].

1.4 Limitations of conventional remediation

Conventional remediation approaches – such as excavation and landfilling, elevated temperature pyrometallurgy, and aggressive chemical leaching – have been applied to some e-waste contaminated areas but are often financially and environmentally unsustainable at the scales required [3, 5]. Excavation and offsite disposal can cost on the order of USD 100,000–500,000 per hectare, generate around 10,000 tons of contaminated soil and debris per hectare, and irreversibly disrupt soil structure, hydrology, and local ecosystems [3]. Pyrometallurgical processing can recover a portion of valuable metals but typically consumes large amounts of energy (approximately 500 kWh/ton of material), emits greenhouse gases, and can release dioxins and furans at levels of 0.1–1 ng toxic equivalent (TEQ) per cubic meter of flue gas if emission control technologies are inadequate [3, 5]. Acid or cyanide-based chemical leaching can mobilize up to 90% of target metals; however, it generates large volumes of highly hazardous liquid waste requiring complex treatment, and any failures in containment can contaminate groundwater and surface waters [5].

These technologies are largely reactive, addressing contamination after it has occurred rather than preventing it, and they rarely restore ecological function or provide cobenefits such as habitat creation or climate mitigation [3, 5, 8]. Furthermore, they do not address the socioeconomic drivers of e-waste generation and informal recycling. Consequently, there is an urgent need for sustainable, solar-powered, and ecosystem restorative alternatives that can be implemented in resource-limited settings while aligning with broader climate and biodiversity goals [5, 8].

1.5 Phytoremediation: a sustainable alternative

Phytoremediation, broadly defined as the engineered use of plants and their associated rhizosphere microorganisms to remove, stabilize, transform, or degrade environmental contaminants, is increasingly recognized as a promising, low-cost strategy for managing metal and organic polluted environments, including e-waste impacted sites [6–8]. For e-waste derived contamination, phytoremediation can be classified into five primary strategies: (a) phytoextraction, where hyperaccumulating species take up metals into harvestable biomass; (b) phytostabilization, in which root systems immobilize contaminants and reduce their bioavailability; (c) phytovolatilization, involving plant mediated transformation of certain elements (e.g., selenium, Hg) into volatile forms; (d) rhizofiltration, where plant roots sorb or precipitate contaminants from aqueous media; and (e) phytodegradation, in which root exudates stimulate microbial communities that biodegrade organic pollutants such as PBDEs and PAHs [6–8]. In practice, these processes often act synergistically; for example, *Brassica juncea* and related species can accumulate Pb in above-ground tissues while simultaneously fostering microbial consortia capable of degrading PBDEs in the rhizosphere, thereby achieving multicontaminant control [3, 6–8].

Economic assessments indicate that phytoremediation typically costs 10–100 times less than excavation-based approaches, with reported project costs in the range of USD 5,000–15,000 per hectare depending on site conditions, plant species, and remediation duration [6–8]. Because phytoremediation leverages solar energy and natural plant growth, operational energy requirements are comparatively low, and the method can often be implemented using locally available species and agricultural practices [7, 8]. In addition to contaminant reduction, phytoremediation delivers cobenefits such as erosion control, increases in plant and microbial biodiversity (reported gains of around 20% species richness in some projects), and carbon sequestration rates of approximately 0.5–2 ton CO_2 equivalent per hectare per year in established vegetative cover [3, 6–8]. When these ecosystem services and potential resource recovery pathways (e.g., phytomining of metals from harvested biomass) are factored into cost-benefit analyses, net present value (NPV) estimates of around USD 30,000 per hectare over a 5-year horizon have been reported for certain scenarios [3, 5, 8].

1.6 Chapter roadmap and objectives

Given the projected rise to 82 Mt of e-waste by 2030 and the mounting health and ecological burdens associated with current disposal practices, this chapter seeks to provide a data-driven framework for scaling phytoremediation as part of an integrated response to the global e-waste challenge [1, 5, 8]. Section 2 will describe the global e-waste burden and its environmental impacts in greater detail. Section 3 will examine the chemical composition and ecotoxicology of e-waste, with emphasis on metals, REEs, and key organic pollutants. Section 4 will outline the fundamental principles and classification of phytoremediation. Section 5 will explore mechanisms and pathways of metal uptake, translocation, and transformation in plant-microbe systems. Section 6 will discuss criteria for plant selection and the role of microbial synergies. Section 7 will evaluate enhancement strategies, including chelators, biochar, CRISPR-based genetic modifications, and nanomaterials. Section 8 will present field applications and case studies from Scopus-indexed research conducted between 2023 and 2025. Section 9 will address practical challenges, limitations, and regulatory considerations. Section 10 will explore circular economy opportunities, such as phytomining and biomass valorization. Section 11 will highlight future research needs, and Section 12 will conclude with policy recommendations. Collectively, this chapter positions phytoremediation as a scalable, sustainable solution for remediating e-waste contaminated environments while contributing to broader planetary health objectives [5–8].

2 Global e-waste burden and environmental impacts

The proliferation of electronic devices in the twenty-first century has catalyzed an unprecedented environmental crisis, with e-waste now recognized as one of the fastest-growing waste streams worldwide [9]. The Global E-waste Monitor 2024 documented a record 62 million Mt of e-waste generated in 2022, representing an 82% increase since 2010 and an average of 7.8 kg per capita globally [9]. Projections indicate that annual generation could reach approximately 82 Mt by 2030 – a further 32% increase that is expected to outpace current formal recycling capacity by roughly a factor of 5 [9]. This widening gap between e-waste generation and environmentally sound management has created a planetary health emergency, particularly in LMICs that receive large volumes of transboundary waste from high-income nations [10, 11]. This section reviews global generation trends, environmental fate and transport of key toxicants, associated health and ecological risks, and illustrative hotspot case studies, thereby establishing the rationale for sustainable remediation options such as phytoremediation.

2.1 Generation trends and regional disparities

E-waste generation is increasing rapidly due to technological diffusion, shortened product lifespans, and rising consumption [9–11]. Between 2010 and 2022, the compound annual growth rate of e-waste was approximately 3.6%, with an average increase of 2.6 Mt per year, a trajectory projected to continue in the coming decade [9]. Key drivers include rapid expansion of the Internet of Things, deployment of 5G infrastructure, proliferation of artificial intelligence hardware, and shrinking replacement cycles for consumer electronics: for example, the typical lifespan of smartphones and laptops has fallen from 4–5 years to around 2–3 years in many markets [10]. The COVID-19 pandemic further accelerated this trend; remote work, online education, and telehealth collectively increased device purchases by an estimated 15% between 2020 and 2022, even as lockdowns disrupted collection networks and formal recycling systems [10, 11].

Marked regional disparities in both e-waste generation and management capacity highlight underlying environmental injustice [9–11]. High-income regions such as the

European Union, the United States, and Japan generate roughly 15–20 kg of e-waste per capita annually, yet formally recycle only about 35–40% within their own borders, with a substantial fraction of the remainder exported to LMICs under the guise of “second-hand goods” or “donations” [9, 10]. Estimates suggest that up to 80% of e-waste originating in some developed economies ultimately flows to informal recycling hubs in Asia and Africa, circumventing or exploiting gaps in Basel Convention enforcement [9, 10]. By contrast, many LMICs generate only 2–5 kg per capita but are forced to process three to five times their domestic generation volume through largely informal, unregulated recycling, often without adequate infrastructure for safe collection, treatment, or disposal [10, 11].

Asia bears an especially heavy e-waste burden [9–11]. In 2022, China generated approximately 10.1 Mt of e-waste, while India produced around 3.2 Mt, with projections suggesting that India could reach 5 Mt by 2030 if current trends continue [10]. Informal recycling sectors in these and other Asian countries collectively employ tens of millions of workers – estimates exceed 18 million people, including nearly 5 million children – who are frequently exposed to hazardous conditions without protective equipment [10, 11]. Formal recycling rates remain low; across Asia, the average documented recycling rate is about 11.9%, and in South Asia, it is closer to 5% [9]. Africa receives an estimated 40% of global e-waste exports, with West African hubs such as Agbogbloshie in Ghana processing roughly 215,000 tons/year and employing about 40,000 informal workers [10]. This pattern of “toxic trade” externalizes environmental and health costs to populations with the least capacity to manage the risks, amplifying intergenerational harm [10, 11].

2.2 Environmental fate and transport of e-waste toxicants

The environmental behavior of e-waste-derived contaminants is governed by complex physicochemical processes that determine their mobility, persistence, and bioavailability [3, 5, 9, 12]. Heavy metals such as Pb, Cd, and Cu exhibit variable mobility depending on soil pH, redox potential, and organic matter content. Pb and Cd often sorb strongly to clay minerals and iron/manganese oxides (distribution coefficients, K_d, >1,000 L/kg), but their mobility increases sharply under acidic conditions (pH < 6) due to proton competition and complexation with low molecular weight organic acids [3, 12]. Cu readily forms stable complexes with humic and fulvic substances, enhancing its transport in dissolved and colloidal forms through preferential flow pathways in structured soils [5, 12]. Hg released from e-waste can volatilize as elemental Hg^0 or undergo microbial methylation to methylmercury (CH_3Hg^+), a neurotoxic species that bioaccumulates and biomagnifies in aquatic food webs [10, 12]. REEs in acidic leachates (pH < 5) may exist as mobile oxyanionic complexes with relatively long environmental half-lives, contributing to persistent contamination in groundwater and sediments [5, 12].

Persistent organic pollutants (POPs) associated with e-waste, including PBDEs and PCBs, display even greater recalcitrance [6, 9, 12]. PBDEs, with log K_ow values between 6 and 8, preferentially partition to organic matter and fine particles, leading to high sediment-associated burdens; it is estimated that up to 80% of PBDEs released from burning or dismantling operations become buried within the upper 10 cm of sediments in receiving waterbodies [6]. PCBs and PBDEs exhibit BCF exceeding 10^5 in fish muscle, enabling marked bioaccumulation [6, 12]. Atmospheric emissions from open burning and low-temperature incineration transport $PM_{2.5}$-bound metals (e.g., Pb, Cd, Cu) and semivolatile organics over hundreds of kilometers, creating diffuse regional contamination plumes [10–12]. Studies from Guiyu, China, indicate that atmospheric deposition contributes roughly 30% of soil Pb within a 5-km radius of major burning sites, while riverine transport carries an estimated 70% of dissolved contaminant loads downstream toward estuarine and coastal zones [10, 11]. Groundwater advection rates on the order of 0.1–1 m/day facilitate long-range subsurface transport, so that contaminant plumes from e-waste dumpsites may extend 3–5 km over a decade if left unmanaged [3, 12].

The persistence of e-waste toxicants underpins the concept of "legacy contamination" [3, 6, 12]. Metals such as Pb often have effective soil half-lives spanning 100–1,000 years, while PBDEs degrade only slowly via photolysis and microbial debromination, with estimated half-lives of 10–20 years in surface soils and sediments [6, 12]. At some legacy industrial and e-waste impacted sites in Europe, PCB concentrations in groundwater remain on the order of 10 µg/L decades after primary sources were removed, illustrating the limitations of natural attenuation alone [3, 12]. Cumulative contaminant loading in major hotspots can exceed 1 kg/m^2/year for metals and 100 g/m^2/year for organic pollutants, overwhelming the assimilative capacity of local ecosystems [10–12].

2.3 Health and ecological risks

Human health impacts associated with e-waste exposure are severe, multisystemic, and concentrated among the most vulnerable [1, 2–4, 11,13]. Pb exposure from contaminated soil, dust, food, and water causes neurodevelopmental deficits, with meta-analyses suggesting IQ reductions of approximately 4–7 points per 10 µg/dL increase in blood Pb levels in children, alongside increased risk of renal tubular dysfunction and hypertension in adults [2, 13]. In Agbogbloshie, mean blood Pb concentrations in children have been reported around 74 µg/dL – roughly 10 times the 5 µg/dL reference value commonly cited by WHO – while adult workers exhibit high prevalence of respiratory symptoms and a threefold increase in spontaneous miscarriage rates compared to nonexposed populations [2, 10]. Cd accumulates in renal cortex tissue, leading to proteinuria and measurable declines in glomerular filtration rate at relatively low soil concentrations (~0.5 mg/kg), whereas chronic Hg exposure via methylmercury bioaccumulation can precipitate Minamata-like neurological syndromes at hair Hg levels above 50 μg/g [10, 13].

PBDEs disrupt thyroid hormone homeostasis, with observational studies in e-waste-affected communities documenting 30–50% reductions in serum thyroxine (T4) and associated growth and developmental delays in children [3, 6, 13]. In areas near Guiyu, reported serum PBDE concentrations of 200–500 ng/g lipid – up to 100 times higher than in reference populations – have been linked to IQ deficits of 10–15 points and attention deficits in school-aged children [3, 6]. Cr(VI) is a recognized human carcinogen, with oral slope factors around 0.5 $(mg/kg/day)^{1}$ and elevated odds ratios for lung and other cancers in chronically exposed workers [10, 13]. Overall, the global economic burden of e-waste-related health impacts is estimated at around USD 15 billion annually, with approximately 70% of costs borne by LMICs despite these countries generating a minority of the waste [10, 11, 13].

Ecological impacts are equally concerning [1, 4–6, 14]. PBDE concentrations in fish collected from rivers and estuaries near e-waste recycling centers often exceed European Union environmental quality standards by up to two orders of magnitude and have been associated with 50% reductions in benthic invertebrate diversity and endocrine disruption phenomena such as intersex conditions in fish [6, 14]. Elevated REE concentrations (e.g., Nd ~ 10 mg/kg in sediments) have been shown to be teratogenic to amphibians, with embryo mortality rates of ~30% in laboratory exposures, while chronic metal contamination reduces soil microbial diversity by up to 70%, impairing key processes such as nitrogen fixation and organic matter decomposition [3, 5, 12]. Bioaccumulation factors for Pb in fish muscle can reach 100–500 times ambient water concentrations, and PBDEs may biomagnify by factors of 10^4 or more in apex predators, resulting in food web scale contamination [6, 12, 14]. These cascading effects reduce ecosystem productivity, compromise fisheries and agriculture, and undermine long term resilience.

2.4 Case studies: pollution hotspots

Several well-documented pollution hotspots illustrate the magnitude and complexity of e-waste contamination [2, 3, 10–12, 14].

Guiyu, China, has been described as one of the world's largest e-waste processing hubs, historically handling on the order of 100,000 tons per year through informal dismantling, open burning, and acid leaching [10, 11]. Soil Pb concentrations in some areas exceed 2,300 mg/kg – more than 20 times local background – while Cd levels surpass 80 mg/kg, and river sediments downstream contain PBDE concentrations of over 20,000 ng/g, among the highest reported globally [10, 11, 14]. Groundwater Cr(VI) levels around 50 μg/L pose chronic exposure risks for an estimated 150,000 residents, and rice grains cultivated near processing areas exceed national Pb food safety limits multiple times [10, 11]. Cancer incidence in surrounding communities has been reported at up to three times the national averages, illustrating the "cocktail effect" of multicontaminant exposure [3, 12].

Agbogbloshie, Ghana, is a ~40-hectare informal e-waste recycling site that receives a large share of West Africa's imported electronics, processing approximately 215,000 tons per year [2, 10]. Soil Pb concentrations frequently exceed 4,000 mg/kg, while ash samples contain PBDE levels of around 9,000 ng/g [2, 10, 14]. Ambient $PM_{2.5}$ concentrations during burning activities can reach 200–500 µg/m^3 – more than 20 times the WHO guidelines – exposing roughly 40,000 workers, including about 10,000 children, to intense inhalation hazards [2, 10]. Blood Pb levels in resident children average ~74 µg/dL, and school dropout rates exceed 40%, reflecting compounded health and social stressors [2, 10, 13].

Bitterfeld, Germany, a legacy electronics and chemical manufacturing region, demonstrates that long-term contamination is not limited to LMICs [3, 12]. Soils contain elevated Cu concentrations (~1,500 mg/kg), and groundwater PCB levels around 10 µg/L persist more than 30 years after primary manufacturing ceased, complicating brownfield redevelopment and depressing property values by up to 50% [3, 12].

Moradabad, India, represents a 10-hectare cluster where traditional brassware production coexists with intensive informal e-waste recycling, processing roughly 50,000 tons per year [3, 11]. Soil Pb levels around 1,200 mg/kg and Cd levels near 45 mg/kg have been reported, while rudimentary smelting operations emit dioxins at concentrations around 0.5 ng TEQ/m^3, affecting an urban population of over 100,000 [3, 11, 14].

Collectively, these hotspots demonstrate that e-waste contamination is a global crisis characterized by transboundary flows, multicontaminant mixtures, and long-lasting environmental and health impacts [2, 3, 9–14]. They underscore the urgent need for scalable, sustainable remediation strategies – among which phytoremediation stands out as a promising, cost-effective, and ecosystem restorative approach.

3 Chemical composition and ecotoxicology of e-waste

The chemical complexity of e-waste poses a formidable challenge for environmental management, as discarded electronics contain heterogeneous mixtures of metallic, metalloid, and organic contaminants with diverse physicochemical properties, toxicity pathways, and environmental fates [3, 5, 9, 15]. Unlike many traditional industrial waste streams, e-waste simultaneously contains high-value materials (e.g., precious metals and REEs) and persistent toxicants at concentrations often 50–100 times higher than natural ore grades, creating unique risk-benefit trade-offs for resource recovery and remediation [5, 9, 15]. This section synthesizes current evidence on the principal chemical constituents of e-waste, their ecotoxicological mechanisms, and the cumulative risks associated with multicontaminant exposure, drawing particularly on recent Scopus-indexed literature published between 2023 and 2025 [5, 6, 12, 15, 16].

3.1 Metallic and metalloid contaminants

Metals and metalloids typically account for 40–50% of total e-waste mass, transforming disposal and recycling sites into anthropogenic ore bodies while simultaneously creating acute and chronic toxicity hazards for humans and ecosystems [3, 5, 9, 15]. The metallic fraction includes base metals (e.g., Fe, Al, Cu), heavy metals (Pb, Cd, Hg, Cr), metalloids such As and Sb, and technologically critical elements including REEs and precious metals [5, 9, 15]. Each element exhibits distinct speciation, mobility, and toxicity characteristics that shape its environmental behavior and biological effects [3, 12, 15].

Pb is often regarded as the dominant contributor to e-waste-related metal toxicity, originating from solder alloys, cathode ray tube glass, and certain PVC stabilizers [3, 10, 15]. Mixed e-waste streams may contain on the order of 5–15 kg Pb/ton, while soils at major informal recycling hubs such as Guiyu (China) and Agbogbloshie (Ghana) have been reported with Pb concentrations exceeding 2,000–4,000 mg/kg – more than an order of magnitude above typical background levels [2, 10, 11, 15]. In soils, Pb occurs in multiple forms, including carbonates, sulfates, and organically complexed species; its solubility and mobility increase markedly under acidic conditions ($pH < 6.5$), enhancing plant uptake and leaching potential [3, 12, 15]. BCF for Pb in plants are generally modest (≈0.1–15), but higher values (≈50–100) have been reported for soil invertebrates such as earthworms, facilitating trophic transfer [3, 15]. Toxicologically, Pb disrupts calcium-dependent signaling and interferes with *N*-methyl-D-aspartate receptor function, leading to neurodevelopmental deficits; epidemiological studies in e-waste-affected communities have documented strong correlations between elevated blood Pb (15–25 µg/dL) and 4–7 point IQ reductions per 10 µg/dL increment in children [2, 10, 13, 15]. Risk quotients (= C_environmental/C_threshold) for Pb frequently exceed 1 in hotspot soils, indicating substantial ecological and human health concern [2, 10, 15].

Cd is concentrated in nickel-cadmium batteries, certain phosphors, and electroplated components, and typically occurs at 0.1–1 kg/ton of mixed e-waste [3, 5, 15]. In soils, Cd^{2+} forms complexes with chloride and sulfate, with distribution coefficients (K_d) around 100–500 L/kg in clay-rich matrices; bioavailability increases sharply under acidic conditions ($pH < 5.5$) [3, 12, 15]. BCFs of 5–30 have been reported for plants and 20–80 for earthworms in contaminated sites, indicating efficient transfer from soil to biota [3, 12, 15]. Cd preferentially accumulates in renal proximal tubules; kidney cortex concentrations above ~50 µg/g wet weight are associated with proteinuria, decreased glomerular filtration rate, and increased risks of osteoporosis and certain cancers [2, 13, 15]. Dietary intake via contaminated rice and vegetables near e-waste sites can account for a substantial share of total Cd exposure in local populations [2, 10, 13, 15].

Hg in e-waste arises mainly from fluorescent lamps, LCD back lights, switches, and certain sensors, typically at 0.01–0.1 kg/ton [3, 15, 16]. Hg occurs as elemental Hg^0 (volatile), inorganic Hg^{2+}, and methylmercury (CH_3Hg^+), the latter formed via microbial methylation in aquatic and wetland environments [10, 12, 16]. Open burning and uncontrolled heating of e-waste can release Hg^0 to the atmosphere, where it may be transported regionally before deposition; subsequent methylation produces CH_3Hg^+, which bioaccumulates and biomagnifies through aquatic food webs, with biomagnification factors up to 10^6 reported in predatory fish [10, 12, 16]. Hg toxicity targets the nervous system, with clinical effects (e.g., tremors, sensory disturbances, memory impairment) observed at Hg concentrations above ~50 µg/g in highly exposed populations [10, 12, 16].

Cr is present in electronic coatings, pigments, and corrosion-resistant alloys, often partly as hexavalent Cr(VI), a highly mobile and carcinogenic oxyanion (CrO_4^{2-}) [3, 12, 16]. Cr(VI) has low sorption affinity (K_d often <10 L/kg), enabling migration in groundwater at tens to hundreds of µg/L for extended periods [3, 12, 16]. It can penetrate cells via sulfate transporters, induce oxidative DNA damage, and is classified as a Group 1 human carcinogen; oral slope factors on the order of 0.5 $(mg/kg/day)^{-1}$ and elevated lung cancer risks have been reported for chronically exposed workers [12, 16].

Cu is abundant in wiring, printed circuit boards, and motors, with hotspot soil concentrations commonly exceeding 1,000–1,500 mg/kg near informal smelting operations [3, 5, 15]. Although Cu is an essential micronutrient, elevated levels can induce oxidative stress via Fenton-like reactions, impair photosynthesis in plants, and inhibit key soil enzymes [3, 5, 15]. At the same time, Cu plays a role in metallothionein (MT) and phytochelatin (PC) systems that underpin plant metal tolerance and can influence phytoextraction potential [7, 8, 15].

Other trace elements – including As from gallium arsenide components, Sb from flame retardant formulations, and Ba from "getter" materials – occur at lower concentrations but can contribute to overall mixture toxicity, particularly where exposures are chronic and multipathway [3, 15, 16]. Cumulative noncancer hazard indices (sum of hazard quotients for multiple metals) frequently exceed unity in e-waste hotspots, and cancer risk estimates for Cr(VI) and As often surpass common regulatory benchmarks of 1×10^{-4}, indicating the need for active intervention [2, 12, 15, 16].

3.2 Persistent organic pollutants (POPs)

POPs, including brominated flame retardants, PCBs, and combustion byproducts, constitute roughly 15–25% of the organic fraction of e-waste and are characterized by long environmental half-lives, high bioaccumulation potential, and significant endocrine, immunological, and carcinogenic effects [4, 6, 9, 12, 16].

PBDEs are the dominant flame retardants historically used in plastic housings, printed circuit boards, and cable insulation, with commercial penta, octa, and deca

BDE mixtures encompassing 209 congeners [6, 16]. PBDE contents of 10–12% by weight in some plastic components correspond to 10–120 kg of PBDEs per ton of e-waste, and soil and sediment concentrations in informal recycling areas have been reported in the range of 9,000–21,000 ng/g [2, 6, 14, 16]. PBDEs have log $K_{o}w$ values of 6–10, BCFs of 10^4–10^6, and environmental half-lives on the order of 10–20 years in soils, rendering them highly persistent and prone to bioaccumulation [6, 12, 16]. They are resistant to biodegradation, with only slow debromination observed under aerobic conditions.

Toxicologically, PBDEs disrupt thyroid hormone regulation by competing with thyroxine (T4) for transport proteins, altering deiodinase activity and triggering compensatory feedback on the hypothalamic-pituitary-thyroid axis [4, 6, 13, 16]. Human studies in e-waste-affected communities have linked elevated serum PBDE levels (≈100–300 ng/g lipid) with 30–50% reductions in T4 and associated neurodevelopmental deficits, including IQ declines of 10–15 points and attention problems in children [3, 6, 13, 16]. Animal models indicate additional effects such as hepatic enzyme induction (CYP1A1, CYP2B), immunotoxicity, and tumor formation at higher doses [6, 16].

PCBs, formerly used in capacitors, transformers, and as plastic additives, can be present at 0.5–5 kg/ton of older e-waste streams and remain a key contaminant in legacy sites [3, 12, 16]. With 209 congeners, including dioxin-like PCBs (dl PCBs) that possess TEFs up to 0.1, these compounds exhibit half-lives of decades and BCFs of 10^5–10^6 in aquatic biota [12, 14, 16]. Chronic PCB exposure is associated with endocrine disruption, immunosuppression, neurobehavioral effects, and increased cancer risk; PCBs are classified as Group 1 carcinogens by IARC [12, 16].

PCDD/Fs (dioxins and furans) are generated unintentionally during low temperature combustion of chlorinated plastics and flame-retardant-containing materials, particularly under oxygen-poor conditions typical of open burning at informal sites [2, 10, 16]. TEQ concentrations in soils and biota near such sites can exceed WHO and EU guideline values by an order of magnitude or more, with 2,3,7,8-TCDD (TEF = 1) dominating the toxicity profile [2, 10, 16]. Dioxins act primarily through binding to the aryl hydrocarbon receptor (AhR), leading to dysregulation of gene expression, immunotoxicity, teratogenesis, and carcinogenesis [2, 12, 16].

Phthalates and PAHs, originating from PVC casings, cable insulation, and combustion residues, further compound the toxic load [3, 16]. Di(2-ethylhexyl) phthalate and related plasticizers exhibit endocrine-disrupting properties, while high molecular weight PAHs such as benzo[a]pyrene are potent genotoxic carcinogens that form DNA adducts and drive lung and skin cancer risk [3, 16]. Mixture studies suggest synergistic interactions, with coexposure to PBDEs and PCBs enhancing oxidative stress and modulating steroid hormone pathways beyond additive expectations [6, 16].

Risk assessments in e-waste hotspots often show hazard quotients >1 for PBDEs and PCBs in soils and sediments, while incremental lifetime cancer risk estimates for dioxins and certain PAHs exceed 1×10^{-4}, underscoring the need for active remediation rather than reliance on natural attenuation processes that would require many decades [2, 4, 6, 12, 16].

3.3 Rare earth elements and precious metals

REEs and precious metals embody the economic paradox of e-waste: they are highly valuable and critical to modern technologies, yet their recovery from complex waste matrices remains technically challenging, leading to low recycling rates and ongoing primary mining pressure [5, 9, 15, 17].

REEs such as Nd, Dy, Eu, Y, In, and Ga occur in magnets, phosphors, and display technologies at concentrations of roughly 50–200 g/ton of e-waste – often 50–100 times greater than typical ore grades [5, 15, 17]. Nd, for example, is a key component of Nd-Fe-B permanent magnets; while relatively immobile at neutral pH, it can be mobilized in acidic leachates and has been shown to accumulate in hepatic tissue and induce developmental abnormalities in aquatic and amphibian models at elevated exposure levels [5, 15, 17]. In, used in indium tin oxide coatings for LCDs and touch screens, has been linked to pulmonary fibrosis in workers exposed to inhalable particles and exhibits embryotoxicity in fish and invertebrate assays [15, 17].

Precious metals, particularly Au, Ag, and Pd, are present at lower absolute concentrations but have extremely high economic value. Typical content estimates include ~0.25 kg Au, 3.6 kg Ag, and 0.1 kg Pd/ton of e-waste, representing tens of billions of dollars in potential value globally each year [5, 9, 17, 18]. However, low solubility at near neutral pH (e.g., exceptionally low solubility products for Au and Ag compounds) limits their bioavailability and direct phyto uptake, often necessitating pretreatment or chelation [7, 17, 18].

Emerging research on phytomining demonstrates that certain plant species, particularly within the Brassicaceae, can accumulate economically significant quantities of Au and Ag under controlled conditions using complexing agents such as thiocyanate or cyanide analogs [7, 17, 18]. Experimental field and mesocosm studies suggest that BCFs for Au in *B. juncea* can reach values above 10 under optimized conditions, with potential recovery of several kilograms of Au per hectare, offering a promising circular economy pathway that could partially offset remediation costs [7, 17, 18].

In contrast, phytoextraction of REEs remains at an early stage; most plant species exhibit low BCFs (< 1) for Nd, Dy, and related elements under typical field conditions, although chelator amendments, microbial consortia, and genetic engineering of metal transporter pathways (e.g., ZIP and NRAMP families) may enhance uptake in future applications [15, 17, 19, 20].

From a risk standpoint, REEs generally present lower acute toxicity than Pb, Cd, Hg, or Cr(VI), and noncancer [15, 17, 19, 20] hazard quotients for REEs in soils often fall below 1 [5, 15, 19]. However, long term accumulation in biomass and food chains, particularly if contaminated biomass is used for fodder or bioenergy, warrants careful evaluation [15, 19]. The circular economy potential is substantial: modeling suggests that recovering even a small fraction of Au and Ag from global e-waste via phytomining and related technologies could meaningfully reduce reliance on high-impact primary mining while simultaneously contributing to site remediation [5, 17, 18].

4 Principles of phytoremediation

Phytoremediation is a suite of plant-based technologies that employ living plants and their associated rhizosphere microbiota to contain, remove, transform, or detoxify contaminants in soils, waters, and sediments [7, 8, 21]. It is particularly attractive for e-waste-contaminated sites because it can treat, diffuse, large area pollution at comparatively low cost, with minimal physical disturbance and important cobenefits for ecosystem structure, function, and landscape aesthetics [7, 8, 21].

4.1 Definition and classification

From a regulatory and engineering perspective, phytoremediation refers to deliberately designed systems in which selected plant species are established, managed, and, where appropriate, harvested to achieve specific remediation endpoints such as compliance with soil quality guidelines or reduction of leachate concentrations below discharge standards [7, 22]. These systems differ from simple "green cover" because plant selection, planting density, harvest frequency, and the use of soil amendments or hydraulic controls are tailored to contaminant chemistry, plant physiology, and site hydrology [7, 22].

For e-waste sites, five core phytoremediation strategies are especially relevant and often operate simultaneously within a single planting [7, 8, 21, 22]:

- **Phytoextraction:** Hyperaccumulator or high biomass species take up metals and metalloids from the root zone, translocate them to shoots, and permit removal through periodic harvesting. In e-waste-impacted soils, species such as *B. juncea* have been shown to accumulate Pb, Cd, and Cu, and under certain conditions, even trace amounts of Au and Ag, at concentrations far exceeding those of typical agronomic crops, enabling both risk reduction and potential phytomining [7, 8, 23].
- **Phytostabilization:** Deep and fibrous-rooted plants immobilize contaminants in situ via adsorption to roots, precipitation as less soluble mineral phases, and enhancement of soil organic matter, thereby reducing leaching, dust emissions, and erosion without necessarily removing contaminant mass from the site [7, 22]. Grasses such as vetiver (*Chrysopogon zizanioides*) are widely recommended around dumps and informal recycling yards to trap Pb and other metals within the rhizosphere and limit their migration into groundwater or adjacent agricultural land [22, 23].
- **Phytovolatilization:** Certain plant species and engineered lines can enzymatically transform inorganic forms of elements such as Hg, Se, and As into volatile species released through stomata, effectively transferring contamination from soil or water to the atmosphere [21, 24]. At e-waste sites, phytovolatilization is mainly considered for Hg originating from lamps and switches, using plants ex-

pressing bacterial *merA*/*merB* genes; however, concerns about atmospheric redistribution and long-range transport mean that this approach is applied only under stringent risk assessment frameworks [21, 24].

- **Rhizofiltration:** Aquatic and semiaquatic plants cultivated hydroponically or in constructed wetlands sorb and absorb dissolved metals and suspended particulates from contaminated waters, including leachates and drainage from e-waste yards [7, 8, 22]. Species such as *Eichhornia crassipes* (water hyacinth), *Typha* spp., and duckweeds (*Lemna* spp.) have demonstrated high removal efficiencies for Pb, Cu, and Cd when hydraulic loading rates, residence times, and harvesting regimes are appropriately designed [7, 22].
- **Rhizodegradation (phytodegradation):** Root exudates (e.g., sugars, organic acids, phenolics) stimulate microbial communities that cometabolize or directly degrade organic contaminants such as PBDEs, PCBs, and PAHs, while plant enzymes contribute to oxidation, reduction, and conjugation reactions within root and shoot tissues [6, 23, 24]. Fast-growing trees (e.g., *Populus* spp.) and perennial grasses have been shown to enhance dissipation of PCB mixtures and brominated flame retardants relative to unplanted controls, primarily through enrichment of dehalogenating bacteria and fungi in the rhizosphere [6, 23, 24].

In practice, these mechanisms overlap: for example, *B. juncea* established on e-waste contaminated soils may simultaneously phytoextract Pb, partially stabilize residual metals in root-associated phases, and support rhizodegradation of co-occurring flame retardants via enriched dehalogenating microbiota [7, 23]. Mixed species plantings and rotational designs are increasingly advocated to integrate deep stabilization, surface layer extraction, and attenuation of organic pollutants within a single, multifunctional phytotechnology system [21–23].

4.2 Solar-powered ecosystem services

A defining feature of phytoremediation is that contaminant capture and transformation are powered primarily by photosynthesis rather than external energy inputs [7, 8, 22]. Compared with excavation, thermal treatment, or aggressive chemical leaching – methods that demand substantial fossil energy, heavy machinery, and often generate secondary waste streams – phytoremediation offers a low energy, low carbon alternative, particularly well suited to large e-waste landscapes and resource-constrained settings [7, 22, 24].

Technoeconomic assessments indicate that well-designed phytoremediation systems for metal-contaminated soils can operate at 10–100 times lower direct cost than excavation and offsite disposal, especially when applied over multihectare areas [7, 8, 22]. Operating expenditures are dominated by site preparation, planting, limited irrigation or fertilization, and periodic harvesting; there is no requirement for intensive soil handling or elevated temperature processing [7, 22]. When potential revenues from phytomining (e.g., Au/Ag recovery) or biomass valorization (bioenergy, biochar, and fiber) are incorporated, NPV analyses over 5–10 years frequently become positive, with estimated benefits of tens of thousands of dollars per hectare at metal-rich sites [7, 17, 18, 23].

Beyond direct contaminant mass reduction, phytoremediation delivers important ecosystem services that conventional remediation approaches may damage or eliminate [8, 21, 23]. Establishing vegetative cover on e-waste dumps markedly reduces wind and water erosion, thereby lowering particulate emissions and runoff-mediated transport of contaminants, while also stabilizing embankments and dump slopes [21, 23]. Root systems enhance soil structure, porosity, and aggregate stability, and litter inputs gradually rebuild soil organic matter, improving cation exchange capacity and further immobilizing metals in less bioavailable pools [22, 23].

Biodiversity enhancement is another key cobenefit. Simple grass-tree mixtures on degraded e-waste land can increase local plant and invertebrate species richness by approximately 20% compared with bare or sparsely vegetated controls, providing habitat for pollinators and natural enemies of crop pests in surrounding agricultural mosaics [21, 23]. At landscape scales, vegetated buffer strips and corridors established around e-waste clusters can function as ecological stepping stones, reconnecting fragmented habitat while intercepting contaminated dust and surface runoff [21, 24].

Phytoremediation also contributes to climate regulation. Typical biomass productivities of 5–15 ton/ha/year for grasses and short rotation trees correspond to carbon sequestration rates on the order of 0.5–2 ton CO_2 equivalent per hectare per year, depending on species, management, and site conditions [8, 21, 24]. When contaminated biomass is directed to controlled combustion with energy recovery and appropriate flue gas cleaning, or to pyrolysis for biochar production with metals sequestered in stable mineral phases, net greenhouse gas balances can compare very favorably with excavation or landfilling scenarios [8, 21, 24]. These combined services – erosion control, biodiversity support, carbon sequestration, and potential resource recovery – offset a substantial share of apparent remediation costs and support framing phytoremediation installations around e-waste sites as multifunctional green infrastructure rather than purely technical cleanup projects [8,17,18, 21-23].

5 Mechanisms and pathways of metal uptake and transformation

The efficacy of phytoremediation for e-waste contaminants depends on a suite of physiological, biochemical, and microbial processes that operate in concert within the rhizosphere and plant tissues [7, 8, 21, 25]. In contrast to conventional physicochemical methods that rely on external energy and chemical reagents, phytoremediation exploits the intrinsic capacity of plants to acquire nutrients, exude metabolites, and host diverse microbial symbionts, thereby driving contaminant mobilization, sequestration, and detoxification through solar-powered pathways [7, 8, 21, 25, 26]. This section examines four core mechanisms – phytoextraction, phytostabilization, phytovolatilization/rhizofiltration, and rhizodegradation – emphasizing molecular underpinnings, kinetic constraints, and quantitative performance metrics reported in recent field and laboratory studies [25–30].

5.1 Phytoextraction and hyperaccumulation

Phytoextraction is the most direct route for removing metallic contaminants from e-waste-impacted soils: plants absorb dissolved metal ions at the root surface, translocate them via the xylem, and sequester them in harvestable shoot biomass [7, 8, 20, 21, 25]. The process is governed by three sequential bottlenecks: (a) rhizosphere bioavailability, (b) root uptake and xylem loading, and (c) shoot detoxification and storage [25–27]. Hyperaccumulator species overcome these bottlenecks through specialized traits that are increasingly being elucidated at transcriptomic and proteomic levels [26, 27].

Rhizosphere acidification and metal mobilization: In neutral to alkaline soils typical of many e-waste dumps (pH 6.5–8.0), Pb, Cd, and Cu are often present in sparingly soluble forms such as carbonates, phosphates, or sorbed phases on Fe/Mn oxides [3, 5, 15, 25]. Hyperaccumulating plants can acidify the rhizosphere by extruding H^+ via plasma membrane H^+ ATPases, lowering pH by approximately 0.5–1.5 units and thereby increasing metal solubility and desorption [25–27]. Organic acid exudation (e.g., citrate, malate, oxalate) further mobilizes metals through complexation, forming soluble metal-ligand species that diffuse toward the root surface [25–27]. Studies with *B. juncea* grown on metal-rich soils show substantial rhizosphere pH decreases accompanied by multifold increases in soil solution metal concentrations and corresponding gains in plant uptake [25, 26].

Root uptake and membrane transport: Mobilized metals are taken up across root plasma membranes by specific and nonspecific transport systems. Members of the ZIP (ZRT/IRT-like Protein) family mediate Cd^{2+} and Zn^{2+} influx, while Ca^{2+} channels and nonselective cation channels contribute to Pb^{2+} entry [26–28]. Once in the symplast, P-type ATPases such as HMA4, localized in pericycle and xylem parenchyma cells, play key roles in loading metals into the xylem stream, underpinning high shoot translocation factors (TF = C_shoot/C_root) observed in hyperaccumulators like *Arabidopsis halleri* and *Thlaspi caerulescens* [26–28]. Experimental overexpression of HMA4 and related transporters in model and crop species has been shown to enhance Cd and Zn translocation several-fold, illustrating the potential of molecular breeding and genetic engineering to improve phytoextraction performance [26–28].

Intracellular detoxification and sequestration: Within root and shoot cells, metals are detoxified primarily via chelation and compartmentation. Glutathione-derived PCs and cysteine-rich MTs bind Cd, Pb, and other metals, forming complexes that are transported into vacuoles by ABC-type transporters [27–29]. This vacuolar sequestration enables tolerant plants to accumulate leaf metal concentrations orders of magnitude higher than those tolerated by nonaccumulators, without severe physiological damage [27–29]. Quantitatively, phytoextraction potential is often expressed through BCF (BCF = C_plant/C_soil) and TF values; hyperaccumulators are typically defined by BCF > 1 and TF > 1, with shoot metal concentrations exceeding 100 µg/g dry weight (DW) for Cd and 1,000 µg/g DW for Pb [26–29]. Field trials on contaminated soils have reported that successive cropping cycles of high biomass accumulators can remove 10–20% of the plant available metal pool per year, though total depletion to regulatory thresholds often requires multiyear programs [7, 8, 25, 29].

5.2 Phytostabilization and immobilization

Phytostabilization is often preferred at e-waste sites where complete contaminant removal is impractical or uneconomical, or where immediate control of erosion and leaching is the primary objective [7, 8, 21, 25]. Instead of extracting metals, this strategy uses plant cover to reduce contaminant mobility and bioavailability through physical, chemical, and biological mechanisms in the rhizosphere [21, 22, 25, 29].

Root adsorption and physical entrapment: Dense fibrous root systems increase soil cohesion and physically trapped particulate bound metals, while root surfaces and cell wall components (e.g., pectins, lignin, and hemicelluloses) provide abundant functional groups (carboxyl, hydroxyl, phenolic) for cation binding [21, 22, 25]. Species such as vetiver grass (*C. zizanioides*) form extensive root networks that enhance slope stability and substantially reduce dust and runoff from tailings and dumps, thereby lowering off-site contaminant fluxes [21, 22, 25, 29].

Precipitation and sorption to new mineral phases: Root-associated microbial communities, including phosphate-solubilizing and sulfate-reducing bacteria, can promote the formation of low solubility metal phosphates and sulfides, effectively locking Pb, Cd, and other metals into mineral forms such as pyromorphite and CdS [25, 29]. Rhizosphere-driven increases in soil organic matter and cation exchange capacity further enhance sorption, shifting metals into less bioavailable pools [21, 22, 25].

Redox modifications: Root oxygen release and exudate-driven microbial activity can alter redox conditions in the rhizosphere, influencing speciation and mobility of redox-sensitive elements such as Fe, Mn, As, and Cr [22, 25, 26]. For example, the formation of Fe plaques on roots can sequester arsenate and chromate, while localized reducing conditions may favor the conversion of more mobile and toxic Cr(VI) to less mobile Cr(III) [22, 25, 26].

Field studies indicate that well-designed phytostabilization systems can reduce metal leaching by 70–90% and decrease measures of bioavailable metal fractions by 50–70% within a few years, without the need for biomass removal [21, 22, 25, 29]. Long-term performance, however, depends on maintaining vegetation cover and monitoring soil chemistry to ensure that immobilized fractions remain stable under changing environmental conditions [25, 29, 30].

5.3 Phytovolatilization and rhizofiltration

Phytovolatilization and rhizofiltration address specific contaminant classes – volatile metalloids and dissolved metals, respectively – and can complement phytoextraction and phytostabilization at e-waste sites [7, 8, 21, 24–26].

Phytovolatilization of Hg and metalloids: In phytovolatilization, plants transform certain inorganic species (e.g., Hg^{2+}, SeO_3^{2-}, and AsO_4^{3-}) into volatile forms that are released through stomata [24–26]. Transgenic plants expressing bacterial *merA* and *merB* genes, for example, can reduce Hg^{2+} to Hg^0 and, in some systems, demethylate CH_3Hg^+, thereby lowering soil and aqueous Hg concentrations [24–26]. While such approaches have demonstrated substantial Hg removal in experimental settings, concerns about atmospheric redistribution and long-range transport have limited their application to carefully controlled sites with robust air quality monitoring [24–26]. Similar processes have been described for Se and, to a lesser extent, As, although volatilization fluxes are generally small relative to total site inventories [24–26].

Rhizofiltration of leachate and runoff: Rhizofiltration uses aquatic or semi-aquatic plants grown in water or saturated substrates to remove dissolved metals and particulates from contaminated effluents, such as leachates from e-waste storage yards or drainage canals [7, 8, 22, 25]. Species such as *Typha* spp., *E. crassipes*, and *Lemna* spp. possess extensive root systems with high sorptive capacity, enabling them to remove large fractions of Pb, Cd, Zn, and Cu from water at relatively low operating cost [7, 22, 25, 26]. Constructed wetland systems incorporating these plants can achieve 80–95% reductions in dissolved metals under appropriate hydraulic loading and residence times, with contaminants captured primarily in root biomass and associated sediments [7, 22, 25, 26]. Periodic harvesting of roots and above-ground biomass is required to prevent release and maintain treatment capacity [22, 25].

5.4 Rhizodegradation of organic contaminants

Rhizodegradation (or plant-assisted biodegradation) is central to addressing organic pollutants associated with e-waste, including PBDEs, PCBs, and PAHs [4, 6, 12, 16, 25].

Table 1: Mechanism performance metrics for e-waste contaminants [4, 6, 12, 16, 25].

Mechanism	Target contaminants	Removal efficiency (%/year)	Key plant species	Limiting factors
Phytoextraction	Pb, Cd, Cu, Au, and Ag	10–20 (metals)	*B. juncea* and *P. vittata*	Slow kinetics and mass transfer
Phytostabilization	Pb, Cr, Cu, and REEs	70–90 (leaching reduction)	*V. zizanioides* and *P. vittata*	Metals remain in situ
Phytovolatilization	Hg, As, and Se	1–3 (volatilization)	Transgenic *Arabidopsis*	Atmospheric redeposition
Rhizofiltration	Cd, Cu, Zn, and Pb	80–95 (aqueous)	*T. latifolia* and *E. crassipes*	Root saturation, area
Rhizodegradation	PBDEs, PCBs, and PAHs	20–35 (organics)	*P. virgatum* and *Populus*	Recalcitrant congeners

Root exudates and microbial stimulation: Plants release a substantial fraction of their photosynthate as root exudates – comprising sugars, amino acids, organic acids, and phenolic compounds – which serve as carbon and energy sources for heterotrophic microbes and can act as cosubstrates or inducers for degradative enzymes [2, 25–27]. These exudates typically increase microbial biomass and activity in the rhizosphere relative to bulk soil, stimulating assemblages of bacteria and fungi capable of reductive dehalogenation, cometabolic oxidation, and ring cleavage reactions [2, 25–27].

Degradation pathways for PBDEs, PCBs, and PAHs: Anaerobic reductive dehalogenation in micro anaerobic niches can remove bromine or chlorine atoms from PBDE and PCB molecules, producing less halogenated intermediates that are more amenable to subsequent aerobic oxidation [16, 25–27]. Poplar and willow plantations, as well as deep-rooted grasses, have been shown to enhance dechlorination and debromination rates by enriching dehalogenating genera such as *Dehalococcoides*, *Dehalobacter*, and various *Pseudomonas* species in the rhizosphere [16, 25–27]. Under aerobic conditions, mono- and dihalogenated intermediates can be further metabolized via dioxygenase and monooxygenase pathways, ultimately yielding CO_2 and inorganic halides [16, 25–27]. For PAHs, rhizosphere stimulation of ring hydroxylating dioxygenases in bacteria and ligninolytic enzymes (laccases, peroxidases) in fungi accelerates degradation relative to unplanted controls [16, 25–27].

Field and mesocosm studies on contaminated soils indicate that rhizodegradation can reduce concentrations of certain PBDE congeners and PCB mixtures by 40–70% over 2–3 years, depending on site conditions and plant-microbe combinations [16, 25–27, 29]. However, highly brominated PBDEs (e.g., BDE 209) and dioxin-like PCBs remain recalcitrant, often requiring sequential anaerobic-aerobic treatments or bioaugmentation with specialized consortia for substantial transformation [16, 25, 27, 29].

5.5 Comparative performance and integration of mechanisms

In practice, effective phytoremediation designs for e-waste sites integrate multiple mechanisms within mixed or sequential plantings [7, 8, 21, 23, 25]. High biomass accumulators such as *B. juncea* (Table 1) may be combined with deep-rooted stabilizers like vetiver and riparian buffers incorporating *Typha* or *Phragmites* for leachate polishing, while tree or grass species that promote rhizodegradation address co-occurring organic pollutants [7, 21–23, 25, 29]. Conceptual performance metrics derived from the literature can be summarized as follows [7, 8, 16, 21–23, 25–29]:

- Phytoextraction: Removal of ≈10–20% of the plant available metal pool per year for Pb, Cd, and Cu under favorable conditions, with >80% of accumulated metals in above-ground biomass.
- Phytostabilization: 70–90% reduction in metal leaching and 50–70% decrease in bioavailable fractions within 1–3 years.
- Rhizofiltration: 80–95% reduction of dissolved metals in low to moderate concentration effluents.
- Rhizodegradation: 20–35% annual reduction of selected POPs, yielding 40–70% decline over 2–3 years for more labile congeners.

Such integrated systems allow practitioners to tailor remediation objectives – risk reduction, resource recovery, or site restoration – to local priorities and constraints, while capitalizing on the synergies among mechanisms [7, 8, 21–23, 25–29].

6 Selection of plant species and microbial synergies

The practical success of phytoremediation at e-waste-contaminated sites depends on deliberate selection of plant genotypes and rhizosphere microbiota that can tolerate extreme metal and metalloid loads while maintaining high biomass, rapid growth, and broad-spectrum contaminant uptake [7, 8, 21, 25, 31]. Unlike conventional agriculture – where yield and nutritional quality dominate breeding targets – phytoremediation requires a multitrait optimization framework that balances hyperaccumulation capacity, agronomic robustness, root architecture, and ecological compatibility [7, 8, 23, 25, 31]. Recent field and greenhouse studies have begun to quantify these traits and to clarify how plant-microbe interactions can amplify remediation efficiency [23, 25, 26, 31, 32].

6.1 Hyperaccumulator criteria: a multitrait optimization framework

Optimal phytoremediation species for e-waste sites must satisfy a set of interrelated criteria that determine both technical performance and economic viability [7, 8, 25, 31]. Five core requirements emerge from mass balance calculations, kinetic models, and field validations:

1. **Hyperaccumulation capacity:** Classical thresholds for hyperaccumulation are shoot concentrations >100 μg/g DW for Cd, >1,000 μg/g DW for Pb, and >10,000 μg/g DW for Zn, achieved without severe phytotoxicity [26, 27, 31]. At e-waste sites where soil Pb frequently exceeds 1,000 mg/kg, plants must reach at least 1,000 μg/g DW Pb in shoots to ensure realistic timeframes for remediation; otherwise, the number of cropping cycles needed to lower soil concentrations to guideline values becomes prohibitive [7, 25, 31]. High accumulation depends on coordinated upregulation of high-affinity transporters (ZIP, NRAMP, HMA families), chelation systems (PCs, MTs), and vacuolar sequestration pathways [26, 27, 31, 32].
2. **High biomass production:** Biomass yield directly scales total contaminant mass removal per unit area. A plant accumulating 1,000 μg/g DW Pb but producing only 2 ton/ha/year removes 2 kg Pb/ha/year, whereas a 15 ton/ha/year crop removes 15 kg Pb/ha/year under the same tissue concentration, dramatically improving areal removal rates [7, 8, 25, 31]. For e-waste sites targeted for redevelopment within 3–5 years, species with annual above-ground biomass yields of >10 ton/ha are generally preferred [7, 8, 23, 25].
3. **Fast growth and short rotation:** Rapid growth and short growth cycles permit multiple harvests per year in suitable climates, accelerating cumulative contaminant removal and shortening exposure windows before canopy closure and full root development [7, 8, 23, 25, 31]. In tropical and subtropical regions, species capable of completing a growth cycle in ≤6 months enable two harvests annually, while in temperate regions, fast early growth remains important for soil stabilization and competitive ability [7, 8, 23, 31].
4. **Deep and extensive root systems:** Rooting depth >1 m and high root length density are crucial for accessing contaminants in subsoil horizons and for providing large contact surfaces for uptake and immobilization [21, 22, 25, 31]. Deep-rooted grasses and trees can intercept leached metals and metalloids that have migrated below the plough layer, a common pattern at legacy dumps and leachate plumes [21, 22, 25, 31, 32].
5. **Multicontaminant tolerance.** E-waste matrices contain complex mixtures of Pb, Cd, Cu, Cr, As, REEs, and organic pollutants (e.g., PBDEs, PCBs, and PAHs); thus, species must tolerate and, ideally, remediate more than one contaminant type [3–5, 9, 15, 31]. Tolerance to combined metal and organic stress is underpinned by robust antioxidant systems, efficient chelation, membrane transport regulation, and, in many cases, beneficial microbial associations [3, 25–27, 32].

Additional agronomic and ecological considerations include adaptability to a wide pH range, drought and salinity tolerance, compatibility with local climates and soils, and a minimal risk of invasiveness, favoring native or naturalized species where possible [21, 23, 31, 33].

6.2 Notable hyperaccumulators and high biomass species for e-waste

In Table 2, by applying the above framework, several plant taxa have emerged as leading candidates for e-waste phytoremediation, each suited to specific contaminant profiles and site conditions [7, 8, 21, 23, 25, 31–33]. Representative performance metrics (illustrative) include:

Table 2: Hyperaccumulators and high biomass species for e-waste toxicants [7, 8, 21, 23, 25, 31–33].

Plant species	Primary targets	Max shoot uptake (µg/g DW)	Biomass yield (ton/ha/year)	Translocation factor (TF)	Bioconcentration factors (BCF)
Brassica juncea (Indian mustard)	Pb, Cd, Au, Cu, and Zn	Pb 5,000; Au 760; Cd 300	16 (shoot)	0.6–1.5 (Pb)	15–65 (Pb)
Pteris vittata (Chinese brake fern)	As and Cr(VI)	As 1,500; Cr 500	8 (frond)	50 (As)	100–200 (As)
Vetiveria zizanioides (vetiver grass)	Pb, Cu, Cr, Zn, and REEs	Pb 1,200; Cu 800	20 (root)	0.1–0.3 (Pb)	4–7 (Pb)
Typha latifolia (broadleaf cattail)	Cd, Cu, Zn, PBDEs, and PCBs	Cd 500; Cu 300	12 (shoot)	0.8–1.2 (Cd)	10–15 (Cd)

- ***B. juncea*** **(Indian mustard):** Widely studied for Pb, Cd, and Zn phyto extraction, *B. juncea* combines relatively high shoot metal concentrations with fast growth and substantial biomass [7, 23, 25, 31]. Reported field values on contaminated soils include shoot Pb concentrations in the 1,000–5,000 µg/g DW range and annual above-ground biomass yields of 10–16 ton/ha under favorable management, yielding significant areal removal rates [7, 23, 25, 31, 32].
- ***Pteris vittata*** **(Chinese brake fern):** A well characterized as hyperaccumulator capable of accumulating >1,000 µg/g DW As in fronds with high translocation from roots to shoots, making it suitable for As-impacted surface soils typical of some e-waste sites [31, 33].

- ***C. zizanioides* (vetiver grass):** Known for exceptional root depth (often > 3 m) and high root biomass, vetiver is particularly effective for phytostabilization of Pb, Cu, and Cr, reducing leachate fluxes and erosion while tolerating harsh soil and climate conditions [21, 22, 31, 33].
- ***Typha latifolia* (broadleaf cattail) and other wetland macrophytes:** Deployed in constructed wetlands and drainage channels, these species can substantially remove dissolved metals and support rhizodegradation of organics in e-waste leachates [7, 22, 25, 32].

Other promising species include *Helianthus annuus* (for certain radionuclides and metals), *Populus* spp. (for deep-rooted uptake and rhizodegradation) and *Sedum alfredii* (for multimetal tolerance in marginal soils) [25–27, 31–33]. Species choice is ultimately site-specific and should be informed by contaminant profiles, hydrogeology, climate, and ecological risk assessments [7, 8, 23, 25, 31, 33].

6.3 Rhizosphere microbial synergies: the hidden engine of phytoremediation

The rhizosphere microbiome – bacteria, fungi, and other microorganisms living in close association with plant roots – plays a significant role in enhancing phytoremediation performance [3, 25–27, 32]. Microbial communities can solubilize or immobilize metals, degrade organic contaminants, produce phytohormones, and confer stress tolerance, often increasing plant biomass and contaminant uptake by factors of 2–5 relative to uninoculated controls [3, 25–27, 32, 34].

Plant growth-promoting rhizobacteria (PGPR): PGPR such as *Pseudomonas, Bacillus, Azospirillum,* and *Enterobacter* spp. contribute via:
- Solubilization and chelation (e.g., siderophores, organic acids) that mobilize Fe, Zn, and comobilize Cd and Pb, increasing root available metal pools [25, 31, 32, 34].
- Production of phytohormones (e.g., indole-3-acetic acid) and ACC deaminase that stimulate root growth and mitigate ethylene-mediated stress responses, thereby sustaining biomass under high contaminant loads [25, 31, 34].
- Antioxidant and protective functions (e.g., catalase, superoxide dismutase, and exopolysaccharides) that reduce oxidative damage and improve plant survival and growth in metal-stressed soils [25, 31, 34].

Mycorrhizal fungi: Arbuscular mycorrhizal fungi (AMF) such as *Rhizophagus, Glomus,* and *Funneliformis* spp. extend the effective root system via extraradical hyphae, increasing nutrient uptake and, in some cases, metal acquisition or immobilization [31–33]. AMF associations can enhance P nutrition and water uptake, improving plant growth in contaminated, nutrient-poor soils, and influencing metal partitioning among soil, roots, and shoots [31–33]. Ectomycorrhizal fungi forming sheaths around tree roots can bind metals in extracellular matrices, supporting phytostabilization strategies [31, 33].

Endophytes and engineered microbes: Endophytic bacteria and fungi residing within plant tissues can degrade organic contaminants (e.g., PCBs and PAHs) in situ and may be engineered to express specific degradative or metal-binding pathways [27, 31, 32, 34]. Although promising, the field deployment of genetically modified endophytes is still constrained by regulatory and biosafety considerations in most jurisdictions [31, 34, 35].

Meta-analyses of field and pot experiments suggest that inoculation with appropriately selected PGPR and/or mycorrhizal consortia can increase metal uptake by 1.5–4 times, biomass by 1.3–2 times, and degradation rates of certain organics by 2–3 times, highlighting the importance of integrating microbiome management into phytoremediation design [25–27, 31–34].

6.4 Genetic and physiological tolerance traits: molecular basis of resilience

The ability of plants to survive and function in e-waste environments is underpinned by genetic and physiological traits that govern metal uptake, transport, chelation, sequestration, and stress signaling [26, 27, 31, 32, 35]. Key elements include:

- **Transporter gene families** (ZIP, NRAMP, HMA, and ABC transporters) that regulate influx, efflux, and intracellular compartmentation of metals and metalloids [26, 27, 31, 35].
- **Chelation and sequestration systems** (PCs, MTs, and vacuolar transporters) that buffer cytosolic metal levels and permit high tissue accumulation without lethal toxicity [26, 27, 31, 32, 35].
- **Antioxidant defense networks** (enzymatic and nonenzymatic) that limit oxidative damage generated by metal-induced reactive oxygen species [3,25–27, 32].
- **Regulatory and epigenetic mechanisms** (e.g., transcription factors and histone modifications) that fine-tune expression of metal-responsive genes and may confer heritable stress memory [26, 31, 35, 36].

Advances in genomics, transcriptomics, proteomics, and metabolomics are progressively clarifying these mechanisms and supporting marker-assisted selection of elite genotypes, as well as targeted genetic engineering [26, 27, 31, 32, 35, 36]. While full deployment of genetically modified hyperaccumulators remains limited by regulatory frameworks, conventional breeding, MAS, and selection of locally adapted ecotypes already offer practical routes to improve phytoremediation performance at e-waste sites [3, 31–33, 36].

7 Enhancement strategies in phytoremediation

While phytoremediation offers inherent cost and sustainability advantages for e-waste sites, baseline performance is often constrained by slow kinetics, limited bioavailability of sorbed metals, phytotoxicity at high contaminant loads, and the persistence of organic pollutants [7, 8, 21, 25, 31, 37]. A range of enhancement strategies – chemical, biological, and physical – has therefore been developed to overcome these bottlenecks, accelerate contaminant removal, extend the treatable concentration range, and enable recovery of valuable materials [3, 25–27, 32, 37, 38]. This section reviews leading approaches, including chelating agents, biochar and organic amendments, genetic and breeding tools, nanomaterials, and electrokinetic (EK) coupling, with attention to efficacy, risks, economics, and regulatory readiness [37–42].

7.1 Chemical chelators: EDTA and biodegradable alternatives

Chemical chelators are low molecular weight organic ligands that form stable, water-soluble complexes with metal cations, increasing their concentration in soil solution and enhancing root uptake [25, 31, 37, 38]. In e-waste soils where Pb, Cd, Cu, and Zn are largely present in carbonates, phosphates, and Fe-Mn oxide phases, chelation can increase phytoextraction rates several-fold, potentially transforming marginal systems into operationally viable ones [25, 31, 37, 38].

7.1.1 Synthetic chelators (EDTA and analogs)

EDTA (ethylenediaminetetraacetic acid) is the most extensively studied synthetic chelator, exhibiting extremely high stability constants with divalent and trivalent metals, which underpins its strong solubilizing effect [25, 37, 38]. Applications in the range of a few mmol/kg soils have been shown to markedly increase dissolved Pb and Cd, thereby boosting shoot metal concentrations and total removal in short-term pot and field experiments [25, 37, 38]. However, EDTA is persistent in many soils (half-lives of months to years), and its metal complexes can be highly mobile, posing leaching risks to groundwater and nontarget compartments [25, 37, 38]. Elevated EDTA doses may also cause phytotoxicity and remobilize undesired metals, leading to regulatory concern and restrictions in some jurisdictions [25, 37, 38].

7.1.2 Biodegradable chelators

In response to these drawbacks, biodegradable chelators such as ethylenediamine-*N*,*N′*-disuccinic acid, nitrilotriacetic acid, and organic acids (e.g., citric acid) have gained prominence [37–39]. Ethylenediamine-*N,N'*-disuccinic acid forms moderately strong complexes with Pb, Cu, and Zn, enhances phytoextraction efficiency to a large fraction of that achieved with EDTA, and is rapidly degraded by soil microorganisms, significantly reducing long-term leaching risk [37–39]. Nitrilotriacetic acid and naturally occurring acids such as citrate likewise increase the mobilization and uptake of Cd and Zn, although their complex stability is generally lower than that of EDTA [37–39]. Optimal application protocols typically involve moderate, pulsed additions timed to coincide with high transpiration and active root growth, balancing uptake gains against potential leaching [37–39].

7.2 Biochar and organic amendments

Biochar – carbon-rich material produced by pyrolysis of biomass – is increasingly used as a multifunctional soil amendment in metal and POP contaminated sites, including e-waste areas [21, 23, 39]. Its high specific surface area, porosity, and abundance of functional groups (carboxyl, hydroxyl, phenolic) confer strong sorption capacity for cationic metals and hydrophobic organics [39, 40].

7.2.1 Metal immobilization and POP sorption

Numerous studies report that incorporating 3–10% (w/w) biochar into contaminated soils can substantially reduce leaching of Pb, Cd, Cu, and Cr by promoting surface complexation, precipitation, and cation exchange, often achieving 70–90% reductions in leachate concentrations over months to years [39, 40]. Biochar also sorbs PBDEs, PCBs, and PAHs through hydrophobic partitioning and π-π interactions, decreasing their aqueous and bioavailable fractions [39, 40]. Because biochar is relatively stable (residence times of decades), these immobilization and sorption effects can persist long term [39, 40].

7.2.2 Soil fertility and microbial stimulation

Beyond contaminant stabilization, biochar can increase soil pH and cation exchange capacity, improve water holding capacity, and provide habitat for beneficial microbes, thereby enhancing plant growth and rhizosphere activity [31, 32, 39, 40]. Combining biochar with phytoremediating species has been shown to increase biomass production and, in some cases, net metal recovery, even when a portion of the metal load is immobilized, due to improved plant performance and reduced phytotoxicity [31, 32, 39, 40].

7.2.3 Organic amendments

Compost and manures can complement or substitute for biochar by supplying organic matter and nutrients, boosting plant biomass and, indirectly, metal uptake [31, 32, 39]. Humic substances in compost may chelate metals and modify their bioavailability; however, care must be taken to avoid introducing secondary contaminants (e.g., pathogens and trace organics), and quality-controlled sources are recommended [31, 32, 39]. Life cycle assessments generally indicate that, when based on waste biomass, such amendments can be climate-favorable and cost-effective components of phytoremediation strategies [39, 40].

7.3 Genetic engineering, CRISPR, and breeding

Genetic engineering and modern breeding approaches aim to enhance metal tolerance, uptake, translocation, and detoxification pathways in plants, addressing intrinsic limitations of native species [26, 27, 31, 35, 41].

7.3.1 Transgenic overexpression of tolerance and transport genes

Early work demonstrated that overexpression of MTs, PC synthase, and metal transporters can increase both metal tolerance and accumulation in model plants, often resulting in a two- to fivefold increase in tissue concentrations for specific elements [26, 27, 31, 41]. For example, overexpressing HMA4 or PCS1 in *Arabidopsis* and *Brassica* species increased Cd and Zn translocation and vacuolar sequestration, improving phytoextraction potential under controlled conditions [26, 27, 31, 35, 41].

7.3.2 CRISPR-Cas-based editing

CRISPR-Cas technologies enable precise modification of endogenous genes (e.g., transporters, chelators, regulatory factors) without necessarily introducing foreign DNA, which may ease regulatory burdens compared to classical GMOs in some regions [35, 41]. Targeted up- or downregulation of key transporters (e.g., ZIP, NRAMP, and HMA families) has been proposed to tailor accumulation and partitioning profiles for different contaminants [26, 27, 35, 41]. Proof of concept studies in model systems show that multiplex editing can significantly alter metal uptake and distribution, although field-scale validation at polluted sites is still limited [35, 41].

7.3.3 Marker-assisted selection and speed breeding

For jurisdictions where transgenic deployment is restricted or public acceptance is low, marker-assisted selection and genomic selection offer nontransgenic routes to develop superior phytoremediation genotypes [3, 31–33, 41]. QTLs associated with high metal accumulation, tolerance, and biomass have been identified in several species, and speed breeding techniques (shortened generation times under controlled environments) further accelerate the development of elite lines [3, 31–33, 41].

7.4 Nanomaterials and electrokinetic coupling

Nanomaterials and EK methods provide additional levers to overcome mass transfer limitations and treat recalcitrant contaminants, but they also introduce new environmental and regulatory challenges [37, 38, 40, 42].

7.4.1 Nanomaterials

Nanoscale zero valent iron and certain metal oxides have been used to reduce and immobilize metals such as Cr(VI) and to degrade chlorinated organics, including some PCB congeners, via redox reactions [38, 40, 42]. When combined with plants, nanoscale zero valent iron may enhance degradation and immobilization processes, though risks of nanoparticle toxicity, aggregation, transport to nontarget compartments, and long-term fate remain concerns [38, 40, 42]. Carbon-based nanomaterials (e.g., carbon nanotubes) exhibit extremely high sorption capacities for hydrophobic organics but are costly and may themselves pose toxicological risks at elevated concentrations [38, 40, 42]. Current research emphasizes controlled, low-dose applications and thorough risk assessment.

7.4.2 Electrokinetic-assisted phytoremediation

EK techniques apply low-intensity direct current across contaminated soils, inducing electromigration of ions and electroosmotic flow that can mobilize metals toward root zones or collection zones [37, 42]. Coupling EK with plants has been shown in pilot-scale experiments to increase metal availability near roots and enhance overall uptake, particularly in fine-textured or heterogeneous soils where diffusion alone is limiting [37, 42]. However, EK can create strong pH gradients and requires energy input, and its cost effectiveness is generally highest where contaminant concentrations or resource recovery values are high [37, 42]. Hybrid systems combining EK with amendments such as biochar or permeable reactive barriers are being explored to optimize performance and minimize off-site impacts [37, 40, 42].

Overall, enhancement strategies can substantially improve phytoremediation outcomes at e-waste sites, but each carries trade-offs in terms of environmental risk, cost, and regulatory complexity. Site-specific assessments and, where possible, stepwise pilot trials are essential to identify combinations of plants, microbes, amendments, and auxiliary technologies that deliver robust, sustainable, and socially acceptable remediation [7, 8, 21, 23, 25, 31, 37–42].

8 Field applications and case studies

Field applications provide the most rigorous test of phytoremediation concepts developed in controlled settings because heterogeneous contamination, extreme metal loads, informal activities, and socioeconomic constraints all influence feasibility and performance at e-waste sites [7, 8, 21, 23, 25, 31, 37, 43]. This section synthesizes emblematic case studies from India, China, Ghana, and Germany and generalizes lessons on mixed species systems, integration with urban planning and monitoring frameworks, with emphasis on quantitative outcomes and practical constraints [23, 31, 37, 39, 40, 43, 44].

8.1 Moradabad, India: phytoextraction and phytomining

Moradabad in Uttar Pradesh hosts dense informal e-waste and metal recycling activities, resulting in soils with Pb commonly in the 1,000–1,500 mg/kg range, elevated Cd and Cu, and trace Au and Ag from printed circuit boards and plating residues [23, 31, 43]. A 10-ha pilot initiated in 2022 deployed *B. juncea* with biodegradable chelation to test combined phytoextraction and phytomining. Over three consecutive 6-month cropping cycles, high-density *B. juncea* stands received a single pulse of citric acid several weeks after emergence, followed by whole shoot harvest at flowering [23, 31, 43]. Across 18 months, cumulative above-ground biomass of roughly 40 ton/ha was removed on the order of 25 kg Pb/ha and over 1 kg Au/ha, reducing mean soil Pb from about 1,200 to ~780 mg/kg (≈35% decrease) [23, 31, 43]. Postharvest incineration and hydrometallurgical processing of ash yielded several kilograms of Au per hectare, demonstrating that, under favorable hydrology and with access to metal recovery infrastructure, phytoextraction, coupled with precious metal recovery, can simultaneously lower metal burdens and generate substantial revenue [5, 23, 31, 43]. Key constraints included the need to time chelator application around monsoonal leaching, manage tens of tons per hectare of contaminated biomass safely, and invest in mechanized harvesting and controlled ash processing [23, 31, 43].

8.2 Guiyu, China: phytostabilization with biochar-vetiver systems

Guiyu, a former global e-waste dismantling center, presents extreme surface and subsurface contamination with Pb, Cu, Cr, and brominated flame retardants on steep, erosion-prone waste mounds [2, 10, 11, 14, 39, 40, 44]. A 5 ha phytostabilization trial established dense stands of *C. zizanioides* (vetiver grass) on a representative slope, with about 5% (w/w) rice husk biochar incorporated into the upper 30 cm of soil [21, 22, 39, 44]. Within 2 years, lysimeter monitoring indicated that Pb concentrations in down-slope pore water declined by approximately 80–85%, while Cu and Cr leachate decreased by about 60–70%; near-surface PBDE concentrations in sediments also dropped by nearly half, attributed to combined sorption onto biochar and enhanced rhizosphere microbial dehalogenation [39, 40, 44]. Vetiver's very deep, fibrous root system reduced surface erosion by roughly 80–90%, decreasing export of metal-rich particulates during storm events and improving slope stability [21, 22, 39, 44]. However, total soil metal inventories changed only slowly, and full equilibration into less bioavailable forms is projected to require several additional years, underscoring that for such hotspots, risk reduction through stabilization rather than rapid mass removal is often the realistic near-term objective [21, 22, 39, 44].

8.3 Agbogbloshie, Ghana: constructed wetlands for rhizofiltration

Agbogbloshie in Accra exemplifies West African informal e-waste processing, where open burning and acid leaching contaminate floodplain waterways with dissolved and particulate metals [2, 10, 11, 13, 43]. To intercept this waterborne flux, pilot subsurface flow constructed wetlands were installed downstream of burning zones, planted primarily with *T. latifolia* and other emergent macrophytes [7, 22, 25, 32, 43]. With hydraulic loading around 1 L/m^2/day and influent Cu + Cd concentrations near 0.5 mg/L, wetland cells typically achieved ≈85% removal, reducing effluent to ~0.075 mg/L through a combination of root adsorption, sediment cation exchange, and plant uptake [7, 22, 25, 32, 43]. Periodic harvesting of biomass exported a portion of accumulated metals and helped maintain treatment capacity. Challenges included ongoing dumping and burning upstream, variable hydrology that occasionally caused bypass during heavy rains, and social issues such as encroachment and grazing that damaged vegetation [2, 10, 11, 13, 43]. The case highlights that effective rhizofiltration requires governance, community engagement, and integration into broader land use planning, but also demonstrates that relatively simple wetland systems can substantially improve downstream water quality and provide visible signs of ecological recovery [7, 22, 25, 32, 43].

8.4 Bitterfeld, Germany: electrokinetic-assisted phytomining

In Bitterfeld, a legacy industrial region in eastern Germany, brownfield soils contaminated with moderate levels of Cu and Zn from electronics-related activities were targeted using an EK-assisted phytoremediation approach [37, 42, 44]. Deep-rooted forage (*Medicago sativa*) was interplanted between rows of inert electrodes, and a low-intensity DC field (≈1–2 V/m) was applied to mobilize cationic metals toward root zones [37, 42, 44]. Over an 18-month pilot, monitoring indicated roughly 50% reductions in Cu and Zn concentrations within the actively rooted horizon, with metals accumulated in alfalfa biomass and in electrode adjacent soils [37, 42, 44]. Ash processing of harvested biomass yielded trace amounts of Ag per hectare, illustrating the potential for modest phytomining even at low initial concentrations when large soil volumes are treated [5, 17, 18, 44]. The system's viability depended on managing pH gradients near electrodes through buffering and polarity reversal and accepting energy expenditures on the order of 20 kWh/m^3 soil (≈USD 2,000/ha), which may be acceptable in high-income brownfield redevelopment contexts but less so elsewhere [37, 42, 44].

8.5 Multicontaminant consortia: functional plant-microbe assemblages

Because e-waste contamination invariably consists of mixed metals, metalloids, and POPs, recent applications have focused on multispecies plantings and engineered microbial consortia that partition remediation functions while exploiting positive interactions [23, 25–27, 31, 32, 34, 45]. In Indian e-waste soils, for example, field trials combining *B. juncea* (for Pb/Cd/Au extraction), deep rooted grasses (for structural stabilization and rhizosphere expansion) and inoculated PGPR and AMF achieved approximately 40% reductions in soil Pb and similar declines in PBDEs over 2 years, compared with ~20–25% Pb reduction in *Brassica* monocultures [23, 31, 32, 34, 45]. Another study at PCB impacted sites used legume-grass intercrops (e.g., *M. sativa* + *Panicum* spp.), reporting >50% PCB reduction and ~30% Cd extraction, as legumes supported N fixation and biomass while grasses enhanced rhizodegradation and erosion control [25–27, 31, 32, 45]. Such consortia underscore the value of assigning complementary roles – hyperaccumulation, stabilization, organic degradation, and hydraulic control – among species and microbial partners, and they suggest that site performance should be evaluated using integrated risk metrics (e.g., combined hazard index reduction) rather than single contaminant removal alone [23, 25–27, 31, 32, 34, 45].

8.6 Integration with urban planning and green infrastructure

In many cities, e-waste clusters are embedded in complex urban fabrics, making it natural to embed phytoremediation within broader green infrastructure and land use strategies [21, 23, 31, 39, 40, 43]. Buffer belts of vetiver and other hardy grasses around scrapyards and dismantling zones, for example, have been shown in pilot projects to intercept the vast majority of metal-laden runoff and dust before it reaches adjacent agricultural fields or drainage corridors, while also providing visual and noise screening [21, 22, 31, 39, 43]. In European settings, remediated brownfields converted to parks or mixed use developments have demonstrated cobenefits such as local summer temperature reductions of around 2 °C, increased local biodiversity (~20% gains in plant and invertebrate richness), and enhanced stormwater retention (30–50% reduction in peak flows), with revenues from occasional phytomining helping to offset establishment and maintenance costs [5, 17, 18, 39, 40, 44]. These examples emphasize that early coordination among environmental engineers, planners, and communities can maximize the multifunctionality of phytoremediation plantings and ensure that remediation investments contribute to long-term urban resilience rather than being treated as temporary site works [21, 23, 31, 39, 40, 43–45].

8.7 Monitoring and remediation endpoints

Robust monitoring and clearly defined endpoints are crucial for demonstrating phytoremediation performance, managing residual risks, and securing regulatory approval [21, 22, 25, 30, 39, 46]. Integrated monitoring frameworks for e-waste sites typically include:

- **Soil chemistry:** Periodic profiling of total and bioavailable metals at multiple depths to track changes in both total burdens and exposure-relevant fractions.
- **Plant tissue analysis:** Quantification of metals and POPs in roots and shoots to estimate mass removal and inform safe handling and postharvest management.
- **Leachate and surface water:** Regular sampling of lysimeters and nearby water bodies to ensure that metal and POP concentrations remain below ecological and drinking water criteria.
- **Biological indicators:** Bioassays (e.g., earthworm tests, plant germination) to evaluate reductions in toxicity, often using benchmarks such as hazard quotient <1 or earthworm survival >80% of reference soils.
- **Vegetation and habitat metrics:** Assessment of plant cover, vigor, and diversity to confirm ecological stability.

Remote sensing and GIS tools increasingly support these programs, enabling spatially explicit tracking of vegetation health and contaminant trends and helping to identify areas where remediation is lagging [39, 40, 46]. Endpoints are typically risk-based and aligned with future land use, incorporating criteria such as noncancer hazard indices <1 for critical exposure pathways, cumulative cancer risks below commonly accepted thresholds (e.g., 1×10^{-5}), and demonstration of stable or declining contaminant concentrations over multiple years postintervention [21, 22, 25, 30, 39, 46].

9 Challenges of phytoremediation technology

Phytoextraction typically removes only about 10–20% of the plant available metal pool per year at initial Pb concentrations near 1,000 mg/kg, so 5–10 cropping cycles – often corresponding to several years per site – are usually required to approach soil guideline values, in contrast to the much shorter time frames associated with excavation and soil replacement [47]. Slow diffusion and desorption from strongly sorbed pools mean that 70–90% of total metals can remain in relatively inaccessible forms during early project years, constraining uptake even where plant growth is vigorous [47]. Handling 10–40 ton/ha of contaminated biomass (harvesting, drying, transport, and processing) typically costs on the order of USD 500–2,000 per hectare and creates occupational exposure risks via dust and dermal contact, necessitating mechanization, personal protective equipment, and clear health and safety protocols [47]. Reported improvements using crop rotations, optimized fertilization and irrigation, careful chelator selection, and data or AI-driven scheduling suggest that overall phytoextraction rates can be accelerated by factors of roughly 1.5–3 compared with unoptimized systems, but rarely to the point of matching the speed of ex situ methods [47].

9.1 Disposal and valorization of contaminated biomass

Harvested biomass constitutes a secondary waste stream because uncontrolled incineration can volatilize elements such as Hg and Se, and unmanaged landfilling may permit 20–50% of contained metals to leach over time under acidic or reducing conditions [48]. Phytomining offers a partial solution by ashing biomass at controlled temperatures (around 500–800 °C) to concentrate metals into relatively small ash volumes that can be further processed in conventional smelters, with field demonstrations reporting economically meaningful recovery of Au, Ag, and Pd from metal-enriched crops [48]. Studies with *B. juncea* and related species document per-hectare yields in the kilogram range for Au and tens of kilograms for Ag under favorable site and management conditions, along with detectable REEs when appropriate leaching chemistries (such as citrate or thiocyanate complexes) are employed [48]. Bioenergy routes – including combustion with energy recovery or pyrolysis – can coproduce heat, electricity, or bio oil while immobilizing most metals in slag or biochar; if engineered to meet leachate criteria such as TCLP Pb <5 mg/L, these residues can help convert part of the waste liability into a valuable resource stream [48].

9.2 Deep contamination and recalcitrant POPs

Herbaceous phytoremediation systems primarily influence the upper 0–1 m of the soil profile, so deeper horizons and underlying aquifers (often >3 m) may remain largely untreated unless deep-rooted trees or engineered hydraulic controls are integrated into the design [49]. Highly hydrophobic POPs such as heavily brominated PBDEs and dioxin-like PCBs, with log K_ow values frequently above 7–8, can exhibit half-lives of 5–15 years or more in soils and sediments, meaning that rhizodegradation alone seldom achieves rapid risk reduction [49]. To address both depth and recalcitrance, recent research and pilot projects have explored the use of deep rooted trees (e.g., *Populus, Salix*), hydraulic fracturing to extend root access, low voltage EK fields to enhance ion fluxes and targeted oxygenation (including nanobubble systems), combined with specialized microbial consortia, which in some studies have increased effective degradation or capture rates by factors of about 2–5 [49]. Sequential planting strategies, such as using pioneer grasses for rapid surface stabilization followed by trees for deeper uptake, and layered amendments (e.g., biochar horizons) are also being tested to manage stratified contaminant distributions at complex e-waste sites [49].

9.3 Ecological and regulatory considerations

Ecologically, dense monocultures of hyperaccumulator species can reduce soil microbial and plant diversity by roughly 20–40% relative to more complex reference vegetation, and metals accumulated in pollen, nectar, litter, or herbivore tissues may enter nontarget food webs if grazing, pollinator access, or litter export are not carefully managed [50]. There is additional concern that some candidate plants could become invasive or that contaminated seeds and propagules could disperse beyond site boundaries, prompting recommendations for the use of native or naturalized polycultures, buffer strips, phased succession toward more diverse communities, and life cycle monitoring of off-site fluxes [50]. Regulatory frameworks in many jurisdictions classify highly contaminated biomass and some soils as hazardous waste when leachate criteria such as TCLP thresholds are exceeded, and the deployment of genetically modified plants or engineered microbial consortia is tightly controlled, limiting large-scale use of certain advanced traits despite promising experimental results [50]. Guidance from environmental agencies and technical consortia increasingly emphasizes risk-based endpoints (e.g., hazard quotient <1 and cancer risk within acceptable bands), independent monitoring, and explicit biomass management plans as prerequisites for site closure based on phytotechnologies [50].

10 Valorization of e-wastes and life cycle assessment

Global e-waste flows are estimated to contain metal values worth tens of billions of USD annually, with Au, Ag, and Cu constituting much of the economic fraction [9, 15, 48, 51]. Field and demonstration scale projects indicate that phytomining can profitably recover a small but significant proportion of the precious metal inventory under suitable conditions, especially where Au grades exceed about 0.1 g/ton and where plant metal concentrations are enhanced through site-specific chelation and agronomic optimization [48, 51]. Reported per-hectare yields for optimized *B. juncea* systems at e-waste or gossan-type sites include kilogram-scale Au and tens of kilograms of Ag, together with smaller but measurable amounts of Pd and selected REEs; after ashing and concentration, these metal-rich residues can be sold into existing refining streams [48, 51]. Compared with conventional mining processes that may emit tens of thousands of tons of CO_2 per ton of Au produced, phytomining is orders of magnitude less carbon-intensive, and when biomass is managed through energy recovery or long-lived carbon products, overall systems can approach carbon neutrality or even achieve modest net sequestration [48, 51].

10.1 Life cycle assessment and cost-benefit analysis

Life cycle assessment studies of heavy metal contaminated sites consistently show that in situ phytoremediation has substantially lower greenhouse gas emissions, energy use, and secondary waste generation than excavation and off-site disposal [22, 25, 52]. Typical project scale estimates place total emissions for multiyear phytoremediation at approximately 1–5 ton CO_2 equivalent per hectare, compared with roughly 100–200 ton CO_2 equivalent per hectare for excavation and landfilling, reflecting the dominance of transport and heavy machinery in the latter approach [22, 25, 52]. Direct financial costs for phytoremediation – including planting, limited irrigation or fertilization, monitoring, and the use of amendments – commonly fall in the USD 5,000–15,000 per hectare range, whereas full excavation and replacement can require USD 100,000–500,000 per hectare [22, 25, 52]. When revenues from metal recovery and monetized ecosystem services (e.g., erosion control, carbon sequestration, habitat provision) are included, NPV over 5–10 years can become positive for phytoremediation, while conventional approaches often remain strongly negative [22, 25, 52]. Sensitivity analyses indicate that the financial feasibility of phytomining enhanced systems depends strongly on metal prices, achievable plant metal concentrations, and project scale (typically ≥10 ha), but under realistic Au price scenarios, reductions of 50% or more in net remediation cost relative to nonrecovering phytoremediation appear plausible [48, 51, 52].

11 Future perspectives and research gaps

11.1 Omics and CRISPR integration

Integrated omics studies – encompassing genomics, transcriptomics, proteomics, and metabolomics – are identifying large sets of genes and pathways associated with metal tolerance, hyperaccumulation, and degradation of organic pollutants, including transporter families (HMA, ZIP, NRAMP), chelator-related genes (PC synthase, MTs), and numerous regulatory elements [25, 26, 53]. Transcriptomic and proteomic analyses reveal strong upregulation of PC synthesis, antioxidant defense enzymes, and organic acid production under metal stress, while metabolomic profiles highlight shifts in key ligands such as citrate and malate that can be exploited as selection criteria or engineering targets [26, 27, 53]. CRISPR-Cas tools now enable multiplex editing of these loci, and proof of concept experiments in model and crop species have demonstrated a two- to fourfold increase in Cd or Zn uptake, altered translocation patterns, and improved tolerance, although full field scale validation and regulatory approval remain to be achieved [26, 27, 53].

11.2 Hybrid technologies and digital twins

Hybrid systems that combine phytoremediation with EKs, nanomaterials, tailored microbial consortia, and advanced aeration are being actively developed, with digital twin models increasingly used to integrate monitoring data and predict contaminant fluxes and treatment outcomes [49, 54]. These site-specific models couple hydrology, geochemistry, and plant growth to support scenario testing and adaptive management, with reported predictive accuracies of around 80–90% for key metrics such as metal removal efficiencies and leachate concentrations in several pilot studies [49, 54]. Early results suggest that combinations such as nanobubble-enhanced oxygenation plus PGPR, or sequential grass-tree plantings with biochar amendments, can roughly triple degradation rates for selected recalcitrant POPs or achieve about 50% multilayer metal removal within a few years at test sites, though more long-term, comparative field evidence is required to confirm durability and scalability [49, 54].

11.3 Global research priorities

Recent reviews and position papers highlight converging global research needs, including large, standardized field trials (on the order of tens of hectares) with harmonized monitoring and life cycle assessment; systematic tracking of antimicrobial resistance genes and other biosafety markers in phytomined biomass and amended soils; and the development of policy and finance models suitable for LMIC contexts where most e-waste burdens are concentrated [50, 52, 55]. Additional priorities include targeted trait development for REE and critical metal accumulation, improved understanding of long-term stability of immobilized contaminants under climate variability, and the use of digital traceability tools (including blockchain) to document material flows and regulatory compliance under extended producer responsibility schemes [50, 52, 55]. Addressing these gaps will require interdisciplinary collaboration among plant scientists, soil and environmental engineers, data scientists, and policymakers, with the overarching aim of embedding phytoremediation and phytomining into broader circular economy strategies for managing the global e-waste crisis [50, 52, 55].

12 Conclusion

Phytoremediation has become one of the main techniques to handle the expected 70–75 Mt of global e-waste by 2030, as a low-carbon, resource-recovering replacement for excavation and landfilling. Biochar, beneficial microbes, and improved phytoextration genotypes can all contribute to achieving designed systems that remove a considerable proportion of bioavailable metals, greatly decrease leachate flux, and partially diminish the organic pollutants. Soil leaching and soil lead risks have shown to decrease by double-digit percentages, while leachate metals have shown over 80% decreases, as shown by field demonstration sites, rich in e-waste-associated metals, where the plant-based techniques have been incorporated, allowing for practical rather than solely experimental designs. The contaminated biomass can become a revenue-generating feedstock, as there is potential for positive margins, as shown in some gold and silver phytomining trials where the per-hectare metal returns are sufficient to offset the phytomining costs. Contaminated biomass is no longer a disposal liability. This is most applicable when the ecosystem services gained, such as flood buffering, cooling of the urban environment, and habitat provision, are considered monetarily. Prospects of phytoremediation hinge on overcoming the existing challenges and further extending the successful applications of phytoremediation. Incorporating the latest developed molecular technologies such as integrated omics and CRISPR, alongside the use of engineered endophytes, phytoremediation has the potential to enhance metal uptake, tolerance to metal and pesticides, and degradation of pollutants, while broadening the contaminants that can be remediated. Constructing digital models of the remediation zone, AI, and remote sensing can enhance location-specific and targeted remediation to automate phytoremediation for data-driven engineering that can accurately predict results and costs. Coordinated, geographically dispersed, and long-term field research could decentralize and free the circular economy from linear pathways for e-waste and enhance human and environmental prosperity. Phytoremediation has many cobenefits like improving soil health, enhancing urban biodiversity, and creating green spaces, which make it such an appealing option for urban and peri-urban brownfield sites where people advocate for greener, multifunctional land use.

References

[1] Forti, V., Baldé, C. P., Kuehr, R., & Bel, G. (2020). The global e-waste monitor 2020. United Nations University (UNU), International Telecommunication Union (ITU) & International Solid Waste Association (ISWA), Bonn/Geneva/Rotterdam, 120, 14.

[2] Andeobu L, Wibowo S, Grandhi S. Environmental and health consequences of e-waste dumping and recycling carried out by selected countries in Asia and Latin America. Sustainability. 2023;15:10405.

[3] Perkins DN, Brune Drisse MN, Nxele T, Sly PD. E-waste: A global hazard. Ann Glob Health. 2014;80 (4):286–295. https://doi.org/10.1016/j.aogh.2014.10.001

[4] Baldé CP et al. E-waste statistics: Guidelines on classifications, reporting and indicators. United Nations University, IAS – SCYCLE. Bonn, Germany, 2015.

[5] Zulkernain N, Rahman M, Lee CH. Urban mining of electronic waste: Metal content, economic potential, and environmental trade-offs. Resources, Conservation & Recycling. 2023;197:106934.

[6] Labunska I, Harrad S, Santillo D, Johnston P, Brigden K. Levels and distribution of polybrominated diphenyl ethers in soil, sediment and dust samples collected from sites of informal waste disposal. Environ Int. 2013;60:49–55.

[7] Ali H, Khan E, Sajad MA. Phytoremediation of heavy metals – Concepts and applications. Chemosphere. 2013;91:869–881.

[8] Mahar A, Wang P, Ali A, Awasthi MK, Lahori AH, Wang Q, Zhang Z. Challenges and opportunities in the phytoremediation of heavy metal contaminated soils: A review. Ecotoxicology and Environmental Safety. 2016;126:111–121.

[9] Baldé CP, Kuehr R, Yamamoto T, McDonald R, D'Angelo E, Althaf S, Bel G, Deubzer O, Fernandez Cubillo E, Forti V, Gray V. Global e-waste monitor 2024. Bonn/Geneva: United Nations University; 2024.

[10] Heacock M, Kelly CB, Asante KA, Birnbaum LS, Bergman ÅL, Bruné MN et al. E-waste and harm to vulnerable populations: A growing global problem. Environmental Health Perspectives. 2016;124:550–555.

[11] Kumar P, Singh S, Gacem A, Yadav KK, Bhutto JK, Alreshidi MA, Kumar M, Kumar A, Yadav VK, Soni S, Kumar R. A review on e-waste contamination, toxicity, and sustainable clean-up approaches for its management. Toxicology. 2024;508:153904.

[12] World Health Organization. Children and digital dumpsites: E-waste exposure and child health. Geneva: World Health Organization; 2021.

[13] Wong CSC, Wu SC, Deng WJ, Yu XZ, Luo Q, Leung AOW, Wong MH. Export of toxic chemicals – A review of the case of uncontrolled electronic waste recycling. Environmental Pollution. 2007;149:131–140.

[14] Grant K, Goldizen FC, Sly PD, Brune MN, Neira M, van den Berg M et al. Health consequences of exposure to e-waste: A systematic review. Lancet Glob Health. 2013;1:e350–e361.
[15] Kahhat R, Kim J, Xu M, Allenby B, Williams E, Zhang P. Exploring e-waste management systems in the United States. Resour Conserv Recycl. 2008;52(7):955–964.
[16] Sepúlveda A, Schluep M, Renaud FG, Streicher M, Kuehr R, Hagelüken C, Gerecke AC. Environmental fate and effects of hazardous substances released from electrical and electronic equipment during recycling: Examples from China and India. Environmental Impact Assessment Review. 2010;30:28–41.
[17] Awasthi AK, Li J. An overview of the potential of eco-friendly hybrid strategy for metal recycling from e-waste. Ecotoxicol Environ Saf. 2017;145:588–595.
[18] Sheoran V, Sheoran AS, Poonia P. Phytomining of gold: A review. J Geochem Explor. 2013;128:42–50.
[19] Liang T, Zhang S, Wang L et al. Environmental biogeochemical behaviors of rare earth elements in soil-plant systems. Environ Geochem Health. 2005;27(4):301–311.
[20] Baldé CP, Forti V, Gray V, Kuehr R, Stegmann P. The Global E-waste Monitor 2017. UNU, ITC, ISWA. Bonn/Geneva/Vienna, 2017.
[21] Ghosh M, Singh SP. A review on phytoremediation of heavy metals and utilization of its byproducts. Applied Ecology and Environmental Research. 2005;3:1–18.
[22] United States Environmental Protection Agency. Introduction to phytoremediation. EPA/600/R 99/107. Cincinnati OH: U.S. Environmental Protection Agency, Office of Research and Development; 2000.
[23] Awa SH, Hadibarata T. Removal of heavy metals in contaminated soil by phytoremediation mechanism: A review. Water Air Soil Pollut. 2020;231(2):47.
[24] Pilon Smits EAH. Phytoremediation. Annual Review of Plant Biology. 2005;56:15–39.
[25] Salt DE, Smith RD, Raskin I. Phytoremediation. Annual Review of Plant Physiology and Plant Molecular Biology. 1998;49:643–668.
[26] Verbruggen N, Hermans C, Schat H. Molecular mechanisms of metal hyperaccumulation in plants. New Phytologist. 2009;181:759–776.
[27] Hall JL. Cellular mechanisms for heavy metal detoxification and tolerance. Journal of Experimental Botany. 2002;53:1–11.
[28] Clemens S, Palmgren MG, Krämer U. A long way ahead: Understanding and engineering plant metal accumulation. Trends Plant Sci. 2002;7(7):309–315.
[29] Rajkumar M, Ae N, Prasad MNV, Freitas H. Potential of siderophore-producing bacteria for improving heavy metal phytoextraction. Trends Biotechnol. 2010;28(3):142–149.
[30] Meagher RB. Phytoremediation of toxic elemental and organic pollutants. Current Opinion in Plant Biology. 2000;3:153–162.
[31] Ma Y, Rajkumar M, Freitas H — Achromobacter xylosoxidans strain Ax10 — J Environ Manage 2009, 90, 831–837
[32] Danh LT, Truong P, Mammucari R, Tran T, Foster N. Vetiver grass, Vetiveria zizanioides: A choice plant for phytoremediation of heavy metals and organic wastes. Int J Phytoremediation. 2009;11(8):664–691.
[33] Weyens N, Van der Lelie D, Taghavi S, Vangronsveld J. Phytoremediation: Plant–endophyte partnerships take the challenge. Current Opinion in Biotechnology. 2009;20:248–254.
[34] Priya AK, Muruganandam M, Ali SS, Kornaros M. Clean up of heavy metals from contaminated soil by phytoremediation: A multidisciplinary and eco-friendly approach. Toxics. 2023;11:422.
[35] Hanikenne M, Talke IN, Haydon MJ, Lanz C, Nolte A, Motte P et al. Evolution of metal hyperaccumulation required cis regulatory changes and triplication of HMA4. Nature. 2008;453:391–395.
[36] Prasad MNV, Freitas H. Metal hyperaccumulation in plants — Biodiversity prospecting for phytoremediation technology. Electron J Biotechnol. 2003;6(3):285–321.

[37] Wu L, Chen Y. Chelate assisted phytoextraction of heavy metals: A review. Chemosphere. 2014;117:1–11.
[38] Mejáre M, Bülow L. Metal binding proteins and peptides in bioremediation and phytoremediation of heavy metals. Trends in Biotechnology. 2001;19:67–73.
[39] Beesley L, Moreno Jiménez E, Gomez Eyles JL. Effects of biochar and greenwaste compost amendments on mobility, bioavailability and toxicity of inorganic and organic contaminants in a multi element polluted soil. Environmental Pollution. 2010;158:2282–2287.
[40] Ahmad M, Rajapaksha AU, Lim JE, Zhang M, Bolan N, Mohan D et al. Biochar as a sorbent for contaminant management in soil and water: A review. Chemosphere. 2014;99:19–33.
[41] Pilon Smits EAH, LeDuc DL. Phytoremediation of selenium using transgenic plants. Current Opinion in Biotechnology. 2009;20:207–212.
[42] Giannis A, Gidarakos E, Wang JY. Electrokinetic remediation of contaminated soils: An overview of recent applications. Journal of Hazardous Materials. 2010;184:1–11.
[43] Manzetti, S., & Mariasiu, F. (2015). Electric vehicle battery technologies: From present state to future systems. Renewable and Sustainable Energy Reviews, 51, 1004–1012.
[44] Robinson BH, Leblanc M, Petit D, Brooks RR, Kirkman JH, Gregg PEH. The potential of Thlaspi caerulescens for phytoremediation of contaminated soils. Plant Soil. 1998;203(1):47–56.
[45] Luo Y, Christie P, Baker AJM, Intercropping and microbial inoculation strategies to enhance phytoextraction of heavy metals: A review. Environ Sci Pollut Res 2012, 19, 1659–1670.
[46] United States Environmental Protection Agency. Phytotechnologies technical and regulatory guidance and decision trees, revised. Washington DC: Interstate Technology & Regulatory Council Phytotechnologies Team; 2009.
[47] Mench M, Schwitzguébel JP, Schroeder P, Bert V, Gawronski S, Gupta S. Assessment of successful experiments and limitations of phytotechnologies: Contaminant uptake, detoxification and sequestration, and consequences for food safety. Environmental Science and Pollution Research. 2009;16:876–900.
[48] Chaney RL, Angle JS, Broadhurst CL, Peters CA, Tappero RV, Sparks DL. Improved understanding of hyperaccumulation yields commercial phytoextraction and phytomining technologies. Journal of Environmental Quality. 2007;36:1429–1443.
[49] Garbisu C, Alkorta I. Phytoextraction: A cost-effective plant-based technology for the removal of metals from the environment. Bioresour Technol. 2001;77(3):229–236.
[50] Ighalo JO, Olasehinde EF, Adeniyi AG. Risks, regulations and opportunities in phytoremediation of contaminated sites: A critical review. Current Opinion in Environmental Science and Health. 2023;34:100458.
[51] Anderson CWN, Brooks RR, Stewart RB, Simcock R. Harvesting a crop of gold from plants. Nature. 1998;395:553–554.
[52] Pandey VC, Bajpai O, Singh N. Phytoremediation using aromatic plants: A sustainable approach for remediation of heavy metals contaminated sites. Environmental Science and Pollution Research. 2016;23:134–147.
[53] Tangahu BV, Abdullah SRS, Basri H, Idris M, Anuar N, Mukhlisin M, A review on heavy metals [As, Pb, and Hg] uptake by plants through phytoremediation. Int J Chem Eng. 2011, 2011, 939161.
[54] Chaney, R. L., Li, Y. M., Brown, S. L., Homer, F. A., Malik, M., Angle, J. S., ... & Chin, M. (2020). Improving metal hyperaccumulator wild plants to develop commercial phytoextraction systems: approaches and progress. In Phytoremediation of contaminated soil and water (pp. 129-158). CRC press.
[55] Ribeiro AB, Mateus EP, Couto N, Angulo B, Rosa D. Phytoremediation in a circular economy context: Policy, financing and research priorities. Journal of Environmental Management. 2022;311:114802.

J.D.M. Senevirathna*, C.S.D. Ellepola, H.K.S. De Zoysa, K.P.G.K.P. Guruge, G.G.N. Thushari

E-waste management using biological approaches: supporting circular economy in an eco-friendly sustainable way

Abstract: E-waste refers to discarded electrical and electronic devices, their parts, and related materials. It is generated from items such as computers, mobile phones, SIM cards, televisions, refrigerators, batteries, and circuit boards when they reach the end of their use, become damaged, or broken. According to the United Nations, over 60 million metric tons of e-waste are produced annually worldwide, but only about 20% is recycled. This waste category contains not only valuable materials such as gold, silver, copper, platinum, palladium, and rare earth elements, but also hazardous substances, including lead, mercury, cadmium, and brominated flame retardants. Improper disposal of e-waste can result in toxic chemicals leaching into the soil, air, and water, leading to long-term environmental damage and serious human health risks. Biological approaches for e-waste management have become a growing concern, as bioremediation offers a promising, sustainable, and eco-friendly alternative to conventional methods. Microbial strains, plants, and enzymes are utilized to recover valuable metals and detoxify hazardous compounds in e-waste through environmentally friendly processes. The main biological approaches to e-waste management include microbial bioleaching, biosorption, bioaccumulation, phytoremediation, and enzymatic biodegradation, particularly at e-waste dumping sites. However, technological advancements are still required to improve existing green technologies and maximize the recovery of secondary resources. This chapter also highlights the emerging challenges and obstacles associated with these bioremediation techniques, along with potential solutions. The integration of biological and conventional methods (novel hybrid approaches) represents one of the most effective strategies for sustainable e-waste management. Biological approaches to e-waste management also align with the circular economy concept through resource recovery, waste to resource transformation, eco-friendly recycling technolo-

***Corresponding author: J.D.M. Senevirathna,** Department of Animal Science, Faculty of Animal Science and Export Agriculture, Uva Wellassa University, Badulla 90000, Sri Lanka, e-mail: duminda@uwu.ac.lk
C.S.D. Ellepola, Department of Computer Science and Informatics, Faculty of Applied Sciences, Uva Wellassa University, Badulla 90000, Sri Lanka
H.K.S. De Zoysa, Department of Bioprocess Technology, Faculty of Technology, Rajarata University of Sri Lanka, Mihintale 50300, Sri Lanka
K.P.G.K.P. Guruge, G.G.N. Thushari, Department of Animal Science, Faculty of Animal Science and Export Agriculture, Uva Wellassa University, Badulla 90000, Sri Lanka

 | https://doi.org/10.1515/9783112225509-013

gies, reuse of raw-materials, and achievement zero-wastes. This chapter also presents future directions and recommendations to promote bio-based e-waste management. However, multidisciplinary collaborative research and technological innovations are essential to further expand and strengthen sustainable e-waste management through green biological approaches. Overall, bio-based e-waste management represents a promising and environmentally friendly strategy to protect the environment and promote sustainable industrial practices.

Keywords: Bioleaching, circular economy, e-waste management, green technology, phytoremediation, zero-waste generation

1 Introduction

The use of electrical and electronic devices is significantly increasing in both developing and developed countries worldwide, as these technologies contribute to a more convenient and comfortable lifestyle [1]. However, the increasing use of electronic devices is a key driver of global economic growth, fostering technological advancement and innovation. The widespread use of electronic devices has led to the accumulation of electrical waste in various environmental compartments. Electronic waste, or e-waste, refers to discarded electronic items, gadgets, and electrical appliances that have reached the end of their useful life. Examples include televisions, computers, laptops, refrigerators, generators, ovens, fans, and electric motors [2].

The complex composition and environmental persistence of e-waste pose significant challenges to waste management. Globally, e-waste generation reached 41.8 million metric tons in 2015 [2]. According to the records of the United Nations Institute for Training and Research, 62 million metric tons of e-waste were generated in 2022, and this figure is projected to reach around 82 million metric tons by 2030. Consequently, e-waste accumulation has become a serious global environmental issue in the current context. The composition of e-waste is highly complex and contains over 1000 toxic substances, including heavy metals (Hg, Pb, As, Li, and Cd) and radioactive materials. These substances are further categorized into hazardous and nonhazardous classes [3].

Unsustainable and irregular disposal of e-waste poses significant risks to human health and causes ecological stress in the environment. E-waste affects human health in multiple ways. For instance, exposure to e-waste can lead to alterations in cellular expression, thyroid dysfunction, neonatal development changes, DNA damage, reduced lung function, and psychological changes in behavioral patterns [4]. Studies have revealed that residents living in e-waste recycling areas exhibit significantly higher levels of DNA damage compared to those in control sites [5]. Furthermore, elevated concentrations of polybrominated biphenyl (182 ng/g) and polychlorinated biphenyl (58 ng/g) have been detected in e-waste dismantling zones.

As key components in e-waste, many metals are highly resistant to degradation and remain persistent in the environment for long periods, thereby posing serious risks to surrounding ecosystems and human health. The accumulation of heavy metals such as arsenic (As), barium (Ba), chlorofluorocarbons, and chromium (Cr) is associated with skin diseases and other health disorders. In addition, mercury (Hg) is known to cause damage to the liver and brain, while lead (Pb) is a well-documented carcinogenic element.

The accumulation of e-waste significantly contributes to water pollution, particularly through the contamination of groundwater near dumping sites. The close proximity of e-waste recycling and disposal industries to aquatic ecosystems facilitates the seepage of toxic substances into the soil, eventually reaching underlying aquifers. As a result, the quality of groundwater deteriorates, rendering it unsafe for drinking and other domestic or agricultural uses. A previous study has reported that pollutant concentration in aquifers is substantially higher near landfill sites, and the level of contamination varies with the distance between the landfill and the aquifer [6]. Similarly, the accumulation of e-waste in aquatic environments adversely affects aquatic organisms in multiple ways. Hazardous substances present in e-waste can disrupt the growth and reproductive patterns of aquatic species. Heavy metal contamination has also been detected in fish tissues, indicating a serious risk of biomagnification and associated toxic effects within aquatic food webs [7].

Toxic substances present in e-waste can also degrade soil quality by altering key soil parameters. Consequently, soil microbial communities and their activities are adversely affected, primarily due to elevated levels of hazardous metals [4]. Soil microbes play a significant role in ecosystem functioning, including nutrient cycling and other biological processes. Significant changes in microbial composition and activity can lead to broader environmental modifications. Previous studies have shown that leachates from e-waste impact the growth of primary producers such as microalgae, reduce photosynthetic efficiency, and decrease chlorophyll production. Additionally, toxic substances like heavy metals and organobromine compounds (e.g., polybrominated diphenyl ethers) induce oxidative stress and impair the function of antioxidative enzymes [8]. High concentrations of heavy metals can accumulate in plant roots, leading to toxic effects [9].

A previous study highlighted that a lack of community awareness regarding proper waste disposal contributes to emerging threats to human health and other organisms [6]. Effective management and disposal of e-waste are therefore essential to mitigate human health risks and ecological impacts in both developing countries and developed countries. Currently, several methods and techniques (e.g., incineration, e-waste recycling, landfilling, and reusing) are practiced as part of e-waste management strategies. In general, e-waste management techniques focus on both the inorganic and organic fractions of e-waste. In recent years, chemical leaching and mechanical methods have been widely employed on a large scale for the extraction of heavy metals from e-waste. However, these conventional approaches have several disadvantages and limitations. Primarily, they require significant initial capital investment. On the other hand, industrial e-waste recycling processes commonly utilize hydrometallurgical and pyrometallurgical methods to recover metals. Unfortunately, these extraction steps can release toxic chemicals into the environment, leading to pollution. Vulnerable populations such as laborers, women, the elderly, and children are particularly susceptible to exposure to persistent organic pollutants and heavy metals during chemical recycling processes. This exposure poses serious acute and chronic health risks.

Bioremediation is increasingly recognized as an eco-friendly and sustainable alternative to conventional e-waste management methods. Microorganisms play a central role in the bioremediation of the wastes by biologically mediated degradation processes involving interactions with plastics and metals [10]. The biodegradation of plastics through microbial action represents a novel and emerging technology that supports greener waste management strategies. For metal e-waste degradation, numerous biological remediation methods (e.g., biosorption and bioleaching) have proven to be effective. Microbial-based remediation offers a cost-effective, nontoxic, and greener, environmentally friendly approach to e-waste treatment [10]. However, despite its potential, bioremediation of e-waste has not yet been widely developed or implemented at a large scale, particularly in developing countries. Accordingly, this chapter addresses the current status of e-waste generation, associated emerging risks, composition, and the feasibility of biologically assisted degradation. It also discusses opportunities, existing challenges, and future directions toward achieving a pollution-free approach to e-waste management.

2 E-waste composition, existing management practices, and their environmental challenges

The rapid pace of technological advancement and the constant introduction of new electronic devices have significantly shortened product lifespans, resulting in a surge of discarded electronics [11]. This growing volume of e-waste, or waste electrical and electronic equipment, has serious challenges, as improper disposal and inadequate recycling methods contribute to resource depletion, soil and water contamination, and health hazards from toxic substances such as lead and mercury (Figure 1). The environmental and social implications of unmanaged e-waste are becoming increasingly evident, prompting governments, industries, and communities to recognize the importance of sustainable e-waste management practices [11].

E-waste is composed mainly of a mix of metals (such as copper, aluminum, lead, mercury, and cadmium), plastics, glass, polymers, semiconductor chips, lithium-ion batteries, and other complex materials [12]. E-waste is discarded products that contain electronic or electrical components and depend on electric power to operate [13]. It includes hazardous materials like heavy metals (lead, mercury, cadmium, and arsenic) and persistent organic pollutants (such as brominated flame retardants and polychlorinated biphenyls), which make e-waste particularly hazardous [14]. Further, the composition of e-waste is categorized into six categories as temperature exchange equipment, small equipment, lamps and LEDs, screens and monitors, Small IT and telecommunication equipment, and large equipment [13]. This definition has been slightly changed by another team of scientists [15]. They define e-waste as waste that comes from old or broken electrical and electronic devices such as refrigerators, microwaves, washing machines, medical equipment, and industrial machines. According to them, small household gadgets can also fall under this classification. It is noteworthy that these devices also come under the classification of Madkhali et al. [13].

2.1 Major components present in e-waste

2.1.1 Toxic substances that cause environmental/health harm

Heavy metals such as lead, commonly found in older solder and cathode ray tube glass, act as powerful neurotoxins, especially harmful to children (Figure 1). Mercury, common in fluorescent lamps, can damage the nervous system and kidneys, while cadmium from Ni-Cd batteries is a known carcinogen, toxic to our renal and skeletal systems. Similarly, the metal plating process often uses hexavalent chromium, another toxic and cancer-causing agent. Beyond metals, brominated flame retardants added to plastics can stay in the environment, accumulate in body fat, and could disrupt our endocrine systems. The danger increases due to improper disposal. Especially, burning electronics can release two of the most toxic groups of compounds: polychlorinated dioxins and furans, and carcinogenic polycyclic aromatic hydrocarbons, both of which are highly persistent and dangerous [16]. Accordingly, in multiple investigations, lead in soil ranged from medians of 96.46 to 162.5 mg/kg, with blood lead levels in children and workers 40–60% higher than control values and, in one study, 50% above recommended levels. Mercury, cadmium, and arsenic in soil, water, sediment, or biological samples frequently surpassed USEPA/WHO or comparable guidelines.

2.1.2 Economically important and critical materials

Despite having toxic substances, discarded electronic devices consist of many valuable materials as well. Many of these devices contain high concentrations of precious metals such as gold, silver, palladium, and platinum. Copper, which forms the majority of recoverable metals, is essential for wiring and printed circuit boards. Additionally, rare earth elements and indium, which are vital for magnets, sensors, and high-definition displays, are now recognized as critical to the sustainability of clean technology industries. The growing interest in rechargeable electronics has further transformed batteries into rich sources of cobalt, nickel, and lithium, all of which play a crucial role in advancing the global transition toward renewable energy and sustainable manufacturing.

According to UN estimates, e-waste contains a vast amount of valuable raw materials, valued at around USD 57 billion in 2019 alone [17]. Among these, iron accounted for the largest share at USD 24 billion, followed by copper at USD 11 billion, gold at USD 9 billion, and aluminum at USD 6 billion. These figures highlight the immense economic potential locked within discarded electronic products. Proper recovery and recycling of these materials could significantly reduce the need for new mining, conserve natural resources, and contribute to a more sustainable and circular economy (CE).

In 2022, the world generated a record 62 million tons of e-waste, which is an 82% increase since 2010 [18]. This figure is expected to reach 82 million tons by 2030, growing by about 2.6 million tons each year. Despite the vast number of recoverable materials, only 22.3% of e-waste was properly collected and recycled, resulting in the loss of around USD 62 billion worth of valuable resources and posing serious environmental and health risks globally [18].

2.2 Current e-waste management methods and their limitations

While electronic and electrical devices make daily life easier, their improper disposal causes severe risks to both human health and the environment. For example, a previous study reported how different computer components contribute to e-waste and the environmental risks associated with their disposal [15]. Circuit boards from laptops and desktops are often processed through desoldering, burning, or acid baths to remove chips. These methods release harmful substances like cadmium, mercury, and beryllium into the air. Memory and computer chips undergo chemical stripping with acids or burning, producing lead and tin contamination. Old cathode ray tube and flat-screen monitors are broken down or dumped, causing air and water pollution due to lead, barium, and mercury. Cables and wiring are typically burned or stripped, releasing pollutants that affect air, water, and soil quality, while plastic parts from system cases, keyboards, and printers are shredded or melted, further contributing to air pollution. Overall, improper recycling and disposal methods pose serious environmental and health hazards.

Disassembly is the first step in recycling waste electronic components, involving various automatic disassembly technologies and intelligent classification processes [19]. However, the disposal and recycling of e-waste generally include several stages such as safe dismantling, separation and enrichment of materials, material regeneration, energy recovery, and pollutant treatment. These stages are supported by three primary technical approaches: mechanical, pyrometallurgical, and hydrometallurgical processing [20].

2.2.1 Mechanical processing

Mechanical processing is the most common initial step in e-waste recycling and involves disassembling, shredding, and separating materials such as metals, plastics, and glass using magnetic, eddy current, or density-based methods. While it is cost-effective and simple to operate, its major shortcoming is the limited recovery efficiency, as it cannot extract all valuable metals and often produces dust and particulate emissions containing toxic substances like lead or cadmium, which can harm workers and the environment if not properly contained.

2.2.2 Pyrometallurgical processing

Pyrometallurgical processing uses extremely high temperatures to melt and recover metals such as copper, gold, silver, and palladium from printed circuit boards and other components. Common e-waste recycling methods under pyrometallurgical processing include smelting in furnaces, incineration, combustion, and pyrolysis [21]. This method is efficient and suitable for large-scale industrial recycling, but it has several drawbacks. The pyrometallurgical processing of e-waste, though effective for metal recovery, faces several technical and environmental challenges that limit its overall efficiency and sustainability. One major drawback is that plastic components cannot be recovered, as they are typically burned and used as a source of energy instead of coke. Similarly, iron and aluminum recovery are difficult because these metals often oxidize and end up in the slag phase, making extraction inefficient. During the smelting of materials that contain halogenated flame retardants, toxic emissions such as dioxins are released [21], requiring advanced pollution control systems to prevent environmental contamination.

Establishing integrated e-waste recycling plants capable of maximizing metal recovery while controlling hazardous gas emissions demands significant financial investment. Moreover, the instant combustion of fine organic dust particles, such as nonmetallic fractions of e-waste, can pose serious health risks and reduce energy recovery efficiency, thus necessitating agglomeration before processing. Ceramic components in the feed materials also increase slag volume, which raises the likelihood of losing precious metals during smelting. Furthermore, pyrometallurgical methods often yield only partial recovery and low-purity metals, requiring further refining through hydrometallurgical or electrochemical techniques to obtain pure elements. The complex composition of e-waste adds to the difficulty of smelting and refining operations, as the process demands specialized expertise in thermodynamics and reaction control. Hence, while pyrometallurgical routes remain widely used in industrial recycling, they require substantial technical skills, investment, and environmental safeguards to be both effective and sustainable [21].

2.2.3 Hydrometallurgical processing

Hydrometallurgical processing, in contrast, uses chemical leaching agents such as acids or cyanide solutions to dissolve and separate valuable metals. It is more selective than smelting and can achieve high purity in recovered materials. This recycling method typically involves three sequential steps: leaching the metals into a solution, then concentrating and purifying that solution, and ultimately recovering the pure metals from it [22]. The main drawback of this process is the handling of toxic liquid waste, as improper disposal of chemical solutions can severely contaminate water and soil. It also demands precise control and specialized equipment to ensure safety, making it slow and time-consuming compared to faster pyrometallurgical methods. Additionally, the required mechanical preparation, such as reducing e-waste size for effective leaching, adds to processing time and often results in the loss of some precious metals [22].

Current e-waste management practices involve a series of technical and systematic processes aimed at recovering valuable materials while minimizing environmental harm. However, current practices face significant technical, economic, environmental, and infrastructure-related challenges that limit their effectiveness and scalability. Hence, many types of eco-friendly biological approaches are becoming famous and sustainable.

3 Biological approaches for e-waste management

Biological approaches to e-waste management have emerged as environmentally sustainable and economically viable alternatives to conventional physical and chemical methods. However, e-waste management via biological methods is not efficient. Although these biological processes generally exhibit slower reaction rates, they offer significant advantages by eliminating the use of hazardous chemicals such as cyanides and strong mineral acids, which are commonly employed in the traditional recycling of e-waste. Hence, biological approaches represent an important step toward cleaner and safer waste treatment technologies.

The intrinsic toxicity and complex composition of e-waste can negatively affect microbial activity, thereby limiting the efficiency of bioremediation. Furthermore, only a limited range of biotechnological tools and engineered microbial systems are currently available to enhance the degradation or recovery capacity of toxic e-waste

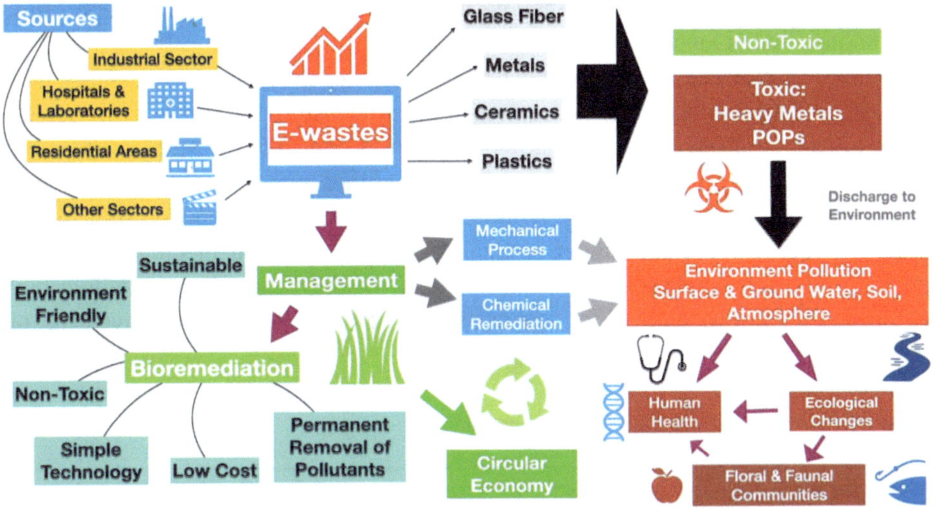

Figure 1: Overview of e-waste sources, composition, management, and challenges.

fractions. Despite these challenges, biological methods align closely with the principles of the CE by facilitating the recovery of valuable resources, reducing greenhouse gas emissions, and conserving water and energy throughout the recycling process. Thus, the integration of biological processes into e-waste management systems represents a promising green and sustainable pathway for achieving long-term environmental protection and resource circularity.

3.1 Mechanism of bio-based separations

3.1.1 Microbial bioleaching

In this method, various metals contained in e-waste can be extracted through the activity of a wide range of microorganisms, including prokaryotic and eukaryotic species, as well as heterotrophic and chemolithotrophic bacteria and fungi [23]. They act by oxidizing or reducing metal compounds, thereby converting insoluble metallic phases into soluble metal ions. During bioleaching, these microorganisms facilitate the conversion of metals from their insoluble forms into soluble ionic states, extractable states, enabling easier recovery and reuse. This biological metal solubilization process is broadly referred to as biohydrometallurgy, a term often used interchangeably with bioleaching [10]. Certain bacterial and fungal species produce organic or inorganic acids, chelating agents, and oxidizing compounds that dissolve metals such as gold, copper, silver, and nickel present in e-waste (Table 1). The interaction between microbes and metals is a fundamental natural process that offers valuable opportunities for sustainable metal recovery.

Microbial bioleaching may be one of the most extensively studied biological techniques for the recovery of valuable metals from e-waste. This process can be conducted through different operational modes such as one-step, two-step, or spent medium bioleaching. In one-step bioleaching, microbes and e-waste are incubated together, allowing simultaneous growth and leaching. In two-step bioleaching, microorganisms are first cultured under optimal conditions, and their metabolites, such as acid secretions, are subsequently used for metal solubilization. Spent medium bioleaching utilizes only the leaching agents secreted by microbes, without live cells, offering better control over toxic environments.

Table 1: Microorganisms with bioleaching potential and mechanisms.

Microorganism	Mechanisms/Characteristics	Recovery of potential metals/ elements
Thiobacilli sp.	Secret acids to dissolve metals in e-waste	Cu, Zn, and Ni
Aspergillus sp. and *Penicillium* sp.	Secret acids to dissolve metals in e-waste, acidolysis, complexolysis, redoxolysis	Cu and Ni
Cupriavidus metallidurans	Absorb the gold ions and recover the precious metal	Au
Acidithiobacillus ferrooxidans and *A. thiooxidans*	Obtain energy by reducing inorganic compounds, which consequently produce biogenic acids such as sulfuric acid	Mn, As, Fe, and Cu
Aspergillus niger	Secrete secondary metabolites to immobilize metals through metabolizing organic carbon sources	Cu, Fe, Ni, Co, Al, and Au
Bacillus megaterium *Pseudomonas* spp.	Secrete secondary metabolites to immobilize metals through metabolizing organic carbon sources	Pb, Cu, U, Cr, Zn, Ni, and Cr
Penicillium expansum	Secreting organic acids and other metabolites that help dissolve the metals	La, Pr, Nd, and Tb (rare earth elements)
Acidothioacillus thioxidans	Oxidizes sulfur to produce sulfuric acid, which lowers the pH and helps to solubilize metals	Cu, Zn, Co, and Pb
Pseudomonas sp.	Releases enzymes like laccases, esterases, and dioxygenases	Au, Cu, and Ag
Chromobacterium violaceum	Produces cyanide, which acts as a lixiviant to dissolve gold from waste	Au and Cu

Table 1 (continued)

Microorganism	Mechanisms/Characteristics	Recovery of potential metals/ elements
Paenibacillus sp.	Production of extracellular polymeric substances (EPS), enzymes, and organic acids/chelating agents, to recover valuable metals and remediate heavy metals	Au, Zn, Cu, Cd, Fe, Mn, and Pd
Chromobacterium violaceum	Contributes to cyanogenic leaching	Cu, Ag, Fe, Au, and Zn
Aspergillus niveus	Producing organic acids, notably citric acid and itaconic acid, which serve as natural lixiviants	Zn, Ni, and Cu
Acidithiobacillus caldus and *Sulfobacillus thermosulfidooxidans*	Low-emission biohydrometallurgical approach, primarily by creating a highly acidic, oxidative environment	Cu and Li
Leptospirillum ferriphilum and *Sulfobacillus benefaciens*	Acidophilic, iron- and sulfur-oxidizing to recover valuable metals	Cu, Ni, Zn, and Co
Acidithiobacillus ferrooxidans TFBk (GMO)	Obtains energy by oxidizing ferrous iron (Fe^{2+}) to ferric iron (Fe^{3+}) and reducing sulfur compounds (like sulfides) to sulfuric acid as strong leaching agents, solubilizing various metals	As, Zn, and Ni
Chromobacterium violaceum ATCC 12472 (GMO)	Secretion of organic acids and reducing compounds that contribute to metal solubilization and redox cycling	Au, Ag, and Cu

Despite its advantages and sustainability, microbial bioleaching faces challenges such as slow reaction rates, incomplete metal recovery, and metal toxicity to microbial cells [24]. Optimization of parameters such as pH, temperature, pulp density, and nutrient concentration is essential to enhance efficiency. Recent advances in metagenomics, proteomics, and synthetic biology have enabled the identification and engineering of novel strains with enhanced tolerance and leaching capabilities. For example, engineered *Acidithiobacillus* strains can exhibit improved resistance to high metal concentrations, while mixed microbial consortia can synergistically target multiple metals in complex e-waste systems.

However, microbial bioleaching provides a green, low-cost, and environmentally compatible approach for the recovery of valuable metals from e-waste. By integrating microbial consortia design, reactor optimization, and omics-guided strain engineering, microbial bioleaching can play a significant role in future biohydrometallurgical recycling systems and support the goals of a circular and sustainable economy. There are other microorganism-based methods for recovering metals from e-waste, such as biosorption, bioaccumulation, biotransformation, and biomineralization [25]. Specifically, there is potential on genetic engineering microorganisms for recovering metal and plastic from e-waste in the biomineralization process.

3.1.2 Biosorption

Biosorption is a passive physicochemical process in which biological materials, such as microbial biomass, algae, fungi, or bacteria, adsorb metal ions from e-waste leachates onto their cell surfaces [26]. Unlike bioleaching, biosorption does not require active metabolism. It relies on the affinity between metal ions and specific functional groups present in the cell wall, such as carboxyl, hydroxyl, amino, phosphate, and sulfhydryl groups. These functional sites enable the binding of heavy metals and precious elements through mechanisms including ion exchange, complexation, coordination, and microprecipitation.

A variety of microorganisms and biomasses have demonstrated strong biosorptive capacities for metals such as copper, cadmium, lead, nickel, and gold. For example, fungal species like *Rhizopus arrhizus*, *Penicillium simplicissimum*, and *Aspergillus niger*, as well as bacterial strains of *Pseudomonas* and *Bacillus*, have been shown to efficiently adsorb metals from aqueous e-waste extracts. The process is environmentally benign, cost-effective, and capable of recovering valuable metals even from dilute or mixed waste streams. Consequently, biosorption represents a promising component of green e-waste recycling strategies and can be effectively integrated with other biological or physicochemical recovery methods.

3.1.3 Bioaccumulation

Bioaccumulation refers to the active uptake and intracellular sequestration of metal ions by living microorganisms. Unlike biosorption, which is a passive surface process, bioaccumulation is metabolically driven and occurs during the growth and metabolic activity of microbial cells. Microorganisms such as bacteria, algae, and yeast absorb metal ions through membrane transport systems, followed by intracellular binding to proteins, peptides, and other biomolecules that serve as detoxification or storage agents [27]. This process enables microbes to tolerate and accumulate substantial quantities of toxic or valuable metals from their surrounding environment.

A diverse range of metal-resistant microorganisms has demonstrated bioaccumulative potential for elements such as copper, cobalt, nickel, zinc, and rare earth elements. For instance, cyanobacteria and green microalgae have been shown to accumulate rare earth and lithium ions, while bacterial species such as *Pseudomonas aeruginosa* and *Bacillus subtilis* can actively uptake copper and cadmium. Although bioaccumulation is limited by microbial tolerance thresholds and slower kinetics, it remains an important mechanism for metal recovery and detoxification. When combined with other biotechnological processes, bioaccumulation contributes to an integrated and sustainable approach to e-waste recycling.

3.1.4 Enzymatic or protein-mediated separation

Enzymatic or protein-mediated separation represents an emerging frontier in bio-based e-waste management. In this process, specific enzymes and metal-binding biomolecules are used to facilitate selective metal recovery and degradation of organic matrices. Certain microbial and fungal enzymes, such as laccases, peroxidases, and hydrolases, are capable of degrading complex polymeric compounds present in PCBs and plastic components of e-waste [28]. This enzymatic degradation enhances metal liberation by breaking down epoxy resins, brominated flame retardants, and other binding materials, thereby improving the accessibility of metallic fractions for subsequent recovery.

Additionally, metal-binding proteins and peptides such as metallothioneins and phytochelatins play a critical role in the selective sequestration of metal ions. These biomolecules exhibit a strong affinity toward heavy metals, including cadmium, zinc, copper, and mercury, and can be harnessed for bio-based metal capture and purification. Advances in molecular biology and synthetic biotechnology have further enabled the engineering of microbial systems to overexpress such proteins, enhancing their binding specificity and capacity through enzyme conjugation [29]. Consequently, enzymatic and protein-mediated strategies provide an innovative, nontoxic, and highly selective approach for the separation and recovery of valuable metals from e-waste, complementing traditional bioleaching and biosorption methods.

3.1.5 Phytoremediation, phytoseparation, and phytomining

Phytoremediation and its specialized branches, such as phytoseparation and phytomining, represent eco-friendly, plant-based strategies for managing e-waste-derived contamination and recovering valuable metals [30]. These processes exploit the natural ability of certain plants to absorb, translocate, and accumulate metals or metalloids from contaminated soils, water, or e-waste leachates. In phytoremediation, plants act as biological filters that stabilize or extract pollutants, thereby reducing environmental toxicity while improving soil and ecosystem health. Phytoseparation specifically refers to the use of plants for selective metal uptake and recovery, where accumulated metals can later be extracted from harvested biomass. Phytomining, on the other hand, focuses on the deliberate cultivation of hyperaccumulator plants capable of extracting economically valuable metals such as nickel, gold, copper, and cadmium from low-grade ores or e-waste contaminated substrates [23]. The metal-rich biomass can then be processed through thermal or chemical methods to recover pure metals, turning waste into a renewable resource.

The success of these plant-based systems depends on factors such as plant species, growth rate, metal tolerance, and soil or substrate composition. Recent studies have explored genetic engineering and symbiotic microbial enhancement to improve metal uptake efficiency and tolerance levels. Overall, phytoremediation, phytoseparation, and phytomining collectively support the CE by transforming e-waste pollutants into recoverable resources through sustainable, green biotechnology. Plants can also be used as a biological component to mitigate e-waste, especially after contamination with water, soil, or air (Table 2). In this process, pollutants from e-waste can be removed, stabilized, and degraded. In many cases, this method is applied to e-waste contaminated sites because of mismanagement and improper discarding. Hence, this method is used to remove, degrade, or contain heavy metals and other contaminants from soil using plants. From e-waste, soil can be contaminated with various types of metals like arsenic, lead, cadmium, zinc, copper, chromium, and mercury. Some plants act as hyperaccumulators and can absorb these metals from the contaminated soil. Sunflowers (*Helianthus annuus*), Indian mustard (*Brassica juncea*), and willow (*Salix* sp.) are famous examples of absorbing metals from the soil and storing them in their plant tissues. Some other plants do phytostabilization, which means that plants can regulate the migration of metals in the contaminated site, and usually prevent contamination of metals with the groundwater. The heavy metals in e-waste can be trapped by some plants like vetiver grass (*Chrysopogon zizanioides*) and certain types of grasses. Some plants may be important for mitigating PCBs in e-waste. Normally, plastics can be subjected to phytodegradation by plants by releasing enzymes to degrade toxic materials into harmless components.

Moreover, aquatic plants like water hyacinth (*Eichhornia crassipes*) and duckweed (*Lemna* sp.) can be utilized to clean contaminated aquatic systems by e-waste components such as heavy metals. Aquatic plants can do rhizofiltration, where their roots can absorb or filter out harmful substances from water. Aquatic macrophytes play an essential role in the bioremediation of e-waste-contaminated water bodies and leachates originating from dumpsites and recycling operations. These plants are highly efficient at absorbing, accumulating, and transforming dissolved metals and organic pollutants due to their rapid growth rates, high surface area, and ability to thrive in nutrient and pollutant-rich environments. Aquatic phytoremediation not only removes contaminants but also improves water quality by reducing turbidity, nutrient load, and toxicity.

Plants use several mechanisms for phytoremediation and phytomining, such as absorption, adsorption, chelation, reduction, and transformation. Metals are initially bound to cell walls or root surfaces via hydroxyl, carboxyl, and amino functional groups, then actively transported into cells and stored in vacuoles or bound with phytochelatins and metallothioneins. Oxidative stress induced by metal exposure triggers the production of antioxidant enzymes (such as superoxide dismutase, catalase, and peroxidase), which protect plant tissues from damage. Also, several factors may influence the efficiency of phytoremediation, such as the type of contaminant, soil condition of the e-waste contaminated site, plant species, and growth period.

Table 2: Plants with phytoremediation potential and mechanisms.

Plant species	Mechanisms/Characteristics	Target metal or e-waste component
Brassica juncea (Indian mustard)	Growing in moderately polluted soils, it is ideal for phytoextraction and phytoaccumulation	Pb, Cd, Cu, Ni, and Zn
Helianthus annuus (sunflower)	High tolerance, used in phytoextraction and rhizofiltration systems	Cd, Pb, and Cr
Vetiveria zizanioides (vetiver grass)	Effective in soil stabilization and phytostabilization	Pb, Hg, and Cu
Phragmites australis (common reed)	Used in constructed wetlands For wastewater and leachate treatment to accumulate metals through rhizofiltration	Cu, Zn, Ni, and Pb
Populus deltoides (poplar) and *Salix viminalis* (willow)	Phytoextraction and stabilization of contaminated soils	Cd, Cu, and Zn
Alyssum murale and *Thlaspi caerulescens*	Hyperaccumulators, exceptional metal uptake efficiency, and use in phytomining applications	Ni and Zn
Acasia sp.	Uptake and accumulation in leaves, stems, and roots	Ni
Lemna minor (duckweed)	Strong metal-binding capacity through ion exchange, surface adsorption, and intracellular sequestration	Cu, Zn, Cd, Pb, Ni, and Cr
Eichhornia crassipes (water hyacinth)	Provide a large surface area for metal adsorption and microbial colonization, supporting synergistic plant-microbe remediation systems	Pb, Hg, Cu, and Cd
Pistia stratiotes (water lettuce)	Accumulate metals both in roots and leaves, showing potential for phytoextraction	Cd, Pb, and Ni
Hydrilla verticillata and *Ceratophyllum demersum*	Absorbing dissolved metals through direct contact with contaminated water, effective in phytofiltration	E-waste wastewater streams

Plants' potential for e-waste management brings opportunities in the CE. Specifically, phytomining plants can absorb a significant number of important metals from e-waste and store them in their tissues. Later, these metals, like gold, lead, and copper, can be extracted and reused in other industries where the CE can be implemented. These potentials can benefit society, providing opportunities for young, innovative groups in e-waste management.

3.2 Integration of bio-based separation in e-waste processing

There are some other hybrid methods as well, such as chemical and biological hybrid methodologies [31]. In these methods, safer chemicals and compatible microbes are encouraged for better and more efficient extraction of metals from the e-waste. The integration of bio-based separation technologies into e-waste management systems represents a transformative step toward sustainable and circular processing. Traditional, mechanical, and chemical methods are often effective but energy-intensive and reliant on toxic reagents such as cyanides, aqua regia, and strong acids. Moreover, there is a potential for hybrid approaches with biocyanidation, which uses cyanogenic bacteria to produce cyanide and efficient recovery of valuable metals like gold from e-waste [32].

Integrating bioleaching with hydrometallurgical or mechanical pretreatment steps is particularly promising. Mechanical dismantling and size reduction increase surface area and improve microbial access to metal-bearing particles. Subsequent bioleaching using acidophilic or metal-tolerant microorganisms such as *Acidithiobacillus ferrooxidans, Leptospirillum ferrooxidans, Chromobacterium violaceum* ATCC 12472, or *Pseudomonas fluorescens* can solubilize metals efficiently. The leachates obtained from bioleaching can then be subjected to biosorption or bioprecipitation for selective metal recovery and purification. Similarly, aquatic phytoremediation systems can be used in the downstream to treat effluents from e-waste facilities. Plants such as *Lemna minor, E. crassipes,* and *B. juncea* absorb residual metals and reduce the toxicity of leachates before discharge or reuse.

3.3 Advantages of bio-based separation

Recent advancements in molecular biology and synthetic biotechnology have opened new possibilities for enhancing the biological management of e-waste through genetically engineered microbes and bionanotechnology. Conventional microorganisms often face limitations such as low tolerance to metal toxicity, slow metabolism, or narrow substrate specificity. Genetic modification enables the development of microbial strains with enhanced capabilities for metal tolerance, biosorption, enzymatic degradation, and bioleaching efficiency. For instance, bacteria and fungi can be engineered to overexpress metal-binding proteins, transporters, or oxidoreductases, improving their capacity to capture and transform toxic metals such as cadmium, lead, chromium, and copper into recoverable or less toxic forms.

In parallel, bionanotechnology integrates biological processes with nanomaterial synthesis to recover and valorize metals from e-waste. Microbes and plants can act as natural nanofactories, producing metal nanoparticles such as gold, silver, or palladium with potential applications in catalysis, electronics, and medicine. This biogenic nanoparticle synthesis not only contributes to e-waste valorization but also reduces reliance on chemical reagents and energy-intensive methods used in traditional nanoparticle fabrication.

Furthermore, integration of genetic engineering with bionanotechnological approaches enhances the precision and efficiency of metal recovery. Engineered microbial consortia and biofilms can be optimized for multimetal systems or complex waste matrices, promoting selective recovery and detoxification. Together, these cutting-edge strategies represent the next generation of bio-based e-waste management tools, bridging green biotechnology and CE principles to create high-value, sustainable recovery pathways.

4 Waste-to-resource transformation in the circular economy (CE) framework

In the digital era, the extensive use of electronic appliances has increased rapidly, leading to a significant rise in e-waste, which is now recognized as one of the largest and fastest-growing sources of pollution worldwide [33]. Global e-waste generation was approximately 53.6 million tons in 2019 and is projected to reach 74 million tons by 2030. E-waste categories include large household appliances, consumer electronics, and information and communication technology devices such as personal computers, monitors, and laptops. E-waste mainly consists of metals (e.g., copper, gold, silver, palladium, aluminum, and iron), plastics, and metal-plastic composites [34], all of which can negatively impact the environment and human health.

Even though e-waste has impacts on the environment and biota, it also holds significant economic potential due to the presence of high-value raw materials such as silver and gold. E-waste dumping and land-filling sites offer significant opportunities for material recovery. The global economic value of e-waste is estimated at approximately USD 62.5 billion annually. Therefore, recovering and reusing materials from e-waste are essential for mitigating environmental impacts and generating financial benefits [35]. Moreover, the massive demand for electronic items may result in a shortage of production resources. Therefore, the implementation of efficient e-waste management practices aligned with the principles of the CE represents an effective approach to reducing environmental pollution and promoting sustainable resource utilization by minimizing the use of virgin materials in production and reducing overall energy consumption [35, 36]. In CE, products and materials are actively repaired and recycled within a closed loop, promoting the principles of reuse, repair, and recycling, opposed to the conventional "take, make, and dispose" model.

The Reduce, Reuse, and Recycle (3Rs) concept forms the foundation of the CE and is regarded as the basis for the development of green manufacturing in the 1990s. In recent decades, the transition from green to sustainable manufacturing has been increasingly emphasized to provide sustainable value to society. Therefore, the 6Rs concept was developed, integrating additional strategies such as Redesign, Remanufacture, and Recovery. Over time, this framework has continued to evolve, broadening the range of strategies available for advancing the CE. Specifically, Redesign has been refined into Refuse and Rethink, while Remanufacturing has been expanded to include Repair, Refurbish, and Repurpose, leading to the establishment of the advanced 10Rs framework within the CE [37].

Accordingly, the 10Rs framework can be further subdivided into three categories based on CE strategies to control the e-waste generation and its accumulation in the environment:

(1) Smarter product use and manufacturing strategy – Refuse, Rethink, and Reduce

(2) Extending the lifespan of products and their components strategy – Reuse, Repair, Refurbish, Remanufacture, and Repurpose

(3) Useful application of materials strategy – Recycle and Recover

4.1 10Rs framework in CE for e-waste management [37]

1) **Refuse:** This concept aims to prevent waste generation at the design and consumption stages. In the context of e-waste management, it can be applied by avoiding short-lived product designs, such as low-cost electronic devices with minimal durability and limited repair options (e.g., low-quality headphones).
2) **Rethink:** This concept is applied at the product design stage to increase durability and enable more efficient use. Designing electronic products for longer lifespans, modular upgrades, or multifunctional use strengthens the CE through the Rethink concept to control e-waste.
3) **Reduce:** The frequent replacement of electronic items can be reduced by designing and using devices with an extended lifespan through repair and reuse, thereby helping to control e-waste generation.
4) **Reuse:** The reuse concept encourages the use of electronic products in good condition for a second or subsequent time.
5) **Repairing:** It helps replace broken parts to restore the proper functioning of the devices, extending the product's lifespan.
6) **Refurbishing:** This concept involves updating and restoring old electronic devices to improve their functionality and align with the latest technology, reducing the need for new products and helping to control e-waste generation.
7) **Remanufacturing:** It involves incorporating components extracted from e-waste into new products to provide the same functionality as the original components.
8) **Repurposing:** It is like remanufacturing, but in this case, the incorporated components provide a different functionality from that of the original devices.
9) **Recycle:** E-waste can be a valuable source of secondary materials. These materials can be upcycled, transforming them into higher-quality materials, or down-

cycled, which produces lower-value outputs. Both upcycled and downcycled components can then be used as raw materials for new electronic devices.
10) **Recovery:** It involves using e-waste as a source of energy or valuable compounds when they cannot be recycled.

Moreover, the World Economic Forum 2019 has further subdivided the CE approach in e-waste management into several sections, including product design, reintegration of manufacturing scrap, repair and durability, higher product collection and return systems with consumer incentives, and advanced recycling and recapture. Products should be designed for durability, reuse, and safe recycling, rather than being discarded after a short lifespan. Priority should be given to developing lighter, more compact devices that incorporate emerging technologies, including advanced digitalization and cloud-based services (Figure 2). Modular designs that enable easy repair and component replacement, together with the use of robust materials and regular software upgrades, can extend product lifespans and promote reuse and refurbishment [35, 38].

Reintegration of manufacturing scrap involves recycling waste materials, such as scrap metal, from production into new components (Figure 2) [38]. GEM Company in China is a leading waste recycling enterprise in the world. Annually, it recycles more than 10% of China's total scrap, including over 10% of waste batteries and more than 30 types of scarce resources such as cobalt, nickel, copper, tungsten, gold, silver, palladium, rhodium, germanium, and rare earth elements. GEM operates six e-waste dismantling enterprises across China to recycle e-waste from products such as televisions, refrigerators, washing machines, air conditioners, computers, telephones, range hoods, water heaters, etc., serving as a pioneer in promoting the CE and low-carbon manufacturing in the country [39].

Repair and durability (life extension) are also concerns in this approach. Advanced repair, refurbishment, and efficient secondary markets ensure the extended lifespans and multiple usage cycles of products. In some cases, products are offered as a service model, where consumers pay for access rather than ownership. This model encourages durable, maintainable design, supports multiple use cycles, and promotes repair, refurbishment, and recycling, thereby advancing CE goals (Figure 2) [38].

Higher product collection and return with incentives for consumers (end of life) are attractive to society. The collection of e-waste and the establishment of take-back systems are essential components of effective e-waste management. Obsolete electrical items are collected from companies under the take-back system for repair, refurbishment, and recycling. In developing countries, the growing involvement of proper e-waste collection systems ensures safer handling, enhanced material recovery, and more efficient integration of discarded products into recycling and refurbishment streams (Figure 2) [38].

Advanced recycling and recapturing are necessary in e-waste management. Rules and regulations that encourage the recycling of e-waste and the use of recycled materials in new products help reduce dependence on virgin resources, thereby strengthening CE initiatives. The implementation of advanced recycling technologies ensures the recovery of valuable materials, such as metals, rare earth elements, and high-grade plastics, while securing their quality for reuse as raw materials in manufacturing (Figure 2) [38].

By 2012, China had established a comprehensive legal and administrative framework to implement CE principles in e-waste management. The ordinance issued by the State Council introduced a producer-funded mechanism to support e-waste treatment and recycling, while the Ministry of Finance developed measures for fund collection and management related to recovery and disposal. This framework facilitated the establishment of 109 standardized e-waste recycling enterprises specializing in the dismantling of televisions, refrigerators, washing machines, air conditioners, and computers. From 2012 to 2020, over 600 million units of e-waste were processed through standard recycling channels, achieving recycling rates of around 94% for televisions and over 77% for refrigerators [39].

The application of microorganisms (i.e., bacteria, algae, and fungi) for metal extraction from e-waste through the biodegradation of metallic ores and electronic components is referred to as an emerging and environmentally sustainable biotechnological approach. This process is known as bioleaching or biohydrometallurgy. Being a low-energy process, it does not require the use of toxic chemicals or other hazardous substances; thus, it is a more economical and environmentally friendly approach. Comprehensive investigations on their industrial-scale applications are recommended to facilitate the implementation of bioleaching in e-waste management [40], as it represents a biological approach to recycling e-waste within the CE framework.

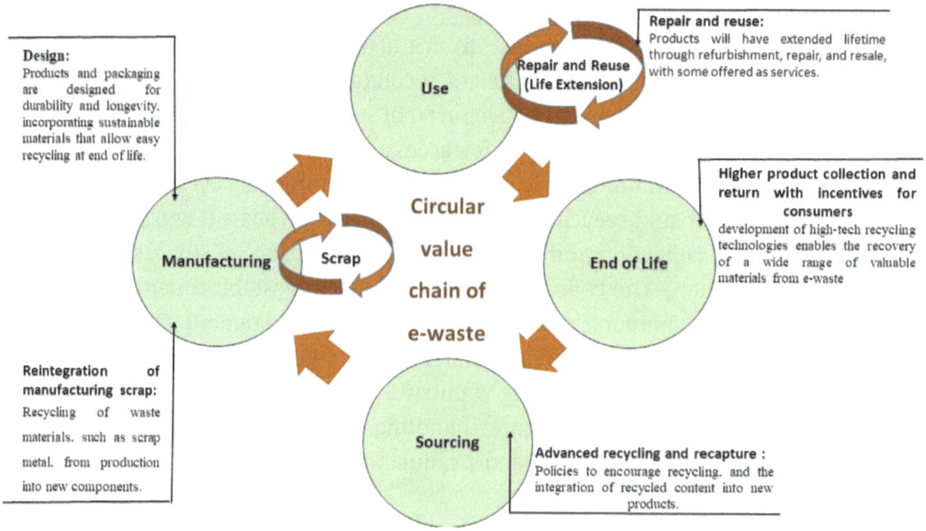

Figure 2: Circular value chain of e-waste [38].

4.2 Hybrid models (bio + conventional methods) in CE

Even though bioleaching is a cost-effective and environmentally friendly approach, it is a time-consuming process. Chemical leaching is a rapid process; however, it has negative effects on the environment and biota. Therefore, combining chemical and biological approaches can provide an efficient and eco-friendly method for metal extraction from e-waste. A hybrid combination of ligands with associated microbes has the potential to efficiently extract metals from e-waste for recycling within the CE framework. For example, Cd, Pb, and Cu extraction from e-waste can be enhanced using oxalate along with ammonium citrate, while the addition of *A. niger* or *Penicillium* sp. can further increase metal extraction efficiency. Moreover, maltol (3-hydroxy -2-methyl-4-pyrone) combined with *Clostridium beijerinckii* shows potential for efficient Fe extraction from e-waste [31].

4.3 Contribution of the CE approach to e-waste management in generating green jobs

Recycling and reusing e-waste not only mitigate their adverse impacts on the environment and living organisms but also contribute to economic development by creating employment opportunities. E-waste recycling promotes the development of "green-collar" employment opportunities that prioritize environmental sustainability. Employers in recycling facilities receive training in environmentally responsible practices, thereby fostering a workforce with a comprehensive understanding of sustainability and its contribution to the global economy. Moreover, integrating new technologies into e-waste recycling is crucial for efficiently and safely recovering valuable materials from e-waste. Therefore, innovations driven by the research and development sector generate high-tech job opportunities across diverse disciplines such as environmental engineering, materials science, and waste management [41].

It is estimated that every 1,000 tons of e-waste can generate approximately 12.6 direct and 20 indirect jobs through processes such as collection, assessment, recycling, refurbishment, and distribution and transportation. Currently, the global e-waste recycling rate is approximately 20%, producing about 0.3 million jobs worldwide; however, it could increase to 1.5 million with the achievement of a 100% recycling rate [42].

4.4 Importance of CE approach in e-waste management for achieving the Sustainable Development Goals (SDGs)

The United Nations Member States adopted the 2030 Agenda for Sustainable Development in 2015, establishing 17 Sustainable Development Goals (SDGs) aimed at eradicating poverty, protecting the planet, and ensuring prosperity for all. However, improper disposal of e-waste and its associated environmental and health effects can negatively affect progress toward achieving several SDGs, particularly those related to environmental sustainability and human well-being [43].

Proper e-waste management is closely linked with SDG 3 (Good Health and Well-being), SDG 6 (Clean Water and Sanitation), SDG 11 (Sustainable Cities and Communities), SDG 12 (Responsible Consumption and Production), and SDG 14 (Life Below Water) [43]. SDG 3 focuses on reducing health impacts and deaths caused by hazardous chemicals, and air, water, and soil pollution. SDG 6 aims to ensure universal and equitable access to safe and affordable drinking water for all while reducing pollution, eliminating dumping, and minimizing the release of hazardous substances. SDG 14 addresses marine pollution and the protection of marine ecosystems [43]. Improper disposal of e-waste can release hazardous substances such as heavy metals, which contaminate freshwater and groundwater, threatening the safety of drinking water and ultimately affecting marine environments. The bioaccumulation of these pollutants along food chains can also cause severe health issues in animals. Therefore, e-waste management through the CE approach represents a promising solution for achieving these SDGs and creating a healthier planet for future generations. SDG 11 aims to reduce the adverse per capita environmental impact of cities by paying special attention to air quality, as well as municipal and other waste management. Since most e-waste is generated in cities, it is crucial to implement effective e-waste management in urban areas, improve collection and recycling rates under the CE approach, and thereby reduce the number of e-waste that ends up in dumpsites. SDG 12 focuses on reducing waste generation through prevention, reduction, repair, recycling, and reuse, which are directly aligned with the principles of the CE [43].

5 Challenges and limitations of biological approaches in e-waste management

The biological approaches of e-waste management offer greater environmental benefits compared to traditional methods. While effective on the laboratory scale, transferring them to the industrial scale is tricky. The key impediments are the heterogeneity of phases of e-waste, slow reaction kinetics that are generally susceptible to inhibition at solid concentrations of relevance, and the technical complexity of operating three-phase reactor systems and treating complicated effluents [10, 28, 44]. In addition, co-dissolution of greater than one metal complicates selective recovery and increases operating costs. Market and policy issues, such as unpredictable supply of feedstock, excessive dependence on backyard recycling, and random regulation, also reduce technoeconomic feasibility [45, 46]. Unless controlled through the acidic/alkaline liquors and metal-rich sludges, these so-called green processes may undermine their environmental benefit through unrecognized risks [25, 47].

One of the ongoing challenges is the heterogeneity of e-waste streams. Printed circuit boards, display screens, batteries, and mixed plastics vary in resin chemistry, solder types, brominated flame retardants, and metal grades. This heterogeneity can cause different issues, such as inhibiting microbial activity, reducing mass transfer rates, leading to passivation, secondary mineral precipitates, and decreasing recovery rates over time [10, 45]. It can also hinder the standardization of operating conditions across batches and complicate the automation of downstream separation technology [10]. To overcome this challenge, front-end liberation and classification, which includes size reduction and density, or eddy-current sorting, should be used to create composition-bounded feeds matched to specific biocatalysts and operating conditions. Quick leachability screening with simple pH stat tests can then direct subfeeds to either acidophilic or fungal lixiviant lines before they enter the reactors.

Kinetic constraints and biocatalyst tolerance are the second major bottleneck. To achieve economically feasible throughputs, higher solids loadings are needed; however, increasing pulp density tends to suppress microbial activity due to metal toxicity, oxygen transfer issues, increased viscosity, and surface precipitates. Conversely, very low solids content reduces contact and extraction efficiency, causing a performance "cliff" [10, 44]. Although aeration can address viscosity problems, excessive gasification may create shear stress and destroy biofilms, while jarosite-type precipitates may reabsorb metals and lead to surface fouling over time [10]. Low-shear aeration in airlift or draft-tube reactors at low pH minimizes corrosion of moving parts and protects biofilms. Speciation control of Fe^{3+} and sulfate (antiscalant strategy) in combination with seeded precipitation off-loop prevents jarosite formation and maintains clean reactor surfaces. Tuning of Fe^{2+}/Fe^{3+} ratios provides in situ redox buffering for maintaining stable oxidation reduction potential (ORP) without overaeration.

Fungal pathways, which are frequently more tolerant of complex matrices, tend to operate at a slower pace. Conversely, acidophilic bacteria and biogenic lixiviants can function more rapidly; however, they produce acidic or alkaline liquors that must be meticulously neutralized and selectively captured to preserve an environmentally sustainable balance [10, 44]. Scaling upward from shake flasks to engineered reactors presents additional challenges, including three-phase (gas-liquid-solid) hydrodynamics, corrosion at low pH levels, solids handling, and the online regulation of pH and ORP, all of which contribute to increased costs and complexity [10]. Methods such as sequential or pulse-plate operations, along with the "grow-then-leach" two-step process, can enhance mass transfer and mitigate toxicity during microbial growth phases in degradation processes. Nonetheless, these techniques also lead to increased energy consumption and media requirements, with their efficacy diminishing at elevated pulsing frequencies. Overall, these methods raise capital costs such as for corrosion-resistant vessels and separation equipment, and operational expenses, including aeration, mixing, acids/bases, nutrients, and effluent treatment. This makes their overall cost profile closer to that of traditional hydro and pyrometallurgical methods, unless heat and reagents are effectively recycled [10].

Achieving high selectivity and purity remains a challenge; the co-dissolution of Cu, Ni, Zn, precious metals, and rare earth elements requires multistep processes such as solvent extraction, ion exchange, or selective precipitation, involving additional chemicals and operational steps [10]. Although biosorption and bioaccumulation can improve selectivity for target ions, robust high-throughput capture methods from authentic bioleach liquors that are resistant to fouling and compatible with regeneration are not yet established at the pilot scale [28, 44]. Inline installation of "toxicity traps" (sacrificial biosorbents or selective precipitants) upstream of the main bioreactor can remove acute inhibitors, for example, Cu^{2+} spikes, and protect the biocatalysts. Staged leaching, where the base metals are liberated first, followed by Ni/Co/REEs, reduces separation complexity further downstream. Terminal polishing via electrowinning or selective sulfide precipitation should be matched to liquor speciation, with barren liquor recycled after pH/ORP trim.

Subsequently, technoeconomic assessments and life cycle analysis are founded on residence time (area and volume of the reactor), energy for mixing and aeration, reagents for ORP and pH adjustment, and downstream separation cost and environmental repercussions [10, 25]. Market dynamics and systemic factors further challenge the viability of business models. In India, despite the existence of hundreds of licensed facilities operating under Extended Producer Responsibility regulations, actual capacity utilization remains low, while a well-entrenched informal sector continues to handle most waste flows. This fragmentation compromises stream integrity and undermines feedstock quality and predictability for formal processing plants [25, 46]. Enhancing stockpiles, enforcement mechanisms, and the effective integration of informal factors are vital prerequisites for guaranteeing a stable supply chain to bio-based processors [46]. Supply stability could be further strengthened through contracted take-back schemes with quality specifications, buffer stockpiles, and price premiums for sorted fractions, alongside the integration of informal collectors through cooperatives or franchise models anchored by safe preprocessing hubs that improve stream integrity for bio-based plants.

Furthermore, environmental, health, and safety measures are critical components of any sustainable "green' waste management strategy. During crude recovery, toxic metal substances like heavy metals, Pb, Cd, Hg, Cr, and persistent organic compounds such as polybrominated diphenyl ethers, along with polluted air, soil, and water, must be regularly monitored when using biodegradable e-waste. Research and field studies revealed that exposure to these substances can lead to endocrine disruption, reduce lung function, adverse birth outcomes, and genetic damage, particularly affecting children and workers [25, 28, 47–49]. Bio-based recovery processes help reduce open burning and acid stripping to some extent; however, these methods still produce metal-rich, acidic, or alkaline effluents that require careful collection and further treatment to prevent and mitigate environmental contamination through sludge and effluents [10]. Further, implementing a closed-loop liquor handling system equipped with secondary containment, lined sumps, and real-time pH/ORP and metal monitoring of discharge lines is essential for ensuring safe and efficient operation. In addition, establishing a well-defined end-of-life pathway for metal-laden biomass or solid residues is critical. Effective options include thermal ashing with appropriate off-gas scrubbing systems or wet-chemical stripping and metal recovery, both of which prevent uncontrolled disposal and minimize environmental impact.

Informal operations have grown faster than formal operations in all but the fastest-growing economies in most Asian countries, including Sri Lanka. Hazardous waste is often not disposed of safely and in an engineered way. Policy design needs more collection networks, better data systems, and their engagement in environmentally friendly management [25, 45, 50]. To make biological processes a worthwhile addition to mechanical and hydrometallurgical processes, there is a necessity of developing basic pretreatment plants, improving data gathering, and implementing controls throughout the boundaries. Such actions must be undertaken to realize real benefits from a CE.

Today, the ceiling is set by true solids heterogeneity, narrow kinetic windows, and effluent/purity control costs. In this case, the advances in process intensification, high solids contact with gentle hydraulics, inline selectivity to blunt toxicity, specification driven feed systems which formalize the informal collector, and policy rewarding verified environmental performance over throughput are fundamental for making bio-based flowsheets a credible complement to mechanical and hydrometallurgical routes in a circular e-waste system [10, 28, 44, 46].

6 Future directions

As e-waste continues to increase over time, environmentally friendly approaches that support a CE offer a promising pathway forward. Hence, the future efforts should focus on technology enhancements, innovative treatment methods, increased public awareness, robust policy formulation, and strengthened interdisciplinary collaborations. These elements are essential for developing sustainable, scalable, and effective e-waste management strategies.

From a sustainability perspective, future directions should emphasize process integration and life cycle assessment to evaluate the economic and environmental feasibility of bioremediation methods. Overall, technological advancements in e-waste management present promising opportunities for the fields of systems biology, materials science, and biotechnology. These developments can support the transition of bioremediation from laboratory-scale research to industrial applications while addressing the current technological gaps.

E-waste consists of heterogeneous materials with highly complex compositions. To enhance the efficiency of the bioremediation process, achieving a more homogeneous material stream is essential. From a materials science perspective, segregation of e-waste components is recommended to facilitate this goal. Bioremediation of e-waste can be further advanced through enhanced microbial-assisted degradation, plant-based remediation techniques, and vermiremediation, each utilizing different biological systems to target specific contaminants. These complementary approaches offer promising pathways for improving the overall effectiveness and scalability of biologically based e-waste management [51]. Biological systems offer valuable opportunities to convert environmental e-waste into renewable and green energy by integrating value addition, metal extraction, and sustainable innovation. Approaches rooted in green chemistry and bionanotechnology are particularly effective for producing nanomaterials while ensuring optimal resource utilization. Combining bioleaching with chemical recovery, bionanotechnology, and green catalysts is recommended to improve cost-effectiveness and increase the overall yield of material recovery. Bioelectrochemical systems, such as bioelectrorefineries and microbial fuel cells, also represent promising technologies for metal recovery without generating secondary pollution. Additionally, the performance of phytoremediation can be enhanced through biochar-assisted biosorption systems and phyto-microbial consortia, which improve metal uptake in polluted leachates and contaminated soil environments.

Advances in omics technologies offer powerful and innovative approaches to address future challenges in bioremediation. Multiomics tools, including metabolomics, genomics, proteomics, and transcriptomics, enable the identification of key proteins, genes, and regulatory pathways involved in metal solubilization, tolerance, and resistance associated with e-waste contaminants. Bioinformatics also serves as a crucial component in e-waste bioremediation research. It can be used to analyze microbial genomes and metabolic pathways to determine the capacity of different microorganisms to degrade pollutants and recover heavy metals from e-waste. These computational approaches improve our understanding of biological mechanisms and support the development of more efficient, targeted bioremediation strategies.

Furthermore, genetic engineering presents a promising avenue to enhance the performance of microbial strains in e-waste bioremediation. Innovative genetically modified microorganisms with engineered metabolic pathways can accelerate the degradation of e-waste by improving bioavailability, tolerance to toxic substances, and overall efficiency. These advancements pave the way for the development of next-generation microbial strains specifically tailored to target e-waste components. Future research directions should also include e-waste-based plastic degradation techniques for various polymer types. According to literature, more than 500 potential PET hydrolase enzymes have been identified from terrestrial and marine metagenomic datasets, highlighting a rich resource for developing effective biodegradation strategies [10]. Innovative approaches utilizing locally available microbial strains from operational environments, polluted sites, and natural areas offer promising solutions for plastic waste degradation [10]. However, the implementation of advanced bioremediation techniques requires sophisticated facilities, skilled personnel, and significant initial capital investment. These financial and technological barriers can pose substantial challenges for companies, particularly in developing countries. Therefore, effective technology transfer mechanisms such as Training of Trainers programs and targeted financial investments are essential to promote the large-scale adoption of bioremediation approaches in low- and middle-income countries.

The future implications should emphasize the expansion of scientific research and the translation of laboratory findings into practical bioremediation technologies. Laboratory findings will pave the pathway to realistic bioremediation techniques by developing existing recycling processes toward a resource-assured and zero-waste future. Also, the interdisciplinary research initiatives must be promoted to drive innovation in e-waste recycling, aligning with principles of the CE. For example, toxicity tolerance capacity and enzymatic degradation need to be addressed by future research studies to improve the performance of bioremediation. Additionally, the development of field-assisted phytoremediation systems and pilot-scale bioreactors are critical priority for achieving scalable, industrial-level solutions. The creation of advanced genetic tools represents another important research avenue for the recovery of resources and improving bioremediation from multiple perspectives (e.g., improvement of the substrate, enhancing enzymatic activities, develop the level of toxicity tolerance). Future research implications should prioritize the discovery and development of microbial strains capable of degrading the complex and composite material compositions of e-waste, rather than focusing solely on individual raw materials such as plastics or metals. The practical application of pilot-scale laboratory findings will be essential for reducing selective e-waste accumulation in the environment. Hence, strong collaboration between the industrial sector and the scientific community is highly recommended to develop feasible and effective solutions for real-world challenges in e-waste management. Furthermore, the establishment of comprehensive national and regional e-waste monitoring systems remains a critical requirement for ensuring effective and sustainable future management strategies.

Additionally, strong collaborations among business leaders, policymakers, scientists, and academicians are essential for the formulation of effective policy frameworks to support the establishment of bioremediation approaches. At the national level, governments should prioritize recycling regulations, promote eco-literacy, and privatize e-waste management. Incorporating the Extended Producer Responsibility concept into the national policies for electronic products is highly recommended. This approach holds manufacturers of electrical and electronic goods accountable for responsible production, marketing, waste handling, and management throughout the entire lifecycle of their products [2]. Effective legal enforcement of the responsible disposal of e-waste is also necessary. For instance, India became the first South Asian country to implement dedicated e-waste legislation in 2011. This legal framework comprehensively addresses various aspects of e-waste management, including storage, transportation, and recycling processes [2].

In addition, public perception and behavior play a crucial role in the responsible disposal and management of all types of waste. Therefore, public education and community awareness programs are vital for fostering responsible e-waste management and encouraging the adoption of bio-based recycling practices. Enhancing eco-literacy among key stakeholder groups (e.g., electronic equipment manufacturers and companies) regarding the health risks and ecological damage caused by accumulated e-waste is particularly important. This approach can drive significant changes in attitudes toward e-waste disposal and promote a shift toward environmentally friendly bioremediation methods.

As a summary, biological approaches have the potential to transition from experimental methods to mainstream sustainable practices for global e-waste recovery and product circularity. Technological innovations should integrate e-waste recycling together with biotechnological solutions to enhance efficiency and sustainability. Additionally, robust environmental regulations and comprehensive e-waste recycling policy frameworks are essential to achieving CE objectives effectively.

7 Conclusion

E-waste generated from multiple sources poses significant public health risks and environmental hazards due to the presence of toxic substances. Conventional e-waste management and disposal practices (physical and chemical methods) result in negative effects and limitations. In contrast, biological approaches offer green alternatives that are pivotal in advancing toward a sustainable CE. Accordingly, bioremediation practices act as safer recycling solutions by preventing environmental degradation and protecting vulnerable populations from exposure to hazardous materials found in e-waste. Common bioremediation practices such as bioleaching, biosorption, bioaccumulation, phytoremediation, and bionanotechnology are widely employed to recover metals and other valuable materials from e-waste with minimal environmental impact. However, biological approaches have some challenges, such as limited scalability, slow kinetics, and inconsistent performance that hinder their large-scale applications. Recent advances in molecular biology, biotechnology, and omics technologies offer promising solutions to overcome these obstacles. In particular, genetically engineered microorganisms, plant-based recovery systems, and enzyme-mediated processes provide effective pathways for resource recovery. Technological innovations integrated with multidisciplinary approaches are suggested as a future implication for an effective e-waste recycling system. Strong interdisciplinary collaborations among research institutions, industries, and policymakers are essential to fully realize the waste management potential of bioremediation techniques. Bioremediation methods are expected to become an integral component of next-generation e-waste recycling processes, combining technological progress with economic sustainability and ecological resilience. Ultimately, responsible e-waste management and recycling are not only an environmental obligation but also a critical responsibility to safeguard human health and well-being.

References

[1] Arya S, Kumar S, E-wastes in India at a glance: current trends, regulations, challenges and management strategies. J Clean Prod 2020, 271, 122707.

[2] Jain M, Kumar D, Chaudhary J, Kumar S, Sharma S, Verma AS, Review on e-wastes management and its impact on the environment and society. Waste Manag Bull 2023, 1(3), 34–44.

[3] Wath SB, Dutt PS, Chakrabarti T, E-wastes scenario in India, its management and implications. Environ Monit Assess 2010, 172, 249–262.

[4] Ankit N, Saha L, Kumar V, Tiwari J, Sweta N, Rawat S, Singh J, Bauddh K, Electronic waste and their leachates impact on human health and environment: global ecological threat and management. Environ Technol Innov 2021, 24, 102049.

[5] Grant K, Goldizen FC, Sly PD, Brune MN, Neira M, van den Berg M, Norman RE, Health consequences of exposure to e-wastes: a systematic review. Lancet Glob Health 2013, 1(6), e350–e361.

[6] Gupta N, Nath M, Groundwater contamination by e-wastes and its remedial measure – a literature review. J Phys Conf Ser 2020, 1531, 012023.

[7] Javed M, Usmani N, Accumulation of heavy metals in fishes: a human health concern. Int J Environ Sci 2011, 2.

[8] Nie X, Fan C, Wang Z, Su T, Liu X, An T, Toxic assessment of the leachates of paddy soils and river sediments from e-wastes dismantling sites to microalga, Pseudokirchneriella subcapitata. Ecotoxicol Environ Saf 2015, 111, 168–176.

[9] Mishra J, Singh R, Arora NK, Alleviation of heavy metal stress in plants and remediation of soil by rhizosphere microorganisms. Front Microbiol 2017, 8, 295763.

[10] Han P, Teo WZ, Yew WS, Biologically engineered microbes for bioremediation of electronic waste: wayposts, challenges and future directions. Eng Biol 2022, 6(1), 23–34.

[11] Goyal S, Gupta S, A comprehensive review of current techniques, issues, and technological advancements in sustainable e-wastes management. e-Prime Adv Electr Eng Electron Energy 2024, 9, 100702.

[12] Cole C, Gnanapragasam A, Cooper T, Singh J, Assessing barriers to reuse of electrical and electronic equipment: a UK perspective. Resour Conserv Recycl X 2019, 1, 100004.

[13] Madkhali H, Duraib S, Nguyen L, Prasad M, Sharma M, Joshi S, A comprehensive review on e-wastes management strategies and prediction methods: a Saudi Arabia perspective. Knowledge 2023, 3(2), 163–179.

[14] Kumar P, Kumar P, Thakur SS, E-wastes and hazardous elements: a comprehensive study of chemical components and environmental threats. Int J Innov Sci Res Technol 2024, 9(10), 1074–1085.

[15] Hsu J, Wang J, Stern M, E-wastes: a global problem, its impacts, and solutions. J Glob Inf Manag 2024, 32(1), 1–28.

[16] Houessionon MGK, Ouendo EMD, Bouland C, Takyi SA, Kedote NM, Fayomi B, Fobil JN, Basu N, Environmental heavy metal contamination from electronic waste (e-wastes) recycling activities worldwide: a systematic review from 2005 to 2017. Int J Environ Res Public Health 2021, 18(7), 3517.

[17] Herat S, E-wastes management in Asia Pacific region: review of issues, challenges and solutions. Nat Environ Pollut Technol 2021, 20(1), 45–53.
[18] Baldé CP, Kuehr R, Yamamoto T, McDonald R, D'Angelo E, Althaf S, Bel G, Deubzer O, Fernandez-Cubillo E, Forti V, Gray V, Herat S, Honda S, Iattoni G, Khetriwal DS, Luda di Cortemiglia V, Lobuntsova Y, Nnorom I, Pralat N, Wagner M, Global e-wastes monitor 2024. Geneva/Bonn: International Telecommunication Union (ITU); United Nations Institute for Training and Research (UNITAR); 2024.
[19] Wu C, Awasthi AK, Qin W, Liu W, Yang C, Recycling value materials from waste PCBs focus on electronic components: technologies, obstruction and prospects. J Environ Chem Eng 2022, 10(5), 108516.
[20] Desmarais M, Pirade F, Zhang J, Rene ER, Biohydrometallurgical processes for the recovery of precious and base metals from waste electrical and electronic equipments: current trends and perspectives. Bioresour Technol Rep 2020, 11, 100526.
[21] Khaliq A, Rhamdhani M, Brooks G, Masood S, Metal extraction processes for electronic waste and existing industrial routes: a review and Australian perspective. Resources 2014, 3(1), 152–179.
[22] Nazari L, Xu C, Ray MB, Advanced and emerging technologies for resource recovery from wastes. In: Green chemistry and sustainable technology. 2021.
[23] Erkmen AN, Ulber R, Jüstel T, Altendorfner M, Towards sustainable recycling of critical metals from e-wastes: bioleaching and phytomining. Resour Conserv Recycl 2024, 215, 108057.
[24] Bhasha S, Thulasi M, Chintada V, Bioleaching: concepts and application in microbial metal mobilization. In: Kiran (ed) Sustainable green technologies. Sustainable economy and ecotechnology. Springer, Cham; 2025.
[25] Agarwal S, Darbar S, Saha S, Choudhury M, Singh RP, E-wastes management using different cost-effective, eco-friendly biological techniques: an overview. In: Waste management and resource recycling in the developing world. Elsevier eBooks; 2023. p. 205–235.
[26] Aslam A, Kanwal F, Javied S, Nisar N, Torriero AAJ, Microbial biosorption: a sustainable approach for metal removal and environmental remediation. Int J Environ Sci Technol 2025, 22(13), 13245–13276.
[27] Ghosh S, Chakraborty S, Bloch K, Microbe mediated metal recovery: a sustainable e-wastes management approach. In: Shah MP (ed) Extraction of precious metals from industrial waste using microbial technology. Environmental science and engineering. Springer, Cham; 2025.
[28] Ferreira-Filipe DA, Duarte AC, Hursthouse AS, Rocha-Santos T, Silva ALP, Biobased strategies for e-wastes metal recovery: a critical overview of recent advances. Environments 2025, 12(1), 26.
[29] Muneer MA, Fatima S, Hussain N, Mashifana T, Sayed A, Boczkaj G, Rajoka MSR, Enzyme conjugation – a promising tool for biocatalytic and biotransformation applications: a review. Top Catal 2024, 68(9–10), 876–892.
[30] Zulkernain NH, Basant N, Ng CC, Kriti N, Salari M, Mallick S, Recovery of precious metals from e-wastes through conventional and phytoremediation treatment methods: a review and prediction. J Mater Cycles Waste Manag 2023, 25(5), 2726–2752.
[31] Pant D, Joshi D, Upreti MK, Kotnala RK, Chemical and biological extraction of metals present in e-waste: a hybrid technology. Waste Manag 2012, 32(5), 979–990.
[32] Tran DT, Lin X, Nguyen TM, Song M, Yun Y, Development of an integrated bio-based cyanide production system for recovery of Au from waste printed circuit boards. Sep Purif Technol 2025, 363, 132332.
[33] Quinto S, Law N, Fletcher C, Le J, Antony Jose S, Menezes PL, Exploring the e-waste crisis: strategies for sustainable recycling and circular economy integration. Recycling 2025, 10(2), 72.
[34] Elgarahy AM, Eloffy M, Priya A, Hammad A, Zahran M, Maged A, Elwakeel KZ, Revitalizing the circular economy: an exploration of e-waste recycling approaches in a technological epoch. Sustain Chem Environ 2024, 7, 100124.

[35] Ali S, Shirazi F, The paradigm of circular economy and an effective electronic waste management. Sustainability 2023, 15(3), 1998.
[36] Yazdi M, Moradi R, Nedjati A, Pirbalouti RG, Li H, E-waste circular economy decision-making: a comprehensive approach for sustainable operation management in the UK. Neural Comput Appl 2024, 36(22), 13551–13577.
[37] Ciano MP, Peron M, Panza L, Pozzi R, Industry 4.0 technologies in support of circular economy: a 10R-based integration framework. Comput Ind Eng 2025, 201, 110867.
[38] World Economic Forum, A new circular vision for electronics: time for a global reboot. World Economic Forum; 2019.
[39] Li K, Qin Y, Zhu D, Zhang S, Upgrading waste electrical and electronic equipment recycling through extended producer responsibility: a case study. Circ Econ 2023, 2(1), 100025.
[40] Harshan K, Rajan AP, Unveiling the potential of microbial biominers in bioleaching for heavy metal recovery from e-waste: a comprehensive review. J Hazard Mater Adv 2025, 100691.
[41] EcoSpot Blog, How e-waste recycling creates jobs and drives the circular economy. EcoSpot Recycling; 2024.
[42] He Y, Kiehbadroudinezhad M, Hosseinzadeh-Bandbafha H, Gupta VK, Peng W, Lam SS, Tabatabaei M, Aghbashlo M, Driving sustainable circular economy in electronics: a comprehensive review on environmental life cycle assessment of e-waste recycling. Environ Pollut 2023, 342, 123081.
[43] Baldé CP, Forti V, Gray V, Kuehr R, Stegmann P, The global e-waste monitor 2017. Bonn/Geneva/Vienna: United Nations University (UNU); International Telecommunication Union (ITU); International Solid Waste Association (ISWA); 2017.
[44] Arya S, Kumar S, Bioleaching: urban mining option to curb the menace of e-waste challenge. Bioengineered 2020, 11(1), 640–660.
[45] Ranasinghe WW, Athapattu BCL, Challenges in e-waste management in Sri Lanka. In: Handbook of electronic waste management: international best practices and case studies. Elsevier; 2019. p. 283–322.
[46] Turaga RMR, Bhaskar K, Sinha S, Hinchliffe D, Hemkhaus M, Arora R, Chatterjee S, Khetriwal DS, Radulovic V, Singhal P, Sharma H, E-wastes management in India: issues and strategies. Vikalpa 2019, 44(3), 127–162.
[47] Naik S, Satya Eswari J, Electrical waste management: recent advances, challenges and future outlook. Total Environ Res Themes 2022, 1–2, 100002.
[48] Omondi EA, Complexity of e-wastes and its management challenges in developing countries – a review. Int J Environ Sci Nat Resour 2022, 31(2).
[49] Ghulam ST, Abushammala H, Challenges and opportunities in the management of electronic waste and its impact on human health and environment. Sustainability 2023, 15(3), 1837.
[50] Gaur TS, Yadav V, Mittal S, Sharma MK, A systematic review on sustainable e-waste management: challenges, circular economy practices, and a conceptual framework. Manag Environ Qual 2024, 35(4), 858–884.
[51] Patel S, Kasture A, E (electronic) waste management using biological systems – overview. Int J Curr Microbiol App Sci 2014, 3(7), 495–504.

Tuyelee Das*, Protha Biswas, Sujata Mandal

Exploring phytoremediation for e-waste-contaminated sites: a promising approach for environmental clean-up

Abstract: Electronic waste (e-waste) relates largely to the unusable materials left behind by different electronic devices. Growing global economic activity has made electronic appliances and devices more popular than they have ever been. Globally, there is a lot of e-waste that is generated, most of which ends up being sent to developing countries. In general, e-Waste is composed of metals, including nickel (Ni), gold (Au), silver (Ag), copper (Cu), mercury (Hg), and other hazardous metals. Additionally, e-waste contains hazardous substances, such as polyvinyl chloride, polychlorinated biphenyls, epoxy resins, thermosetting plastics, fiberglass, and persistent organic pollutants. In most cases, heavy metals and organic pollutants adversely affect ecosystems including flora, fauna, and human health. In order to prevent serious risks to human health, it is imperative that e-waste be managed, treated, and recycled to mitigate serious environmental complications. As one of many options available for treating E-waste, phytoremediation can be considered for its eco-friendliness and cost-effectiveness, as well as the fact that it is generally socially acceptable. Our objective for this chapter is to describe in detail phytoremediation as an effective and sustainable way to remediate e-waste contaminants.

Keywords: E-waste, phytoremediation, heavy metals, environmental pollution, contamination

1 E-waste production and global scenario

The electrical industry must develop products that are easier to update since the electric and electronic equipment (EEE) market demand is steadily increasing. This revolution has led to the obsolescence of electronic products. Reports and inspections classify electronic waste (e-waste) as a problematic waste that contributes to pollution on a global, regional, and national scale. The recyclable materials contained in e-waste can be used to generate economic value. The majority of e-waste being recycled today comes from the informal sector, which exposes the recyclers to toxic materials. Most recycling is done by women and children [1].

*Corresponding author: Tuyelee Das, University Centre for Research and Development, Chandigarh University, Mohali 140413, Punjab, India, e-mail:tuyeleedas25@gmail.com
Protha Biswas, Sujata Mandal, Department of Life Sciences, Presidency University, Kolkata, India

 | https://doi.org/10.1515/9783112225509-014

The quantity of e-waste has been increasing over time, despite the fact that different studies provide different estimates. This increase in e-waste production has been attributed to several factors, including increased consumer demand and high rates of obsolescence. These factors have caused an increase in the production of electronic equipment and an increase in the amount of e-waste. In addition, most of the electronic equipment we use today has a short life cycle/replacement interval [2]. E-waste inventories grow because recycling rates are low.

E-waste generated worldwide in 2014 and 2016 amounted to 5.9 kg per person [3]. According to the report, the global e-waste market size grew by 17.6% (93.45 million tons of e-waste in 2016), well above the 41.5 million tons reported in 2011 (93.45 million tons of e-waste in 2016). As of 2015, 9.8 million tons of e-waste were generated from computers, cell phones, and televisions. In 2010, 5.3 million tons of e-waste were generated. Widmer et al. [4] estimate that e-waste represents approximately 8% of municipal garbage in rich countries. A total of 5.5 million tons of e-waste is produced in the EU-15 and 8.3–9.1 million tons in the EU-27 [5] per year in Europe [6]. United Nations Environment Programme (UNEP) estimates that the production of e-waste worldwide ranged from 20 to 50 million tons in 2006 [7], an estimated 1–3% of the world's 1,636 million tons of municipal waste [8]. India today produces 18 times more mobile phones than it did in 2007, according to the UNEP. It will therefore result in a higher rate of e-waste generation [9]. E-waste mainly originates from developed countries with high electricity and electronic consumption. The "hidden flow" of products that is untraceable and undocumented across international borders is facilitated by regulatory loopholes from the perspective of developing and underdeveloped countries. E-waste is transported across international borders primarily as a result of low-cost workers and substandard environmental policies in these countries, which aim for a cheap disposal of e-waste while ignoring their negative effects on the environment [10].

E-waste is illegally exported to India and China from several developed countries [11]. Due to both their domestic production and their illegitimate imports from developing countries, India and China, for instance, are burdened with e-waste on a double basis. A new report by the United Nations University (UNU) predicts that between 2020 and 2030, EEW will increase by 38%. The Chinese will have disposed of USD23.8 billion of metals as e-waste by 2030 (http://www.greenpeace.org/eastasia/press/1397/chinas-e-waste-worth-23-8-billion-by-2030-2/). Among all states in India, Maharashtra produces the most e-waste [9]. Approximately, 66% of the global community is covered by e-waste legislation, according to the UNU's E-waste Monitor 2017 [12]. It is estimated that e-waste accounts for 8% of the total solid waste generated in the European Union. Awasthi [13] estimates that this type of waste represents only 2% of the total municipal waste generated in developing countries. Consequently, government and the public face challenges relating to safely managing and processing e-waste these days. The management and regulation of e-waste through the use of unconventional methods to extract valuable metals is a cause for concern, but many of these methods are damaging to human health and the environment. To address fast-growing e-waste and take advantage of its high value, several regulations and legislation have been enacted in several countries. A few of these regulations address trans-boundary movement, the use of low hazardous components, and the formal disposition of e-waste generated. Recent concerns around the world have prompted a great deal of research on e-waste. In 2017, the International Telecommunications Union determined that its volume needed to be reduced by 50% by 2020. According to the Global E-waste Monitor 2017 [14], only 41 countries of the world had statistical data on e-waste in 2017. Only 20% of e-waste was properly recycled. In addition to implementing adequate techniques for managing e-wastes that are currently present, it is also crucial to remediate sites that have been contaminated by e-wastes in the past.

2 E-waste composition and its toxicity

There are more than 1,000 different substances in used waste electronic and electrical equipment (WEEEs) classified as hazardous and not hazardous. WEEEs are composed of organic, metal, as well as ceramic materials. Vats and Singh [9] estimate that half of WEEEs are ferrous, while 21% are plastic, and 13% are nonferrous. Electronic and electrical waste streams are highly complex waste streams made up of many sources. The scientific literature reports that e-waste contains more than 1,000 different types of inorganic and organic pollutants. Metals such as copper, zinc, aluminum, gold, platinum, and palladium that are not ferrous are classified as nonferrous metals. Due to its composition varying according to equipment type, e-waste is difficult to quantify precisely. Mercury, lithium, beryllium, and several rare earth elements can be traced to toxic metals, which are used in switches and relays, batteries, and contact materials [15]. E-waste can produce a variety of organic pollutants, including polycyclic aromatic hydrocarbons from computer cases and circuit boards, polychlorinated dibenzodioxins and dibenzofurans (PCDD/DF) from flame retardants, and chlorofluorocarbons from refrigerators and air conditioners.

According to reports, e-waste has caused severe impacts because of ineffective waste management methods. Solid waste is generally disposed of in landfills, which are either poorly managed or operate as open dumps. In addition to open burning, toner sweeping, circuit board recycling, acid extraction of chips, plastic fragmentation, and melting, several e-waste recycling sites are reported to use primitive practices, so that toxic substances can escape [16]. When e-waste is disposed of in landfills, lead is released to groundwater, where it becomes bioaccumulated and biomagnified in food chains, posing a significant threat to ecosystems [17].

When waste is dumped, dismantled, burned, and leached, toxic leachates, particulate matter, effluents, and fumes are produced [18]. A number of environmental and health problems have been caused by the pollutants mentioned above. Toxic compounds accumulate in animal tissues and are found in food chains, which has severely impacted the natural ecosystem in South China. Earthworms, soils, and wormcast were positively correlated with PCDD/DFs, polychlorinated biphenyls, and PBDE concentrations, suggesting that earthworm activity has little impact on the distribution of persistent organic pollutants [19]. The gastrointestinal tract, respiratory system, and other organs of humans are adversely affected by exposure to these chemicals. There are several potential consequences of whooping cough, including coughing, choking, difficulty breathing, irritation of the eyes, skin disease, convulsions, and even death [20]. The bioaccumulation of short-chain chlorinated paraffins differs among aquatic species living in ponds contaminated with e-waste [21]. The management of municipal solid waste remains neglected in many countries.

3 E-waste management and treatment: traditional methods

Multiple management techniques are employed in order to manage daily generated e-waste loads. E-waste may be recycled, thermally treated, disposed of in landfills, or disposed into dump sites. The following are the main methods that are widely utilized to handle e-waste.

3.1 Acid bath

Acid bathing is the most common method for recovering precious metals from electronic circuit boards, and involves submerging them in sulfuric, hydrochloric, or nitric acid solutions and letting them soak for a specified period of time [22]. The extraction of copper, silver, and gold from the solution is accomplished by soaking the metals in concentrated acid solution for a certain time period and then precipitating them. Metallic precipitate is recovered and used to manufacture other products, while acid waste is discharged into the environment. Groundwater becomes a source of hazardous acid wastes and chemicals. Acid baths have been reportedly taken on the banks of the Yamuna River in India. In addition, exposing humans to acidic fumes containing hazardous compounds can put their health at risk.

3.2 Burning

The most common method of treating e-waste involves open burning, which is highly invasive to the environment. In the absence of any pollution control measures, open burning is not accompanied by any pollution control measures to prevent pollutants from escaping into the environment. This method is used in most countries because it is a less expensive way to dispose of e-waste. It is common practice in developing countries to burn waste in open air to separate valuable metals from combustible materials, possibly releasing toxic metals and organic compounds into the environment [23].

3.3 Landfills and dumping

E-waste is most commonly disposed of in sanitary landfills, which are designed to minimize the harmful effects on the environment and human health. In the event of e-waste disposal, contaminants could be leached into the groundwater and soil. Consequently, landfilling and dumping are not environmentally friendly practices [24]. There have been a number of reports that demonstrate a significant increase in the contamination of leachate from e-waste sites in terms of the level of trace metals, as well as organic and inorganic substances that are found dissolved and floating in the leachate. Specifically, Alabi et al. found that in soil and surrounding vegetation, e-waste components and constituents may accumulate to toxic levels and can be genotoxic, resulting in adverse health effects for individuals exposed to them [25]. Kumar and Fulekar have recently concluded that unregulated dumping of e-waste leads to a significant increase in the concentration of heavy metals (Cd, Cr, Cu, Fe, Ni, Pb, and Zn) in the soil of dumping site [26].

3.4 Recycling

In the early stages of e-waste recycling, methods of recycling were overly simplified, and recovery methods were too extensive. About 95% of India's recycling is informal, making it an unorganized business that employs thousands of underprivileged populations living in the country's slums and semi-urban areas. In informal recycling, the materials are collected, transported, processed, segregated, repaired, refurbished, and dismantled. Waste is collected by waste pickers from household. About a quarter to a third of the e-waste currently disposed of is recycled; the rest is dumped into rivers, streams, and waterways. Recyclable electrical goods include CPUs, cables, keyboards, laptops, mobile phones, monitors, and connecting wires, which can be used either manually or mechanically [27]. During recycling, obsolete electrical and electronic equipment is disassembled and disassembled in addition to their eventual reprocessing. Recycling includes the removal of hazardous substances (cathode ray tubes and printed circuit boards), as well as the separation of ferrous and nonferrous metals), according to Asante et al. [28]. An efficient recycler can make it possible to recover valuable materials from e-waste, such as gold, copper, and lead.

Among the most vulnerable populations are children, women, and laborers from rural areas, who are exposed to backyard recycling that poses a health risk [29]. A high concentration of Cu, Mo, Ag, Cd, In, Sb, Tl, and Pb was found in hair of male workers at an e-waste recycling site [30]. In women and children, negative neonatal outcomes have been observed, as well as issues related to thyroid functioning, hormonal imbalances, cancers, weak immunizations, and reproductive systems, leading to occupational health hazards and environmental degradation [31].

4 Phytoremediation: eco-friendly option for e-waste-contaminated sites

Phytoremediation describes the process of treating e-waste by using the plants in the ecosystem to remove the contaminants from e-waste. This method is one of the most important treatments for e-waste. Green technologies such as phytoremediation, which produces optimal results onsite, are generally considered to be environmentally friendly. It is possible to achieve successful phytoremediation technology through plants that possess certain select capabilities, including their ability to absorb metals through their roots, translocate through stems, and accumulate metals in leaves. Several plants have the natural capability to accumulate e-waste metals, which is known as hyperaccumulation [32]. The purpose of phytoremediation is to treat contaminated lands by using a plant bioreactor, to minimize risk and costs, and to ensure long-term environmental sustainability. There are a variety of plants that can be used in phytoremediation, including those that can survive in various climates, have roots that protect the soil, can degrade contaminants, and have a high growth rate that is resistant to contamination. Numerous studies have demonstrated that plant species can be used in the treatment of e-waste, including heavy metals and other hazardous substances. The plants used for phytoremediation have specialized physiological characteristics that distinguish them from other plants. Phytoremediation is the removal of pollutant through a variety of mechanisms, such as phytoextraction, phytostabilization, phytostimulation, phytostabilization, phytovolatilization, rhizofiltration, and phytodegradation (Figure 1). In comparison with alternative remediation methods, phytoremediation offers many advantages, including being relatively safe; upon completion of phytoremediation, the plants are harvested, the contaminants are recovered, and the leftover biomass can be applied in a variety of ways.

5 Techniques of phytoremediation

5.1 Phytoextraction

Phytoextraction is the involvement of plants to remove dangerous materials from soil or water. Generally heavy metals and/or hazardous substances are subjected to phytoextraction. A number of plant species possess the ability to flourish in heavy metallic environment; these are known as metallophytes. Metal content of these plants often outstrip the surrounding soil level. Plants having the ability to accumulate high number of heavy metals are known as "hyperaccumulator." Agglomeration of elements such as manganese, lead, cobalt, cadmium, nickel, selenium, and arsenic in plants has been reported. Many metallophytes have the probability to mineral deposition. The success rate of phytoextraction relies on a number of external factors such as high biomass of metallophyte species, metal bioavailability for plant uptake, soil factors (pH, soil type, soli texture, redox potential, etc.), and plant-associated factors (root exudates, area of rhizosphere, microorganisms of rhizosphere, etc.) [31]. Under ideal phytoextraction condition, plants can remove 50% of metal from soil in 25 years [33].

5.2 Phytofiltration

Removal of pollutants from aqueous effluents with the help of aquatic and terrestrial metallophytes is termed as phytofiltration. Mechanism of phytofiltration involves surface assimilation, complexation, adsorption by pore, exchange of ions, precipitation, and adsorption at surface [34]. It is a cost-effective, efficient biological process to clean up the water bodies. There are lots of terrestrial (like *Pteris vittata, Pteris cretica* cv. Mayii, *Medicago sativa*, and *Brassica juncea* L.) and aquatic (viz. *Hydrilla verticillata, Limnocharis flava* L., *Micranthemum umbrosum, Warnstorfia fluitans*, and *Spirodela polyrhiza* L.) plants showing the activity of phytofiltration [35–37].

5.3 Phytostimulation

In phytostimulation process, plants are stimulated by microbes. Hence, an increase in antimicrobial activity of plants takes place. It includes rhizosphere biodegradation or bioremediation assisted by plants. Microbes that help to improve the microbial activity of crop plants are known as biofertilizers. Phytostimulation helps in induced systemic resistance of plants. Using the technique of phytostimulation, multiple heavy-metal-tolerant plants are being produced by researchers. As, for example, *Brassica campestris* L. was stimulated by endophytic fungi *Mucor* sp. MHR-7 strain to develop its tolerance for manganese, chromium, copper, zinc, and cobalt. Thus, phytostimulation is a remarkable step in phytoremediation.

5.4 Phytostabilization

Phytostabilization is the reduction in mobility of heavy metals or hazardous substances by plants. With the help of roots plants create a cover on the soil surface of contaminated zones. Root growth and transpiration play a major role by diminishing leaching, managing erosion, and preparing aerobic environment in root zone. Sometimes plants also add organic matters to chelating compounds of pollutants. Secreted plant compounds, such as alkalizing agents and phosphates, sometimes decrease the solubility of metals, thereby reduce leaching at ground water. Phytostabilization is an important process of phytoremediation to manage heavy metals efficiently.

5.5 Phytovolatilization

The uptake of contaminants by plants, their conversion into gaseous substances and releasing them into environment is termed as phytovolatilization. Volatile plant compounds sometimes volatilized from leaf or stem are known as direct phytovolatilization. In indirect phytovolatilization, compounds volatilized from soil due to activity of plant roots [38]. Some metal pollutants, such as mercury (Hg), selenium (Se), and Arsenic (As), can also exist in gaseous form. Some plants possess the mechanism to convert these heavy metal contaminants into their gaseous form. Hence phytovolatilization contribute in the process of phytoremediation. It was deemed not to be a complete method of remediation due to the possibility of redeposition of the released materials in nearby lands.

5.6 Phytodegradation

Phytodegradation is the breakdown of absorbed pollutants, occurring either inside plants or in the rhizosphere. It is a metabolic process, also referred to as phytotransformation. A variety of contaminants are removed from environment in this process. Enzymes, produced by plants, catalyze and enhance the degradation process. Thus, organic contaminants, heavy metallic compounds are broken down into simpler ones and are subjected to promote plant growth [39]. Sometimes microorganisms associated with plants help in the degradation process. Phytodegradation involve the chemical changes of contaminants; hence decrease the toxicity of environment.

5.7 Rhizofiltration

Rhizofiltration involves the filtration of polluted ground water, surface water, and also waste water by root mass. In this process, plant removes toxic material even excess metals from water. Adsorption and precipitation are two important mechanisms here. Using plants with the ability to translocate heavy metals and to be resistant to toxic heavy metals, such as hyperaccumulators, is the most effective way to perform rhizofiltration [40].

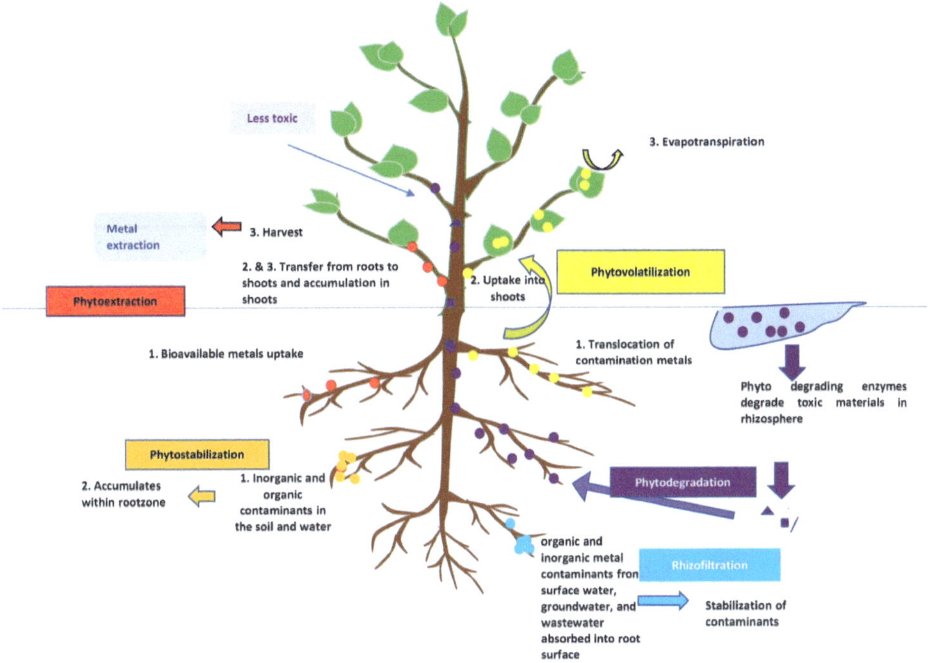

Figure 1: An illustration of the different steps involved in the phytoremediation of contaminated soil containing e-waste.

6 Phytoremediation of e-waste using plant secondary metabolites

Industrialization has exacerbated the problem of heavy metal pollution, which must be addressed seriously. Agricultural production and productivity are adversely affected by heavy metal contamination of water and soil. Remediation of heavy metal has become very much crucial to maintain the health of ecosystem. Phytoremediation of these heavy metals is most effective treatment among all. In this procedure, plants are involved in management and recycling of these metals. Mechanisms of heavy metal processing by plants recruit cell organelles such as mitochondria, plastids, and peroxisomes. This ultimately result in the production of reactive oxygen species and cell damage in plants. Secondary metabolite plays a major role here to prevent cell damage. Metallophytes contain high amount of some secondary metabolites. Even for the conversion of high amount of metals, plants required more amount of defense enzymes associating with it. Furthermore, the reduction of metals and neutralization of toxics also demand high amount of defense enzymes in plant cells. It has been suggested that secondary metabolites and antioxidants may play a role in the decontamination of heavy metals found in e-waste disposal sites. Numerous studies have shown that there is a high probability that many plant species are capable of reclaiming several types of e-waste, removing heavy metals, and other hazardous substances (Table 1).

Table 1: Plant species and their secondary metabolites used in phytoremediation of e-waste metals.

S. no.	Pollutants from e-waste	Plant species	Secondary metabolites	References
1	Barium (Ba)	*Calotropis procera* Aiton.	Tannins, saponins, andflavonoids	[37]
		Helianthus annuus L.	Tannins, saponins, andflavonoids	[41]
		Lactuca sativa L.	Carotenoids and phenolic acids	[42]
		Medicago sativa L.	Phenols, terpenoids, and flavonoids	[37]
2	Cadmium (Cd)	*Sorghum bicolor* L.	Tannins, saponins, andflavonoids	[43]
		Hordeum vulgare L.	Phenolics, flavonoids, andtannins	[44]
		Pyxine cocoes (Sw.)	Tannins, saponins, andflavonoids	[43]
		Oryza sativa L.	Saponins, terpenoids, and tannins	[45]
		Brassica juncea L.	Phytoanticipins andphytoprotectants	[44]
		Spinacia oleracea L. (spinach)	Rutin, quercitin, and gallic acid	[46]
		Eucalyptus camaldulensis Dehnh.	Tannins, saponins, andflavonoids	[46]
		Tagetes patula L.	Phenolic compounds	[47]
3	Beryllium (Be)	*Brassica napus* L.	caffeic acid, chlorogenic acid, quercetin, and kaempferol	[48]

Table 1 (continued)

S. no.	Pollutants from e-waste	Plant species	Secondary metabolites	References
4	Nickel (Ni)	*Hypogymnia physodes* L.	Atranorin, chloroatranorin, and usnic acid	[43]
		Sorghum bicolor L.	Phenolics, flavonoids, and tannins	[43]
5	Mercury (Hg)	*Oryza sativa* L.	Saponins, terpenoids, and tannins	[45]
		Eichhornia crassipes Mart.	Phenolics, flavonoids, and tannins	[49]
		Helianthus tuberosus L.	Chlorogenic acids, phenolic compounds	[50]
		Brassica juncea L.	phytoanticipins, andphytoprotectants	[51]
		Colocasia esculenta L.	Phenolics, flavonoids, and tannins	[49]
6	Lead (Pb)	*Brassica campestris* L.	Flavonoids and anthocyanins	[32]
		Brassica nigra (L.) Koch	Phenolics, flavonoids, and tannins	[52]
		Brassica juncea (L.) Czern.	Phytoanticipins and phytoprotectants	[51]
		Amaranthus hybridus L.	Flavonoids, steroids, andterpenoids	[32]
		Alternanthera philoxeroides (Mart.) Griseb.	Phenols and cardiac Glycosides	[52]
		Impatiens balsamina L.	Phenolic compounds	[53]
7	Arsenic (As)	*Oryza sativa* L.	Saponins, terpenoids, and tannins	[54]
		Populus alba L.	Flavonoids and polyphenols	[55]
		Malus domestica L.	Quercetin and chlorogenic acid	[56]
		Daucus carota L.	Phenolic compounds and ascorbic acid	[57]
8	Antimony (Sb)	*Pteris fauriei* Hieron.	Flavonoids, phenols, andtannins	[58]
		Pteris cretica L.	Alkaloids, flavonoids, and saponins	[59]
		Pteris vittata L.	Flavonoids, titerpenoids, and phenolics	[60]
		Matthiola farinosa Aiton.	Flavonoids	[61]

Table 1 (continued)

S. no.	Pollutants from e-waste	Plant species	Secondary metabolites	References
9	Hexavalent chromium/chromium (VI) (Cr(VI))	*Brassica oleracea* L. var. botrytis (cauliflower)		[44]
		Nicotiana tabacum L.	Alkaloids, steroids, and phenols	[62]
		Lycopersicon esculentum L.	Rutin, quercitin, and gallic acid	[44]
		Calotropis procera Aiton.	Cardenolides, flavonoids, and saponins	[62]
		Convolvulus arvensis L.	Phenolic compounds and ascorbic acid	[63]

In the study by Agwaramgbo [64], for example, plants phytoremediation capability was correlated with their antioxidant capacities. Their study demonstrated that a plant containing the strongest concentration of secondary metabolites, such as carotene and vitamin C, possesses the greatest antioxidant capacity and has the greatest ability to eliminate 2,4,6-trinitrotoluene. An additional study by Smeets et al. [65] examines the cadmium's effects on phenolic compounds in bean plants. When exposed to cadmium, plants accumulated more phenolic compounds. This suggests that phenolic compounds are involved directly in heavy metal detoxification.

In order to not be affected by heavy metals, plants produce phytochelatin synthase as a defense mechanism. Recent research has shown that many yeast and plant species contain gene sequences that encode the phytochelatin synthase enzyme [66]. Metallothionein is a protein synthesized by plants, and it also has affinity for heavy metals. Phytochelatin and metallothionein are technically two different types of cysteine-rich, heavy metal-binding proteins, but it is important to remember that the proteins are related [67]. The detoxification of metals accumulated over time as a result of the improper disposal of e-waste is accomplished by phytochemicals, especially secondary metabolites. It is extremely important to fully understand plants' secondary metabolic response to metal accumulation in order to fully appreciate how plants are able to remove metals from the environment. Further, metabolically engineering plants can also be a good way to boost the generation of metabolites that are necessary for a specific plant to detoxify metals, which are essential in the metabolism of plants. It is anticipated that this revolutionary technology will enable metals to be removed at their very entry point, thus preventing their bioaccumulation and eventual release in the environment by means of phytovolatilization.

7 Conclusion

In recent years, phytoremediation has become a popular method for controlling and managing e-waste contamination. Human consumption of electronic products is becoming increasingly complex, resulting in a drastic increase in the volume and scope of e-waste around the globe. Thus, phytoremediation is continuing to be noticed by the scientific community as being a good way to rehabilitate sites contaminated with e-waste that is cost-effective and environmentally friendly. Not only can this technology be used to remediate a site, but it also provides additional benefits to the environment, such as erosion control, biodiversity preservation by providing habitats for animals and birds, and the enhancement of the beauty of the surrounding environment. Furthermore, a positive outlook is also shown by the public when it comes to phytoremediation. The principal importance of secondary metabolites in plants for phytoremediation of e-waste contaminants cannot be overlooked, even though many pathways are involved in phytoremediation of e-waste contaminants. In the course of phytoremediation, secondary metabolites can play a very important role in interacting with and responding to toxic environmental contaminants. It would be ideal to conduct more research into the existing plants in order to discover their potential for properly removing heavy metals from contaminated environments in order to get even more benefits from the phytoremediation technology. Moreover, further study is necessary to determine whether there are plants with more promising characteristics as well, such as those that can accumulate and remove metals in an aggressive manner. Additionally, genetic engineering techniques can be used to engineer new plant cultivars that can efficiently purify heavy metals.

References

[1] Awasthi AK, Wang M, Wang Z, Awasthi MK, Li J. E-waste management in India: a mini-review. Waste Management & Research. 2018;36:408–414.

[2] Deng WJ, Louie PKK, Liu WK, Bi XH, Fu JM, Wong MH. Atmospheric levels and cytotoxicity of PAHs and heavy metals in TSP and PM2.5 at an electronic waste recycling site in southeast China. Atmospheric Environment. 2006;40:6945–6955.

[3] Baldé CP, Forti V, Gray V, Kuehr R, Stegmann P. The global e-waste monitor. Bonn/Geneva/Vienna: United Nations University (UNU), International Telecommunication Union (ITU), International Solid Waste Association (ISWA); 2017. p. 1–109.

[4] Widmer R, Oswald-Krapf H, Sinha-Khetriwal D, Schnellmann M, Böni H. Global perspectives on e-waste. Environmental Impact Assessment Review. 2005;25:436–458.

[5] Huisman J, Magalini F. Where are WEEE now? Lessons from WEEE: Will EPR work for the US? In Proceedings of the IEEE International Symposium on Electronics and the Environment. 2007. pp. 149–154.

[6] Goosey M. End-of-life electronics legislation: an industry perspective. Circuit World. 2004;30:41–45.

[7] United Nations Environment Programme (UNEP). Call for global action on e-waste. Nairobi; 2006.

[8] Organisation for Economic Co-operation and Development (OECD). OECD environmental outlook to 2030. Paris; 2009.

[9] Vats M, Singh SK. Status of e-waste in India: a review. International Journal of Innovative Research in Science, Engineering and Technology. 2014;3:16917–16931.

[10] Perkins DN, Drisse MNB, Nxele T, Sly PD. E-waste: a global hazard. Annals of Global Health. 2014;80:286–295.

[11] Priya A, Hait S. Comparative assessment of metallurgical recovery of metals from electronic waste with special emphasis on bioleaching. Environmental Science and Pollution Research. 2017;24:6989–7008.

[12] Patil RA, Ramakrishna S. A comprehensive analysis of e-waste legislation worldwide. Environmental Science and Pollution Research International. 2020;27:14412–14431.

[13] Awasthi AK, Zeng X, Li J. Comparative examining and analysis of e-waste recycling in typical developing and developed countries. Procedia Environmental Science. 2016;35:676–680.

[14] Barba-Gutiérrez Y, Adenso-Díaz B, Hopp M. Environmental consequences of European electrical and electronic waste regulation. Resources, Conservation and Recycling. 2008;52:481–495.

[15] Kiddee P, Naidu R, Wong MH. Electronic waste management approaches: an overview. Waste Management. 2013;33:1237–1250.

[16] Robinson BH. E-waste: global production and environmental impacts. Science of the Total Environment. 2009;408:183–191.

[17] Duan H, Hou K, Li J, Zhu X. Systematic characterization of generation and management of e-waste in China. Environmental Science and Pollution Research. 2016;23:1929–1943.

[18] Frenk J, Chen L, Bhutta ZA, Cohen J, Crisp N, Evans T et al. Health professionals for a new century. Lancet. 2010;376:1923–1958.

[19] Shang H, Li Y, Zhang Q, Sun Y, Chen J. Bioaccumulation of PCDD/Fs, PCBs and PBDEs by earthworms in e-waste dismantling areas. Environment International. 2013;54:50–58.

[20] Nordbrand S. Out of control: e-waste trade flows from the EU to developing countries. Stockholm: SwedWatch.

[21] Sun R, Luo X, Tang B, Mai B. Bioaccumulation of short-chain chlorinated paraffins in freshwater food webs near e-waste sites. Environmental pollution. 2017;222:165–174.

[22] Sivaramanan S, Cooray Mawatha EA E-waste management, disposal and environmental impacts.

[23] Stewart ES, Lemieux PM. Emissions from incineration of electronics industry waste. Proceedings of the IEEE International Symposium on Electronics and the Environment. 2003;271–275.

[24] Kasassi A, Rakimbei P, Karagiannidis A, Zabaniotou A, Tsiouvaras K, Nastis A et al. Soil contamination by heavy metals. Bioresource Technology. 2008;99:8578–8584.

[25] Alabi OA, Bakare AA, Xu X, Li B, Zhang Y, Huo X. Environmental contamination and DNA damage from e-waste. Science of the Total Environment. 2012;423:62–72.

[26] Kumar P, Fulekar MH. Multivariate statistical evaluation of heavy metals at e-waste sites. SN Applied Sciences. 2019;1.

[27] Heacock M, Kelly CB, Asante KA, Birnbaum LS, Bergman ÅL, Bruné MN et al. E-waste and harm to vulnerable populations. Environmental Health Perspectives. 2016;124:550–555.

[28] Asante KA, Agusa T, Biney CA, Agyekum WA, Bello M, Otsuka M et al. Trace elements in urine of e-waste workers. Science of the Total Environment. 2012;424:63–73.

[29] Rao LN. Environmental impact of uncontrolled e-waste disposal. International Journal of ChemTech Research. 2014;6:1343–1353.

[30] Ha NN, Agusa T, Ramu K, Npc T, Murata S, Bulbule KA et al. Trace element contamination at e-waste recycling sites in India. Chemosphere. 2009;76:9–15.

[31] Sheoran V, Sheoran AS, Poonia P. Factors affecting phytoextraction. Pedosphere. 2016;26:148–166.

[32] Tangahu BV, Abdullah SRS, Basri H, Idris M, Anuar N, Mukhlisin M. Heavy metal uptake by plants. International Journal of Chemical Engineering. 2011;2011:939161.

[33] Robinson BH, Anderson CWN, Dickinson NM. Phytoextraction: where's the action?. Journal of Geochemical Exploration. 2015;151:34–40.

[34] Gardea-Torresdey JL, De La Rosa G, Peralta-Videa JR. Phytofiltration technologies for heavy metal removal. Pure and Applied Chemistry. 2004;76:801–813.

[35] Sandhi A, Landberg T, Greger M. Phytofiltration of arsenic by aquatic moss. Environmental pollution. 2018;237:1098–1105.

[36] Abhilash PC, Pandey VC, Srivastava P, Rakesh PS, Chandran S, Singh N et al. Cadmium phytofiltration by Limnocharis flava. Journal of Hazardous Materials. 2009;170:791–797.

[37] Gardea-Torresdey JL, Gonzalez JH, Tiemann KJ, Rodriguez O, Gamez G. Phytofiltration of Cd, Cr, Pb and Zn. Journal of Hazardous Materials. 1998;57:29–39.

[38] Limmer M, Burken J. Phytovolatilization of organic contaminants. Environmental Science and Technology. 2016;50:6632–6643.

[39] Wang X, Ye J, Zhang H, Wang L. Phytodegradation by transgenic plants. Biochemical and Biophysical Research Communications. 2008;365:453–458.

[40] Tak HI, Ahmad F, Babalola OO. PGPR-assisted phytoremediation. Reviews of Environmental Contamination and Toxicology. 2012;223:33–52.

[41] Sampaio J, Amaral NMB, Zonta E, Magalhães MOL. Barium and sodium uptake in sunflower. Revista Brasileira de Engenharia Agrícola e Ambiental. 2015;19:1100–1106.

[42] Lamb DT, Matanitobua VP, Palanisami T, Megharaj M, Naidu R. Bioavailability of barium. Environmental Science and Technology. 2013;47:4670–4676.

[43] Muranyi A, Ködöböcz L. Heavy metal uptake by plants. Cereal Research Communications. 2008;36:387–390.

[44] Peralta-Videa JR, Lopez ML, Narayan M, Saupe G, Gardea-Torresdey JL. Heavy metal uptake biochemistry. International Journal of Biochemistry & Cell Biology. 2009;41:1665–1677.

[45] Liu WX, Shen LF, Liu JW, Wang YW, Li SR. Heavy metal uptake by rice. Bulletin of Environmental Contamination and Toxicology. 2007;79:209–213.

[46] Intawongse M, Dean JR. Heavy metal uptake by vegetables. Food Additives and Contaminants. 2006;23:36–48.

[47] Miao X, Wang X, Liu J, Chen H. Phytoremediation of Cd and PCBs. Bulletin of Environmental Contamination and Toxicology. 2022;108:129–135.

[48] Ali S, Abbas Z, Rizwan M, Zaheer IE, Yavaş İ, Ünay A et al. Beryllium stress in oilseed rape. BioMed Research International. 2018;2018:1615968.

[49] Skinner K, Wright N, Porter-Goff E. Mercury uptake by aquatic plants. Environmental pollution. 2007;145:234–237.

[50] Sas-Nowosielska A, Kucharski R, Małkowski E, Pogrzeba M, Kuperberg JM, Kryński K. Mercury remediation and rhizosphere microbes. Environmental Monitoring and Assessment. 2008;137:101–109.

[51] Van Ginneken L, Meers E, Guisson R, Ruttens A, Elst K, Tack FMG et al. Phytoremediation and bioenergy. Journal of Environmental Engineering and Landscape Management. 2007;15:227–236.

[52] Cho-Ruk K, Kurukote J, Supprung P, Vetayasuporn S. Perennial plants in phytoremediation of lead. World Academy of Science, Engineering and Technology. 2006.

[53] Liu W, Zhou Q, Zhang Y, Wei S. Impatiens balsamina for lead remediation. Land Degradation & Development. 2021;32:3773–3784.

[54] Heitkemper DT, Vela NP, Stewart KR, Westphal CS. Arsenic speciation in rice. Journal of Analytical Atomic Spectrometry. 2001;16:299–306.

[55] Vamerali T, Bandiera M, Mosca G. Phytoremediation of pyrite wastes. Environmental pollution. 2009;157:887–894.

[56] Caruso JA, Heitkemper DT, B'Hymer C. Arsenic extraction techniques. Analyst. 2001;126:136–140.

[57] Helgesen H, Larsen EH. Arsenic bioavailability in carrots. Analyst. 1998;123:791–796.

[58] Feng R, Wang X, Wei C, Tu S. Arsenic and antimony accumulation in ferns. International Journal of Phytoremediation. 2015;17:348–354.
[59] Feng R, Wei C, Tu S, Tang S, Wu F. Hyperaccumulation in brake fern. Microchemical Journal. 2011;97:38–43.
[60] Müller K, Daus B, Mattusch J, Stärk HJ, Wennrich R. Arsenic impact on antimony uptake. Environmental pollution. 2013;174:128–133.
[61] Jamali Hajiani N, Ghaderian SM, Karimi N, Schat H. Antimony accumulation in plants. Environmental Science and Pollution Research. 2015;22:16542–16553.
[62] Kim YJ, Kim SH, Kim HJ, Lee SY, Choi JK. Chromium accumulation via MSN1. FEBS Letters. 2006;580:206–210.
[63] Montes-Holguin MO, Peralta-Videa JR, Gardea-Torresdey JL. Chromium stress in plants. Environmental Toxicology and Chemistry. 2006;25:220–226.
[64] Agwaramgbo L. Screening plants for antioxidant and phytoremediation potential. New Orleans: Dillard University; 2005.
[65] Smeets K, Cuypers A, Lambrechts A, Semane B, Hoet P, Van Laere A et al. Oxidative stress in Phaseolus vulgaris. Plant Physiology and Biochemistry. 2005;43:437–444.
[66] Cobbett CS. Phytochelatin biosynthesis and function. Current Opinion in Plant Biology. 2000;3:211–216.
[67] Guo J, Xu W, Ma M, Zhao Y, Wang Y. Metallothionein gene in sugarcane. BioMed Research International. 2013;2013:904769.

N. Sharmila Devi, Tijo Cherian*

E-waste recycling: biological strategies driving the circular economy

Abstract: Electronic waste (E-waste) poses a catastrophic issue for humans because of its complicated constitution. Despite the fact that E-waste contains some hazardous materials, it is also a potentially lucrative industry. Recycling E-waste to extract metallic materials of value and other materials has created commercial opportunities and paved the way for the shift from a linear to a circular economy. The E-waste recycling industry is dominated by chemical, physical, and conventional technologies; nevertheless, sustainability in terms of cost and environmental concerns is a significant challenge related to these technologies. Profitable, sustainable, and eco-friendly technologies must be suggested in order to close these gaps. By taking socioeconomic and environmental factors into account, biological techniques may be a clean and green way to manage E-waste in a sustainable and economical manner. This chapter explains biological methods for managing E-waste and advances in the field. In order to create a sustainable recycling process, more research and development is required in this area. The uniqueness addresses the environmental and socioeconomic effects of E-waste, its solutions, and the potential of biological techniques.

Keywords: Biological approaches, circular economy, E-waste, environmental indicator, socioeconomic indicator

1 Introduction

In every period, technological growth has a significant impact on human life. Human existence grows more accessible with each advancement in technology. The creation of electronic devices is one example of such technical involvement. In addition to making our lives easier, electronic devices are now an integral element of contemporary society [1]. Electronic devices permeate every aspect of our lives, including entertainment, communication, and daily necessities. Apart from this, personal computers and mobile devices are today's essential electrical devices. Without computers and mobile devices, today's world would be like "a fish without water." The market for

***Corresponding author: Tijo Cherian**, Post Graduate Department of Food and Industrial Microbiology, Department of Botany, Bishop Kurialacherry College for Women, Amalagiri 686561, Kottayam, Kerala, India, e-mail: tvarghese891@gmail.com
N. Sharmila Devi, Post Graduate Department of Food and Industrial Microbiology, Department of Botany, Bishop Kurialacherry College for Women, Amalagiri 686561, Kottayam, Kerala, India

 | https://doi.org/10.1515/9783112225509-015

electronic equipment is expanding due to technological advancements. Expanding innovations with improved features draw consumers, which results in the production of electronic waste (E-waste) when obsolete technology is replaced by novel versions. Technology obsolescence is another factor contributing to the production of E-waste. As a result, a large amount of E-waste is being dumped [2]. Lack of understanding about the proper management of E-waste led to a major environmental danger and worldwide stress. Depending on how E-waste is organized, there are a wide range of environmental issues associated with it. Nonmetallic and metallic materials, chemical compounds, alloys, and plastic make up the fundamental components of E-waste [3]. By releasing hazardous gases into the atmosphere during incineration and leaching hazardous anthropogenic pollutants that degrade soil and ground water quality, the disposal of E-waste in incineration and landfills causes adverse ecological effects. The landscape and extent of the ecological effects of E-waste content vary with different ecological conditions, creating new challenges and compelling researchers to develop more effective and innovative technologies to overcome the constraints associated with existing ones for better outcomes. E-waste can be treated in a number of ways, including recycling, incineration, and landfilling. As previously witnessed, the negative impacts of incineration and landfills are not economical or environmentally beneficial. Nowadays, eco-friendly and sustainable recycling is a pressing requirement for handling E-waste. However, like all coins, E-waste has two sides: on the one hand, it includes harmful anthropogenic contaminants, but on the other, recycling offers a plethora of commercial prospects [3]. Therefore, recycling may be the greatest way to eliminate harmful impacts of incineration and large landfills, including health and pollution risks brought on by E-waste. The process of recycling has emerged as one of the best economic prospects in terms of profiting from scrap and lowering the burden of metals recovered from virgin ore, hence strengthening the nation's circular economy model (Figure 1).

An estimated 53.6 million tons of E-waste were produced worldwide in 2019. Only 9.3 of the 53.6 million tons were recycled by the official recycling industry. Potential raw materials worth US$57 billion were extracted through the recycling of gold, copper, and iron from E-waste [4]. Collecting, storing, classifying, disassembling, tearing, mechanical sorting, and recovering precious resources are the steps involved in recycling E-waste [5]. Biological, chemical, and physical methods are used to extract metals from E-waste. Although physical and chemical recycling demonstrated a high rate of metal recovery, it also releases expensive and hazardous products [6]. Despite its slowness, biological approaches are cost-effective and environmentally beneficial for recovering metals. Numerous biological advancements are being made to recycle E-waste in an environmentally friendly and ecologically sound fashion. In order to promote the circular economy and highlight current trends in E-waste management, this chapter discusses potential biological recycling techniques for recovering metals in a more environmentally friendly manner.

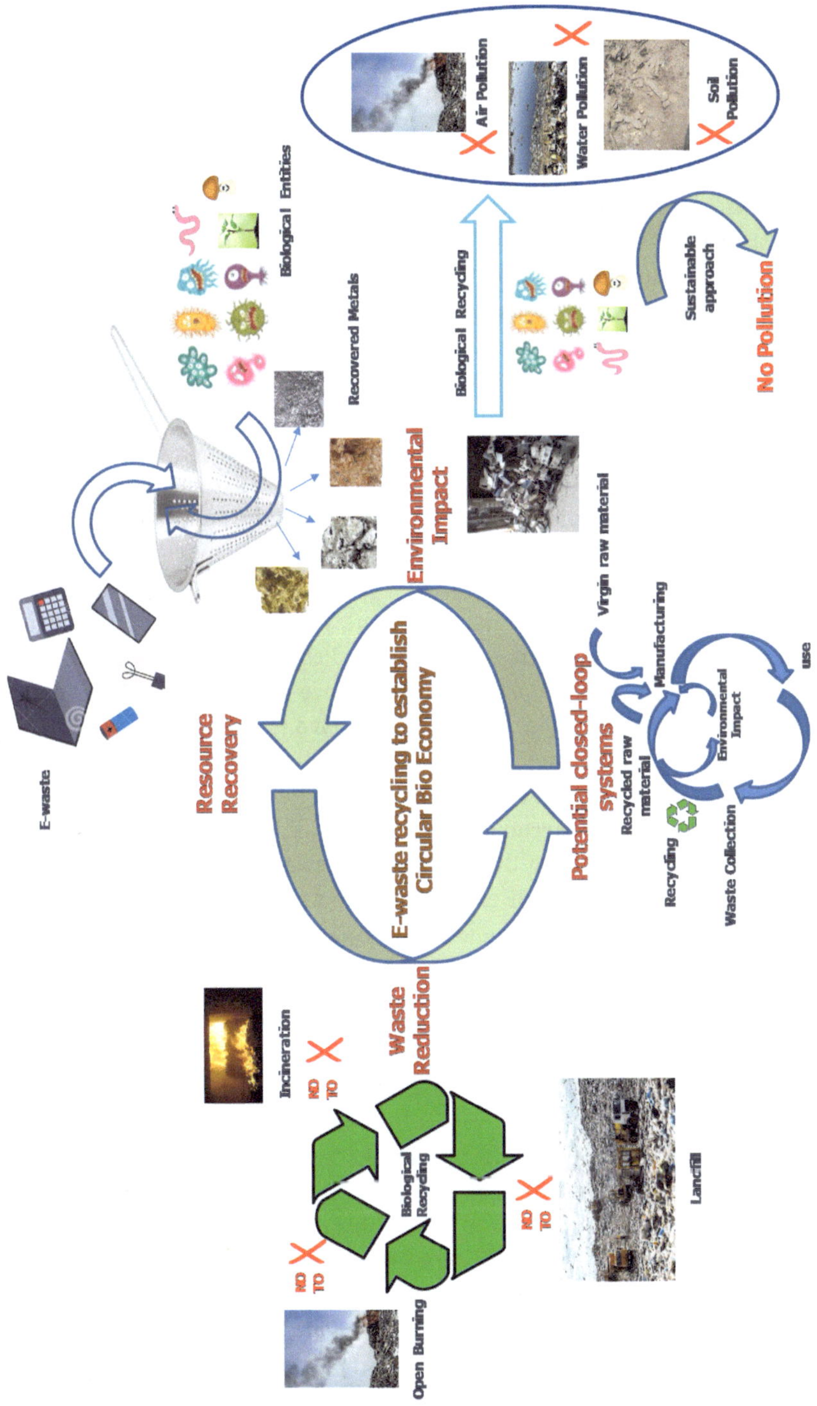

Figure 1: Overall goal of ci-cular bio-economy (reproduced from [1]).

2 Products classified as E-waste and their components

All electrical and electronic equipment components that have been disposed of as waste sans prioritizing repurposing are referred to as electronic waste, or E-waste or also known as waste electrical and electronic equipment. A vast array of products is included in E-waste and is divided into six categories, each of which includes a variety of goods. Small IT and telecommunication equipment, small equipment, large equipment, lamps, monitors/screens, and temperature exchange equipment are the six categories. The items that fall under these categories differ in terms of their configuration, the amount of garbage produce, the effects on the people and environment, and the economic benefits of recovering precious materials from waste through recycle [7]. Plastics, metal-plastic mixtures, CRT and LCD monitors, cables, printed circuit boards, metals, and other components are all included in E-waste [8]. Due to improper waste management and reservations about hazardous metals leaking into multiple ecological categories and endangering human health and the environment, metals hold a prominent position in the composition of E-waste and are therefore a point of interest with regard to the wastage of valuable metals in E-waste. Zinc, tin, silicon, nickel, lithium, germanium, aluminum, silver, gold, and copper are the most often utilized valuable metals in E-waste. To prevent the voidance in the future, these metals could be repurposed responsibly. Since these metals have lower sources of value, E-waste may serve as a center for their recycling in order to transition from linear to circular economy. Lead causes neurological diseases in humans; mercury damages senses, weakens muscles, causes dermal problems, and affects both human and animal reproductive systems. Cadmium disrupts the human respiratory system, which has an impact on the neurological and urological systems. Chromium can cause cancer. In addition to its carcinogenic effects, arsenic has been linked to skin and respiratory disorders.

3 Socioeconomic and ecological effects of E-waste

When E-waste is improperly disposed of at landfills, hazardous substances are released into the soil, water, and, air endangering the environment. Toxic pollutants like dioxins, which are added to the air and have been shown to have detrimental effects on health, are found in the dust particles emitted by E-waste. Toxic vapors were emitted during the unofficial method of extracting metals using different chemicals, harming the environment and posing a number of health risks. Toxic heavy metals, flame retardants, and other hazardous contaminants penetrate into the soil and harm soil biodiversity. They also biomagnify and enter the food chain, which has a significant negative impact on the human health and ecosystem. The ecology is negatively impacted by the informal sector's processing of E-waste to metal recovery by employing chemicals and the unlawful release of treated water-based solutions outlets into water streams [9]. The nation's economy and society are related to the socioeconomic element and the former is significantly impacted by these environmental damages, which in turn pose health risks to people. This could contribute to the establishment of a linear economy in the nation if E-waste is not effectively addressed by tracking the end life of potentially reused materials from E-waste to landfills. When it comes to managing E-waste, the linear economy will put strain on virgin ore mining, which will have an impact on the nation's economy and ecology. An effective gap evaluation and a durable remedy are urgently needed to solve this problem.

4 Remedy for the effects of E-waste

In order to mitigate the effects of E-waste, appropriate schemes for collection and segregation must be established to prevent E-waste from being mixed with municipal solid waste and dumped in landfills. E-waste draws concern as it contains hazardous materials that can harm both the environment and human health. The 3Rs – Reduce, Reuse, and Recycle – are the practical answer. The world's increasing needs and technological advancements make it appear difficult to reduce the production of E-waste. Reusing and recycling electronic devices is therefore the best option. Electronic devices can be repurposed by refurbishing nonworking devices or giving working devices to charitable causes. Recycling may be a strong way to reduce the negative effects of E-waste scrap on both the ecosystem and the economy. An efficient way to solve E-waste issues and support the nation's circular economy is through scientific E-waste recycling.

The World Economic Forum's 2019 Global E-waste Report estimates that the material worth of e-devices worldwide is US$62.5 billion. Large landfills are needed to dispose of garbage if the nation uses a linear economy for managing E-waste, and it would be extremely expensive to appropriately handle the harmful consequences of toxic substances in E-waste to prevent threats to the ecosystem and people. A steady supply of raw materials is also necessary for the production of new technologies in a linear economy. The circular economy, on the other hand, depends on the repurposing of resources from waste and stimulates the economy by generating income and jobs. As the term implies, "circular economy" refers to the preservation of circularity. In contrast to the linear economic notion of take-make-use-dispose, it refers to take-make-use-reuse/recycle-make-take in the context of E-waste. Whereas circular economy is predicated on resource regeneration and restoration, linear economy adheres to the idea of end-life. An industrial system known as the "circular economy" bases its economic structure on the efficient and maximal reuse, refurbishment, recovery, remanufacture, and regeneration of goods and materials before they reach the end of their useful lives [10]. Mining E-waste for valuable metals has the potential to be a profitable venture. According to reports, there is more gold in one ton of E-waste than in one ton of gold ore. Therefore, mining E-waste instead of virgin ore could be an initial move toward sustainability.

E-waste is mined using biological, chemical, and physical methods. CirBioeco (Circular Bioeconomy) is a biological strategy to create a circular economy for resource regeneration and converting waste materials into commodities with added value [11]. Through restoration and regeneration, biological methods for recycling E-waste are a green way to improve the circular economy model's context. By lowering CO_2 emissions, lessening the burden on natural resources, and creating business and employment possibilities, bio-recycling is a process that promotes the Sustainable Development Goals and strengthens the three pillars of sustainability – environment, society, and economy. By facilitating the creation of future closed-loop systems, environmental impact, waste reduction, and resource recovery, the use of biological entities for metal recovery from E-waste promotes the circular economy. It helps achieve the overarching objectives of sustainable resource management and resource conservation by providing an effective and sustainable method of handling E-waste.

5 Recycling

Recycling is the efficient disposal and utilization of waste materials for use in the production of new products. By supporting socioeconomic indicators, recycling can assist the environment and our community in terms of employment, earnings, and tax revenue creation. The adoption of green recycling and sustainable procedures will provide low carbon footprints that support sustainability metrics. Therefore, employing environmentally friendly and sustainable waste recycling techniques may be the greatest way to eliminate large landfills and incinerators, which are contributing to health and pollution risks. Recycling has developed into one of the best ventures for making revenue off of waste while addressing fiscal problems and advancing the nation's economic circularity.

Nonmetallic and metallic elements, plastic, alloys, and other substances, along with a few hazardous contaminants, constitute E-waste. Since there are many different types of E-waste, the main work that needs to be concentrated on during E-waste recycling should be the correct differentiation of the components based on the product. By using E-waste recycling and metal recovery as an alternate method, the strain on natural resources caused by virgin ore mining will be lessened [12]. Biological recycling, traditional recycling, chemical recycling and physical recycling, are popular approaches for recycling E-waste (Figure 2). When it comes to recycling E-waste, traditional recycling is primarily used by the unorganized sector. Chemical and physical recycling procedures are expensive, and the byproducts produced are hazardous to human health and the environment. The biological recycling process is a more economical and environmentally beneficial way to metal recovery and other raw materials from E-waste.

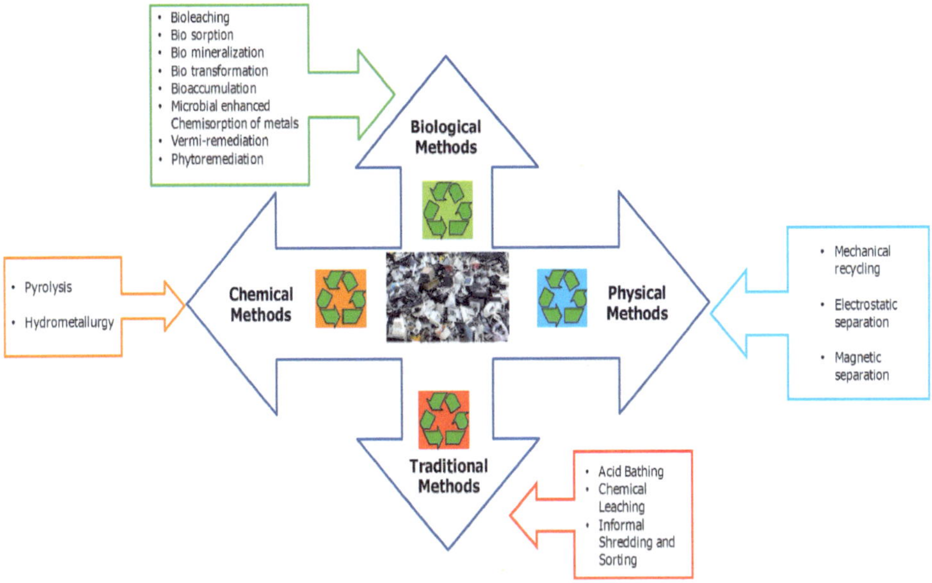

Figure 2: Methods for E-waste recycling (reproduced from [1]).

5.1 Recycling approaches for E-waste

5.1.1 Traditional recycling

5.1.1.1 Informal shredding and sorting

Without proper dust control measures, informal recyclers may use rudimentary shredding procedures, which could spread harmful fibers and dust that contain toxic compounds. Without the proper safety gear, workers who manually sort through the shredded waste are exposed to hazardous compounds, which raises the risk to their safety [13].

5.1.1.2 Chemical leaching

This technique dissolves metals from E-waste using chemicals like strong acids or cyanide. Due to the possibility of hazardous compounds being released into the immediate surroundings, partial treatment of these chemicals, poor storage, and improper handling can provide significant dangers to human health and the environment [14].

5.1.1.3 Acid bathing

To remove precious metals from E-waste, some unauthorized recycling facilities use acid baths, frequently using strong acids like sulfuric acid. This procedure offers serious risks since it introduces employees to caustic materials, releases poisonous gases, and often leads to the erroneous handling of acid waste, which pollutes the ecosystem [15].

Because of their negative consequences on human health, the environment, and workplace safety, these activities are strongly prohibited. Prioritizing sanitary, ecologically conscious recycling methods that adhere to set rules and regulations is essential. Traditional recycling is mostly practiced in developing nations, but the harmful pollutants released by these practices have detrimental effects on human health and the ecosystem and are very responsible.

5.1.2 Chemical recycling

5.1.2.1 Hydrometallurgy

In hydrometallurgical processing of E-waste, metals are leached using various acids and the recovered metals are then concentrated using ion exchange, adsorption, and additional extractions. Chemical recycling techniques have significant negative effects on the ecosystem and human health and demand large capital investments.

5.1.2.2 Pyrolysis

The thermal breakdown of E-waste in anoxic conditions is called pyrolysis. Based on laboratory results, pyrolysis is thought to be the sustainable method of treating E-waste, but it has not yet gained traction in the commercial settings. The benefits and drawbacks of plasma pyrolysis, microwave, co-pyrolysis, catalytic, and vacuum pyrolysis for E-waste were examined by Andooz et al. [16].

5.1.3 Physical recycling

5.1.3.1 Electrostatic separation

The section of E-waste scrap is separated by applying an electrostatic charge. Nonconductor and conductor components of E-waste are often separated via electrostatic separation. Using an electrostatic separation technique, Xue et al. [17] investigated the recycling of E-waste and were able to separate semiconductors up to 88% and conductors up to 82.5%. Although physical recycling techniques are commonly employed in the field, they are not financially viable.

5.1.3.2 Magnetic separation

A common technique for separating waste components, particularly metals, is magnetic separation of E-waste, which uses magnetic disk or plates. This expands the industrial sector's ability to create equipment for recovering metals from E-waste. Woynarowska and Zukowski [18] investigated the magnetic metal separation from mobile E-waste pre-prepared with fluidized bed reactor, inferring separation rates of neodymium (98.2%), iron (90.5%), and cobalt (80.9%).

5.1.3.3 Mechanical recycling

Steps in mechanical recycling include gathering waste from the source, sorting, and grinding it to separate its constituent parts for additional treatment. In order to recover metals from E-waste, it is shredded and put through magnetic separation.

5.1.4 Biological recycling

In order to recover the product from E-waste as metals along with additional important raw materials, biological recycling employs bacteria, plants, and animals (earthworms) [19]. Metals, which are extremely precious, make up the majority of E-waste. In addition to lessening the burden of mining, the metal extraction, particularly using economical and sustainable biological methods increases the flow of national circular economy [20]. The following aspects influence the process of biological recycling: easy availability of compounds for microorganisms to transform waste components, microbial tolerance capability, bacteriostatic effects of the process environment, secondary reactions, and their products. The pros of biological E-waste recycling are that it is economical, it is environmentally friendly, and it does not produce harmful byproducts. The cons include that the process is slow, certain microflora are limited to specific metals, and effectiveness depends on certain environmental conditions. Numerous studies have been conducted and are still being conducted to close the discrepancies. In order to support the circular economy, new methods for efficiently recycling metal mining E-waste will develop as biological recycling technologies advance. Metals can be extracted from solid or liquid waste using a variety of biological techniques. These techniques are also demonstrated to have a substantial removal capability and to be an eco-friendly, clean, and green method for metal removal from E-waste (Figure 3).

The techniques used in the recovery of metals are mentioned below.

5.1.4.1 Phytoremediation

The technique of using plants to eliminate pollutants from a variety of biological matrices, including harmful organic and inorganic chemicals and trace metals, is known as phytoremediation [21]. Phytovolatilization, phytostabilization, rhizofiltration, phy-

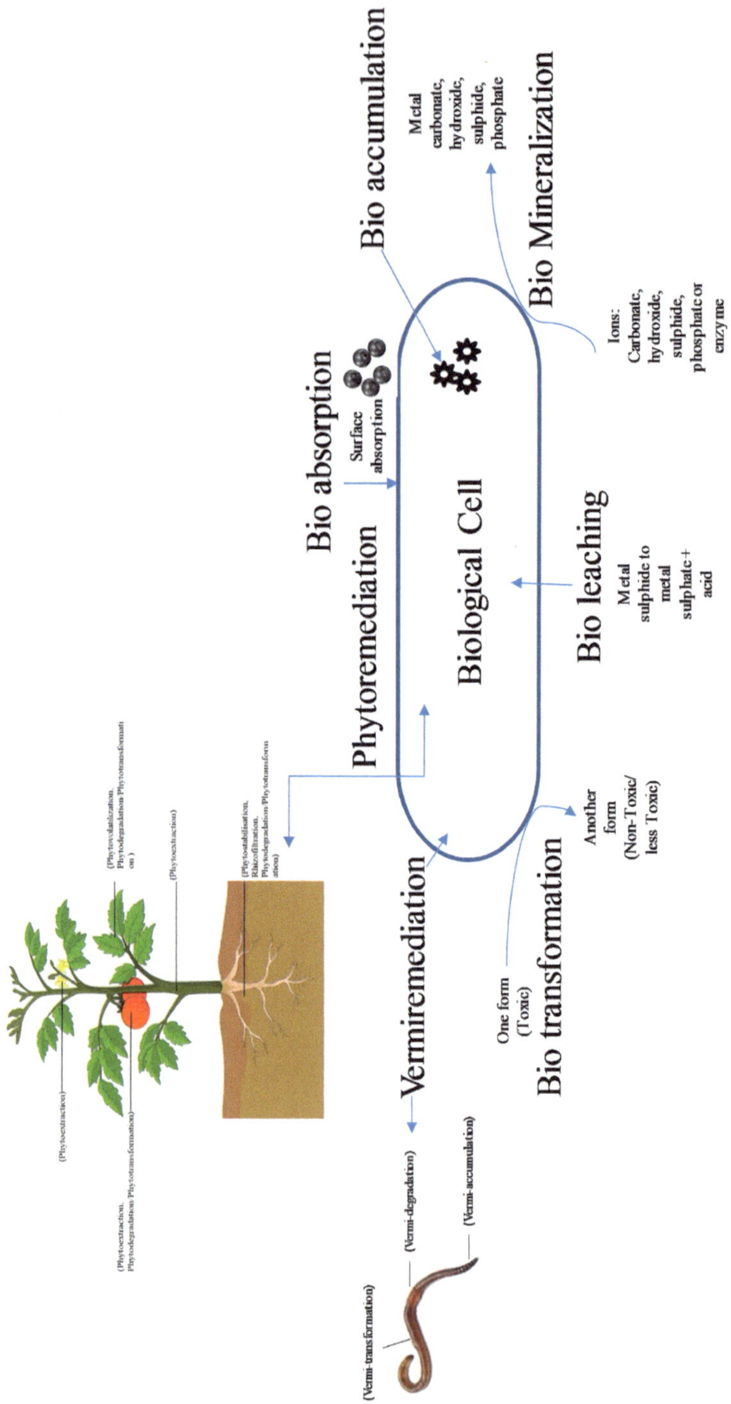

Figure 3: Biological approaches for E-waste recycling (reproduced from [1]).

toextraction, and phytotransformation/phytodegradation are different plant processes used to eliminate pollutants from polluted areas. The concentration of contaminants (metals), the characteristics of the soil, and the species of plant employed in research all have an impact on phytoremediation. It is an eco-friendly method that lessens soil trace element leakage. In addition to the landscape and pollution mitigation, it is a cost-effective method for aesthetic value, biodiversity protection, biomass production, soil weathering management, and carbon sequestration [22]. Concerns with the implementation of phytoremediation in the recovery of metals from E-waste include the bioaccumulation of metals in the food chain and the disposing of plants that are used to absorb metals from contaminated sites [22]. Plants are thought to be the hyperaccumulator of metals. Numerous species of plants have been shown to accumulate metals in different parts of their bodies with remarkable efficiency [21]. One of the more affordable and environmentally beneficial methods for cleaning up an E-waste-contaminated site is phytoremediation. Therefore, phytoremediation may be an eco-friendly way to clean up the site polluted by E-waste [23]. An ecological method for stabilizing metals in the ecosystem to lessen their adverse impacts and for rehabilitation is called phytoremediation.

5.1.4.2 Vermiremediation

Earthworms are used in the vermiremediation technology to absorb heavy metals from soil contaminated by E-waste [24]. By turning E-waste into riches, vermiremediation has been demonstrated to uphold the “circular economy” and “global human sustainability cycle” models. It is the procedure of removing the metal pollution burden from ecological entities with respective to heavy metal accumulation. A number of variables, including excessive worm population expansion in the bed, metal concentration, aeration, temperature, moisture, bed pH, etc., might affect earthworms [25]. Vermicompost setups are simple to build, involve minimal planning, inexpensive and cost-effective – a “one-time investment.” According to Pandit et al. [26], the method is odorous, the worms need to be properly cared for since they are susceptible to a variety of ecological conditions, and the initial establishment is time-consuming and demands additional workspace.

5.1.4.3 Microbial enhancements in metal chemisorption

Similar to bioadsorption, microbiologically enhanced metal chemisorption is a process in which microorganisms play a major role in precipitating a biomineral of a nontarget metal and creating priming deposits. The targeted metals are subsequently deposited using these priming deposits [27]. The parameters of biomass concentration, metal bioavailability, humic acid, metal concentration, temperature, and pH, all have an impact on microbial enhanced chemisorption. According to Ahluwalia and Goyal [28], it is an environmentally acceptable, economical method of recovering metals without the use of chemicals or the production of hazardous bioproducts, although it is less successful for heavy metals.

5.1.4.4 Bioaccumulation

The process by which organisms bioaggregate pollutants from the environmental matrix, such as air, soil, and water, in their body parts is known as bioaccumulation. Environmental parameters such as pressure, temperature, and pH, along with metal dosage and the microbial species involved, all impact bioaccumulation. High concentrations of metal accumulation in living tissues have been found, making bioaccumulation an economical and environmentally acceptable method of recovering metals. However, the process is intensive, and recovering metals from cell structure is a laborious effort. Toxicity, accessibility, ecological shifts in metal speciation, genetic and physical reconciliation, inherent skeletal and biochemical characteristics all contribute to the metal accumulation in the cellular structures of many species. The bioaccumulation process, which essentially relies on the microbial growth in the current environment during the process, is how contaminants (metals) are absorbed by living organisms and transported into the cells [29]. In order to prevent the harmful effects of heavy metals on the human health and environment, bioaccumulation is a technique that uses biological entities, particularly bacteria, for stabilizing the metals in nature. It also suggests that precious metals can be recovered from E-waste and ore mining.

5.1.4.5 Biotransformation

The process known as "biotransformation" uses biological entities to convert hazardous contaminants, such as heavy metals and anthropogenic chemicals, into less hazardous ones. There are two kinds of biotransformation: Ex-situ biotransformation necessitates the excavation of the contaminated matrix; in situ biotransformation is carried out on site [30]. The factors influencing biotransformation include moisture content, temperature, pH, toxicity of the medium, process viability in relation to the target components, and microbial species involved [31]. Although it is a pragmatic and environmentally acceptable method of removing hazardous metals from the ecosystem and recovering precious metals, the process is thought to be slow, and one of the problems with different process conditions is enzyme stability [32]. Heavy metals undergo transformation rather than degradation. One of the best options for removing metals from polluted soil may be the transition of heavy metals by biotransformation employing microbial physiological processes [33]. One of the finest methods for cleaning up soil polluted by E-waste is biotransformation.

5.1.4.6 Biomineralization

Microbes play a crucial role in the process of biomineralization, which converts aqueous metal ions into crystalline or amorphous precipitates [34]. One potential, economical, and environmentally beneficial method of metal recovery from E-waste is biomineralization. The pH of the process and ion inhibition from coexisting metals are the parameters that affect biomineralization. According to Dhami et al. [35], it is a facile, economical, and ecologically sound method of metal recycling; nevertheless, the process is quite complex and difficult to scale up. In terms of electron acceptors, enzymatic co-factors, and microbial pathways, microbes have a robust and intricate relationship with metals. The majority of microbial processes, including co-factorial enzymatic machinery, energy production via dissimilatory reduction in biomineralization and anaerobic respiration, depend on metal features for efficient operation [36]. Microbes have an interface for employing subtle changes in redox conditions to dissociate metals, mainly under stress conditions or high-affinity chelaters (or metallophores), since many metals possess low bioavailability [37]. In order to combat metal stress circumstances, such as toxicity of metals to cells, microbial generated biominerals. The redox reactions, sequestration of weakly soluble minerals, and ionic efflux pumps could all be used as resistance mechanisms to combat metal stress.

5.1.4.7 Biosorption

According to Kumar and Ngueagni [38], biosorption is an energy-free technique that uses biopolymers or biological materials as sorbents to remove contaminants from waste streams through physicochemical or metabolically mediated pathways. The parameters affecting the process of biosorption include affinity of biomass to metals, process agitation speed, absorbent and contaminant contact time, dose of bioabsorbent, metal concentration, temperature, and pH [38]. The process of biosorption is inexpensive, requires fewer chemicals, eliminates sludge, and utilizes regenerative bioabsorbents [39]. However, it takes a lot of time and has a high risk of saturation of the active sites of metal-binding ligands [40], which reduces the efficiency of the process.

5.1.4.8 Bioleaching

The method known as “bioleaching” harnesses microbes to transform insoluble metals into soluble ones in order to recover them from their source. Non-sulfide metals are recovered by fungi whereas metal sulfides are converted into metal sulfate by bacteria, particularly chemolithotrophic bacteria. There are two kinds of bioleaching processes:

- Indirect bioleaching: In this method, microorganisms produce a leaching agent that is used on the metal source to recover the metals, but they are not physically attached to the metal source.

- Direct bioleaching: This method uses minerals that are easily accessible to oxidized metals and enzyme activation to extract metals from the source through introducing microorganisms in immediate proximity with the metal source.

The industrial processing of metals is highly dependent on bioleaching. In situ heap, and slope bioleaching are three economically employed methods. Process variables that impact bioleaching include mineral substrate, concentrations of carbon dioxide and oxygen, nutrient content, temperature, and pH. According to Zhao et al. [41], bioleaching is an economical method of treating complicated and low-grade ores that is also environmentally favorable because it stabilizes sulfate toxicity. However, the procedure takes a long time, occasionally harmful chemicals are produced during it, and once it is started, it cannot be stopped. Bioleaching is a cost-effective and environmentally benign method of extracting metals from their ore or source. One of the greatest methods for recovering metals from E-waste may be bioleaching. The metal extraction from E-waste has been the subject of numerous investigations. The most often utilized microorganisms in metal bioleaching are fungus and bacteria. The position is dominated by bacterial species. Bioleaching uses microorganisms, including acidophilic bacteria, cyanogenic bacteria, thermophilic archaea, and mesophilic bacteria. Precious metals, which include silver, gold, aluminum, and copper are carefully considered, and metal recovery is prioritized [42].

6 Comparative evaluation of biological alternatives for metal recovery

The recovery of metals from E-waste has created new business prospects. The socioeconomic and environmental elements are impacted by the recovery option selection. Numerous studies have shown that chemical and physical methods of recovering metals have negative effects on environmental matrices, including soil, water, and air, and that these methods are expensive in terms of processing costs and harmful chemical emissions. Similar to this, physical methods show promise for recovering metals, but the dust produced during the procedure comprises hazardous substances that lower air quality and endanger the health of processing workers. Physical methods use a lot of energy, require a lot of labor, and have a substantial initial cost. The biological approach is a superior choice to mitigate the negative economic and environmental repercussions of chemical and physical technologies.

A gradient differentiation is illustrated based on each approach's contribution to related environmental and socioeconomic parameters according to the accessible researches. It has been noted that even though biological approaches take a lot of time, their effects on both the economy and the environment are very realistic. Although the term "biological method" refers to biological things including fungi, bacteria, mushrooms, yeast and algae, the majority of metal recovery research uses fungal and bacterial species. In contrast, phytoremediation provides a better alternative for environmental clearing, particularly for soil matrix harmed by concentrations of hazardous metals, whilst bioleaching is the greatest option for recovering metals through a greener approach. Here are some current developments in the area of biological recycling:

7 Modifications of bioleaching method

Numerous researches have indicated that process modification improves the recovery of metals from waste. Mesophilic bacteria can remove copper from E-waste by 24–84% when pyrite is used as a source of iron and sulfur [43]. When combined with *Acidithiobacillus ferrooxidans*, nitrogen-doped carbon nanotubes achieved 80–90% copper bioleaching [44]. Copper recovery via bioleaching with *A. ferrooxidans* was 74%; however, the recovery rate increased to 84% when graphene was added [45].

7.1 Usage of bioreactors

About 23% copper was recovered from E-waste Printed Circuit Board (PCB) using a pulsed plate bioreactor with *Acinetobacter* sp. [46]. According to reports, a common technique for recovering metals is the membrane bioreactor. About 72% of copper was recovered using a membrane bioreactor [47]. To improve the recovery of metals by bioleaching, numerous genetic and molecular advancements have been made; nevertheless, more scope and limitations need to be filled through deployment of recombinant strains, synthetic biology, proteomics, genomics, etc. [48]. On a reactive absorbent, metals are efficiently absorbed and readily recovered. About 85% of gold was absorbed in a study that used the chemical-assisted microbial biosorption (hybrid) technique for gold recovery. In this instance, *Lactobacillus acidophilus* (LA) and ammonium thiosulfate were combined to boost LA's bioabsorptive activity [49]. The enzyme copper oxidase demonstrated efficient gold adsorption for gold recovery or potential usage as a biosensor. To investigate its use for gold adsorption, as a biosensor, or for the bioremediation of E-waste, the copper oxidase gene was sequenced and expressed in *E. coli* [50].

7.2 Hybrid technology

To extract metals from outdated mobile SIM cards, hybrid technologies like chemo-biohydrometallurgy are used. Usage of *Chromobacterium violaceum* at an alkaline pH demonstrated efficient performance in recovering silver, gold, and copper from

SIM [51]. The use of microbiological consortiums in the form of culture seeds or biofilm formation demonstrated an efficient conclusion for co-culturing effects because it has been noted that manufacturing operations are non-sterile. The bacteria improved the metal recovery by adjusting the process conditions.

7.3 Biodismantling

The innovative method of employing bacteria to extract different components from E-waste is called "biodismantling." Four distinct products – depopulated boards, segregated ECs, leaching solutions, and precipitates – are created during this procedure. These products are then progressively processed for the biorecovery of designated metals using different methodologies. The technique known as "biodismantling" improves metal recovery while maintaining a grade akin to that of industrial ores. Recovering metals is therefore economically feasible [52].

7.4 Biohydrometallurgy

One alternative that is becoming more and more important from a business standpoint is biohydrometallurgy. An established business that specializes in the gold bioleaching from ores is Mintek [53]. One metric ton of circuit boards can yield 150 g of gold in a single week, according to an E-waste company called mint innovation [54].

8 Utilizing nanotechnology to recycle E-waste

Waste recycling is one of the many fields where nanotechnology is gaining momentum. If the cost of producing the nanomaterials could be efficiently controlled, this promising technique might fill the environmental and economic gaps uncovered by traditional approaches [55]. As a result, while biological recycling has advanced significantly, it has struggled to gain traction in the recycling industry due to its tedious procedures. If these requirements are satisfied, biological methods will be safe, eco-friendly, and efficient means of resolving the financial and ecological challenges related to physical, chemical, and conventional recycling methods.

9 Innovations in phytoremediation methods for treating sites affected by E-waste

The bioavailability of the pollutant is thought to be the main limiting element in phytoremediation. This issue may be resolved by technological advancements like enzymatic stimulation, improvement of soil microflora, processes revised with electrokinetic applications, surfactant, chelator application, and genetic modification. One promising method for addressing sites affected by E-waste is the utilization of invasive species of plants for phyto-treatment [21]. Invasive plants can thrive on soils with minimal nutrition and exhibit rapid development under unfavorable and adverse circumstances. Certain plant species have demonstrated successful metal removal from contaminated sites, including *Praxelis clematidea, Bidens pilosa, Chromolaena odorata*, and *Amaranthus spinosus*. However, because of their fast growth rate and accompanying effects on humans and the environment (habitat), they should be treated with tremendous attentiveness [21]. The use of transgenic plants to treat sites affected by E-waste may be a preferable choice. Genetic alteration, such as endogenic or exogenic genetic modification, could solve the issue with the wild variety in transgenic plants. According to numerous studies, using genetically modified plants for phytoremediation of sites contaminated by E-waste is a feasible strategy. Another development in this field to speed up process performance is microbe assisted phytoremediation. Biosurfactant is used to make metals more bioavailable to bacteria for subsequent processes. The two biological surfactants that are typically employed to remove metals from polluted soil are rhamnolipids and sophorolipids [56]. Another efficient method for recovering metals is membrane filtering technology [57]. Studies at the molecular level, such as gene array, proteomics, metagenomics, and metallomics, are potential methods to provide a deeper comprehension of the field and may provide different ways to metal recovery from E-waste to support the national circular economy [57].

Future perspectives: Innovations in hybrid and biological technologies that adapt to changing climatic conditions and other factors are highly sought after in biorecovery industry. A collaborative engineering and scientific participation in these domains may result in an efficient process or technology with all practical qualities that can form a genuine practical platform to increase the nation's economy in a circular shape by gaining out of waste materials.

10 Conclusion

One environmentally friendly, resilient, and clean method of managing waste, particularly E-waste, is the biological approach. While phytoremediation, vermicomposting, bioaccumulation, biotransformation, and their advancements are the greener alternatives for enhancing ecological metrics by zero or low greenhouse emissions, which in turn leads to low carbon footprint production, biological approaches like biosorption and bioleaching and their advancements are among the best options for boosting the national economic and socioeconomic metrics. These technologies are becoming more and more popular since they are cost-effective, eco-friendly, efficient, and very promising. These innovations are not generating commercial value because of lengthy consuming procedures and different environmental repercussions. It is necessary to evaluate an equitable experimental interventional viability for field implementation in order to plug these discrepancies.

References

[1] Dixit R, Kumar S, Pandey G. Biological approaches for E-waste management: A green-go to boost circular economy. Chemosphere. 2023;336:139177.

[2] Liu K, Tan Q, Yu J, Wang M. A global perspective on e-waste recycling. Circ Econ. 2023;2(1): 100028.

[3] Shahabuddin M, Uddin MN, Chowdhury JI, Ahmed SF, Uddin MN, Mofijur M, Uddin MA. A review of the recent development, challenges, and opportunities of electronic waste (e-waste). International Journal of Environmental Science and Technology. 2023;20(4):4513–4520.

[4] Forti V, Baldé CP, Kuehr R, Bel G. The global e-waste monitor 2020. United Nations University (UNU), International Telecommunication Union (ITU) & International Solid Waste Association (ISWA. Bonn/Geneva/Rotterdam. 2020;120:14.

[5] Ahirwar R, Tripathi AK. E-waste management: A review of recycling process, environmental and occupational health hazards, and potential solutions. Environmental Nanotechnology, Monitoring & Management. 2021;15:100409.

[6] Kiddee P, Pradhan JK, Mandal S, Biswas JK, Sarkar B. An overview of treatment technologies of e-waste. Handbook of electronic waste management. 2020;1: 38–55.

[7] Baldé C, Wang F, Kuehr R, Huisman J. The Global E-Waste Monitor–2014. Bonn, Germany: United Nations University, IAS–SCYCLE, 2015, pp. 978–992.

[8] Valix M. Bioleaching of Electronic Waste: Milestones and Challenges. Current Developments in Biotechnology and Bioengineering. Elsevier. 2017, pp. 407–442.

[9] Elytus L. E-waste & its Negative Effects on the Environment. 2019. https://elytus.com/blog/e-waste-and-its-negative-effects-on-the-environment.html.

[10] India. Circular Economy in Electronics and Electrical Sector. 2021.

[11] Ilyas S, Srivastava RR, Kim H, Das S, Singh VK. Circular bioeconomy and environmental benignness through microbial recycling of e-waste: A case study on copper and gold restoration. Waste Management. 2021;121:175–185.

[12] Ramanayaka S, Keerthanan S, Vithanage M. Urban mining of E-waste: treasure hunting for precious nanometals. In Handbook of Electronic Waste Management 2020 (pp. 19–54). Majeti Narasimha Vara Prasad, Meththika Vithanage and Anwesha Borthakur (Eds.) Elsevier. Butterworth-Heinemann.

[13] Tran CD, Salhofer SP. Processes in informal end-processing of e-waste generated from personal computers in Vietnam. Journal of Material Cycles and Waste Management. 2018;20:1154–1178. https://doi.org/10.1007/s10163-017-0678-1.

[14] Cortes M, Patiño M. Recycling of electronic waste, using basic and acid leaching. Open Journal of Applied Sciences. 2016;6:169–176. https://doi.org/10.4236/ojapps.2016.63018.

[15] Sivaramanan S E-waste management, disposal and its impacts on the environment. Universal Journal of Environmental Research and Technology. 2013;3(Issue 5):531–537.

[16] Andooz A, Eqbalpour M, Elaheh Kowsari E, Ramakrishna S, Cheshmeh ZZ. A comprehensive review on pyrolysis of E-waste and its sustainability. The Journal of Cleaner Production. 2022;333:130191.
[17] Xue M, Yan G, Li J, Xu Z. 2012. Electrostatic separation for recycling conductors, semiconductors, and nonconductors from electronic waste. Environmental Science & Technology. 2012;46(19):10556–10563.
[18] Woynarowska A, Zukowski W. Magnetic separation of electronic waste after the combustion process in the fluidized bed. Proceedings of ECOpole. 2014;8(1):052.
[19] Patel S, Kasture A. E (electronic) waste management using biological systems- overview. International Journal of Current Microbiology and Applied Sciences. 2014;3:495–504.
[20] Van Yken J, Boxall NJ, Cheng KY, Nikoloski AN, Moheimani NR, Kaksonen AH. E-waste recycling and resource recovery: A review on technologies, barriers and enablers with a focus on oceania. Metals. 2021;11(8):1313.
[21] Gunarathne V, Gunatilake SR, Wanasinghe ST, Atugoda T, Wijekoon P, Biswas JK, Vithanage M. Phytoremediation for E-waste contaminated sites. In Handbook of Electronic Waste Management 2020 (pp. 141–170). Elsevier, Butterworth-Heinemann.
[22] Pandey VC, Pandey DN, Singh N. Sustainable phytoremediation based on naturally colonizing and economically valuable plants. The Journal of Cleaner Production. 2015;86:37–39.
[23] Lukose N. A review on E-waste management and recycling challenges in India. International Journal of Science and Engineering Research. 2015;6:18–29.
[24] Ashwini M, RdKP, Nandhini S, Srinivasan J. Degradation of e-waste using biological techniques. Waffen-und Kostumkunde Journal. 2020;XI:79–96.
[25] Kaur T. Vermicomposting: an effective option for recycling organic wastes. Organic Agriculture. 2020;1–10, IntechOpen.
[26] Pandit NP, Ahmad N, Maheshwari SK. Vermicomposting biotechnology an eco-loving approach for recycling of solid organic wastes into valuable biofertilizers. Journal of Biofertilizers and Biopesticides. 2012;3:1–8.
[27] Marappa N, Dharumadurai D, Thajuddin N. Biological recovery of metals from electronic waste polluted environment using microorganisms. Microbes for restoration of degraded ecosystems New India Publishing Agency, New Delhi, India. 2017;149–172.
[28] Ahluwalia SS, Goyal D. Microbial and plant derived biomass for removal of heavy metals from wastewater. Bioresource Technology. 2007;98:2243–2257.
[29] Aderonke AK, Oladimeji OO, Shittu OB, Okeyode IC, Taiwo MO. Bioaccumulation of heavy metals using selected organisms isolated from electronic waste dumpsite of two south-western states in Nigeria. Applied Environmental Research. 2017;39:29–40.
[30] Speight JG. Chapter 8 – biological transformations. In Speight JG (ed.). Reaction Mechanisms in Environmental Engineering. 2018 (pp. 269–306). Elsevier, Butterworth-Heinemann .
[31] Emenike CU, Jayanthi B, Agamuthu P, Fauziah SH. Biotransformation and removal of heavy metals: a review of phytoremediation and microbial remediation assessment on contaminated soil. Environmental Reviews. 2018;26:156–168.
[32] Hegazy ME, Mohamed TA, ElShamy AI, Mohamed AE, Mahalel UA, Reda EH, Shaheen AM, Tawfik WA, Shahat AA, Shams KA, Abdel-Azim NS. Microbial biotransformation as a tool for drug development based on natural products from mevalonic acid pathway: A review. Journal of Advanced Research. 2015;6(1):17–33.
[33] Narayanasamy M, Dhanasekaran D, Thajuddin N. Biological recovery of metals from electronic waste polluted environment using microorganisms. Microbes for restoration of degraded ecosystems New India Publishing Agency, New Delhi, India. 2017;1:149–172.
[34] Cheng Y, Ke Z. Bio-mineralization and Remediation of Heavy Metals: from Microbe Cellular Level to Molecular Level. Archives of Clinical Microbiology. In 10th Edition of International Conference on Advanced Microbiology Education, London, UK 2018.
[35] Dhami NK, Reddy MS, Mukherjee A. Biomineralization of calcium carbonates and their engineered applications: a review. Frontiers in Microbiology. 2013;4:314.

[36] Staicu LC, Stolz JF. Editorial: microbes vs. metals: harvest and recycle. FEMS Microbiology Ecology. 2021;97.

[37] Wells M, Kanmanii NJ, Al Zadjali AM, Janecka JE, Basu P, Oremland RS, Stolz JF. Methane, arsenic, selenium and the origins of the DMSO reductase family. Scientific Reports. 2020;10(1):10946.

[38] Maulin Shah, Susana Rodriguez-Couto and Jayanta Biswas. Removal of volatile organic carbon and heavy metals through microbial approach. In An Innovative Role of Biofiltration in Wastewater Treatment Plants (WWTPs). Maulin Shah, Susana Rodriguez-Couto and Jayanta Biswas (Eds.). 2022 (pp. 285–308). Elsevier.

[39] Liu Y, Serrano A, Villa-Gomez D. Biological treatment of mine-impacted waters on the context of metal recovery. Future Effluent Treatment Plants. 2021;1:499–522.

[40] Abdi O, Kazemi M. A review study of biosorption of heavy metals and comparison between different biosorbents. Journal of Materials and Environmental Science. 2015;6:1386–1399.

[41] Zhao H, Yang C, Zhang X, Zhang Y, Qiu G. Industrial Application for Chalcopyrite Bioleaching. Biohydrometallurgy of Chalcopyrite. 2021;1:211–231.

[42] Das S, Natarajan G, Ting Y-P.2017. Bio-extraction of precious metals from urban solid waste. AIP Conference Proceedings. 1805(1); 020004. AIP Publishing LLC.

[43] Bas AD, Deveci H, Yazici EY. Bioleaching of copper from low grade scrap TV circuit boards using mesophilic bacteria. Hydrometallurgy. 2013;138:65–70.

[44] Bai J, Gu W, Dai J, Zhang C, Yuan W, Deng M, Luo X, Wang J. The catalytic role of nitrogen-doped carbon Nanotubes in bioleaching copper from waste printed circuit boards. Polish Journal of Environmental Studies. 2016;25:951–957.

[45] Gu W, Bai J, Dong B, Zhuang X, Zhao J, Zhang C, Wang J, Shih K. Catalytic effect of graphene in bioleaching copper from waste printed circuit boards by *Acidithiobacillus ferrooxidans*. Hydrometallurgy. 2017;171:172–178.

[46] Jagannath A, Shetty KV, Saidutta MB. Bioleaching of copper from electronic waste using *Acinetobacter* sp. Cr B2 in a pulsed plate column operated in batch and sequential batch mode. Journal of Environmental Chemical Engineering. 2017;5:1599–1607.

[47] Wang G, Fan Z, Wu D, Qin L, Zhang G, Gao C, Meng Q. Anoxic/aerobic granular active carbon assisted MBR integrated with nanofiltration and reverse osmosis for advanced treatment of municipal landfill leachate. Desalination. 2014;349:136–144.

[48] Pattanaik A, Samal DPK, Sukla LB, Pradhan D. Advancements and use of omic technologies in the field of bioleaching: a review. Biointerface Research in Applied Chemistry. 2021;11(3):10185–10204.

[49] Sheel A, Pant D. Recovery of gold from electronic waste using chemical assisted microbial biosorption (hybrid) technique. Bioresource Technology. 2018;247:1189–1192.

[50] Tan S-I, Ng IS, Yu Y-J. Heterologous expression of an acidophilic multicopper oxidase in *Escherichia coli* and its applications in biorecovery of gold. Bioresources and Bioprocessing. 2017;4:20.

[51] Sahni A, Kumar A, Kumar S. Chemo-biohydrometallurgy – a hybrid technology to recover metals from obsolete mobile SIM cards. Environmental Nanotechnology, Monitoring & Management. 2016;6:130–133.

[52] Monneron-Enaud B, Wiche O, Schlömann M. Biodismantling, a novel application of bioleaching in recycling of electronic wastes. Recycling. 2020;5:22.

[53] Gericke M, Neale JW, Van staden PJ. A Mintek perspective of the past 25 years in minerals bioleaching. Journal of the Southern African Institute of Mining and Metallurgy. 2009;109:567–585.

[54] Han P, Teo WZ, Yew WS. Biologically engineered microbes for bioremediation of electronic waste: wayposts, challenges and future directions. Engineering Biology. 2022;6:23–34.

[55] Dermatas D, Mpouras T, Panagiotakis I. Application of Nanotechnology for waste management: challenges and limitations. Waste Management & Research. 2018;36:197–199.

[56] Mulligan CN, Wang S. Remediation of a heavy metal-contaminated soil by a rhamnolipid foam. Engineering Geology. 2006;85:75–81.

[57] Wang G, Fan Z, Wu D, Qin L, Zhang G, Gao C, Meng Q. Anoxic/aerobic granular active carbon assisted MBR integrated with nanofiltration and reverse osmosis for advanced treatment of municipal landfill leachate. Desalination. 2014;349:136–144.

Tusheema Dutta*, Tisha Gupta, Mayukh Chakraborty, Medha Gupta, Nishita Mathur

Microalgae-mediated remediation of e-waste pollutants: mechanisms, challenges, and sustainable pathways

Abstract: Electronic waste has emerged as one of the fastest growing and most environmentally hazardous waste streams worldwide. E-waste-derived pollutants, particularly toxic heavy metals and persistent organic compounds, accumulate in ecosystems, contaminate water bodies, and pose serious health risks to human and ecological systems. Microalgae offer a sustainable and circular biotechnological pathway for remediating such pollutants through biosorption, bioaccumulation, biotransformation, and biomass-mediated precipitation mechanisms. This chapter reviews the mechanisms underpinning microalgal pollutant sequestration, evaluates key cultivation and reactor technologies, and assesses techno-economic feasibility and environmental performance. It further explores opportunities for integrating microalgal systems into circular economy models, including biomass valorization and metal recovery from contaminated streams. Finally, policy, social equity, and environmental governance considerations are discussed, highlighting microalgae as a transformative yet under-utilized technology for e-waste pollution management aligned with sustainable development goals.

Keywords: Microalgae, e-waste, heavy metals, biosorption, circular economy, bioremediation, sustainable technology

1 Introduction

The global digital economy has accelerated the production, consumption, and discarding of electronic devices at an unprecedented pace [1]. Smartphones, laptops, tablets, household appliances, communication systems, industrial control technologies, and renewable energy components have become essential to modern life. However, the same products that enable innovation and development eventually enter the waste stream as electronic waste (e-waste) – a complex, hazardous, and rapidly expanding

***Corresponding author: Tusheema Dutta**, Center for Research and Innovation in Frugal Technology Management, FORE School of Management, New Delhi 110016, India, e-mail: 183113@fsm.ac.in
Tisha Gupta, Mayukh Chakraborty, Medha Gupta, Nishita Mathur, Center for Research and Innovation in Frugal Technology Management, FORE School of Management, New Delhi 110016

 | https://doi.org/10.1515/9783112225509-016

category of solid waste [2]. Over the past two decades, global e-waste generation has increased dramatically and continues to rise annually. Projections indicate that future volumes will grow even faster as populations urbanize, devices become more affordable, and upgrading cycles shorten [1]. Yet, despite the significant economic value embedded in discarded electronics – such as gold, copper, silver, rare earth elements, cobalt, and high-performance plastics – only a fraction of e-waste is processed through formal recycling systems [3]. Much of it is handled informally, stored in households, shipped across borders, dumped in landfills, or openly burned, particularly in low- and middle-income regions [2]. The environmental consequences are profound. Printed circuit boards, batteries, wiring, solder, coatings, and casings contain hazardous substances including lead, cadmium, chromium, mercury, nickel, antimony, brominated flame retardants, per- and polyfluoroalkyl substances (PFASs), and other persistent organic pollutants [4, 2]. When released into soil, groundwater, rivers, and air, these contaminants accumulate and disperse widely across ecosystems, resulting in long-term ecological degradation and bioaccumulation in food chains [4]. Informal recycling practices such as open burning, acid leaching, and crude dismantling further intensify pollutant release and disproportionately expose workers and surrounding communities to severe health risks [2]. Conventional treatment approaches – including chemical precipitation, activated carbon adsorption, ion-exchange resins, oxidation processes, and membrane filtration – can remove heavy metals and organic contaminants but often involve high energy consumption, costly reagents, operational complexity, and secondary waste generation [5,6]. These limitations have prompted growing interest in biological remediation technologies, particularly approaches that operate in synergy with natural ecological and biochemical processes [7]. Among available biological strategies, microalgae offer a particularly promising platform for the remediation of e-waste-derived pollutants [8, 9]. Microalgae are photosynthetic microorganisms capable of rapid growth in aquatic environments using sunlight, carbon dioxide, and nutrients from wastewater streams [7]. Their cell walls contain polysaccharides and proteins with functional groups that bind metal ions, while intracellular metabolism and extracellular polymeric substances (EPSs) contribute to pollutant sequestration, transformation, and immobilization [5, 8]. Unlike conventional treatment systems, microalgae-based processes simultaneously support biomass production, nutrient recovery, CO_2 utilization, and downstream valorization pathways, thereby aligning remediation with circular economy objectives [3, 10]. This chapter examines – critically and comprehensively – the potential of microalgae-mediated remediation as a sustainable response to e-waste pollution. It explores pollutant characteristics, biological sequestration mechanisms, cultivation and reactor systems, techno-economic feasibility, environmental trade-offs, social equity considerations, and future innovation pathways [11, 12]. The objective is not only to present microalgae as an environmentally friendly biotechnology, but also to position it within the broader transition toward circular, regenerative, and resource-efficient industrial systems aligned with Sustainable Development Goals 6 and 12 [3, 6].

2 Nature and toxicity of e-waste-derived pollutants

2.1 A complex and persistent pollutant mixture

E-waste is chemically diverse, with more than 1,000 different substances identified in discarded electronic equipment [8, 10]. Two groups dominate environmental risk: heavy metals and persistent organic pollutants (POPs) [1, 8]. When electronic devices deteriorate or are dismantled, metals leach into surrounding environments through rainfall runoff, landfill leachate, evaporative deposition, and wastewater discharge pathways [1, 6]. Open-air combustion of electronic components generates toxic fumes and fine particulate matter that remain airborne and subsequently deposit onto soils, vegetation, and built surfaces [8]. Informal recycling hubs are frequently located near residential settlements, agricultural land, and surface water bodies, thereby amplifying pollutant exposure and associated public health risks among workers and neighboring communities [8].

2.2 Heavy metals and human health

Heavy metals are of particular concern due to their persistence, bioaccumulation potential, and systemic toxicity in environmental and biological systems [1, 5]. Key examples include lead (Pb) which damages the nervous system, reduces cognitive performance and IQ, and causes behavioral and developmental disorders, particularly in children [1]; cadmium (Cd) which accumulates in the kidneys and liver; prolonged exposure weakens bone structure and disrupts essential metabolic pathways [5, 6]; chromium (Cr), particularly hexavalent chromium (Cr^{6+}), which is a confirmed human carcinogen that adversely affects the respiratory system and compromises DNA integrity [1, 12]; mercury (Hg) which exhibits pronounced neurotoxicity in its methylated form and biomagnifies through aquatic food webs, posing long-term ecological and health risks [1, 8]; and nickel, copper, zinc, and cobalt which are essential trace elements at low concentrations but capable of interfering with enzyme systems and cellular homeostasis when present above physiological tolerance limits [5]. Field investigations consistently report elevated concentrations of lead, cadmium, chromium, and copper in soils and sediments adjacent to informal and formal e-waste recycling regions [8]. Agricultural crops irrigated with contaminated surface water or groundwater frequently accumulate metals in edible tissues, thereby directly linking environmental contamination from e-waste activities to food safety concerns and chronic human exposure pathways [1, 6].

Table 1: Major heavy metals in e-waste and their environmental and health impacts.

Heavy metal	Major sources in e-waste	Environmental behavior	Key human health impacts	References
Lead (Pb)	Solder, cathode ray tubes, batteries, and cables	Persistent; accumulates in soils and sediments; low mobility once deposited	Neurotoxicity, reduced IQ, behavioral and developmental disorders, and cardiovascular effects	[1, 8]
Cadmium (Cd)	Rechargeable batteries, semiconductors, and circuit boards	Highly mobile in soils; readily taken up by plants; bioaccumulative	Kidney and liver damage, bone demineralization, and metabolic disruption	[5, 6]
Chromium (Cr^{6+})	Metal coatings, anticorrosive treatments, and pigments	Highly soluble and mobile; strong oxidizing agent	Carcinogenicity, respiratory toxicity, and DNA damage	[1, 12]
Mercury (Hg)	Switches, relays, fluorescent lamps, and batteries	Volatile; methylates in aquatic systems; biomagnifies in food webs	Neurotoxicity, impaired cognitive and motor function, and developmental toxicity	[1, 8]
Nickel (Ni)	Alloys, rechargeable batteries, and electronic components	Moderately mobile; accumulates near recycling sites	Enzyme inhibition, allergic reactions, and respiratory effects	[5]
Copper (Cu)	Wiring, printed circuit boards, and motors	Strong soil binding; elevated concentrations near recycling hubs	Oxidative stress, and liver and kidney toxicity at high doses	[5, 8]
Zinc (Zn)	Circuit boards, coatings, and alloys	Mobile under acidic conditions; phytotoxic at high levels	Gastrointestinal distress and interference with micronutrient balance	[5]
Cobalt (Co)	Lithium-ion batteries and magnets	Bioaccumulative at elevated levels	Cardiovascular and endocrine disruption, and oxidative stress	[5, 11]

2.3 Organic pollutants and endocrine disruption

Beyond metals, e-waste contains brominated flame retardants, phthalates, PFAS and PAHs, and dioxins (from burning plastics). These substances resist environmental degradation, act as endocrine disruptors, impair reproductive and developmental systems, and significantly increase long-term cancer risk in exposed populations [1, 8]. In summary, e-waste pollution is chemically complex, chronic in nature, global in scale, and socially inequitable – disproportionately affecting vulnerable populations such as informal recycling workers, children, and low-income communities [8, 11]. This context underscores the urgent need for sustainable, low-impact, and socially inclusive remediation platforms capable of addressing both environmental contamination and public health protection [3, 12].

3 Microalgae as a biological remediation platform

Microalgae encompass a diverse group of unicellular photosynthetic organisms including *Chlorella, Scenedesmus, Chlamydomonas, Spirulina, Nannochloropsis*, and *Ankistrodesmus* [12]. Their key features relevant to remediation include rapid biomass productivity under appropriate light and nutrient conditions, high surface-area-to-volume ratio enabling efficient contact with pollutants, biochemically active cell surfaces containing carboxyl, hydroxyl, phosphate, and amino groups, production of extracellular polymeric substances (EPSs) that bind pollutants, and ability to acclimate to variable environmental stress. Unlike many microorganisms, microalgae generate oxygen via photosynthesis, supporting mixed microbial communities that further assist pollutant transformation [4]. They also assimilate nitrogen and phosphorus from wastewater, meaning remediation and nutrient recycling occur simultaneously.

4 Mechanisms of pollutant removal

Microalgae remove pollutants through multiple, overlapping processes. Understanding these mechanisms is essential to designing efficient systems.

4.1 Biosorption (passive binding)

Biosorption occurs when dissolved metals bind to negatively charged functional groups on the algal cell wall. It is rapid, metabolism-independent, and reversible in some cases. This means even **inactive or dried algal biomass remains effective** as a biosorbent material.

4.2 Bioaccumulation (active uptake)

Metals may cross cell membranes and enter intracellular compartments. They are often chelated by metalloproteins or stored in organelles to minimize toxicity. Bioaccumulation requires metabolic activity and energy expenditure.

4.3 Biotransformation and reduction

Some algae reduce highly toxic metal ions to less toxic or less soluble forms – for example, chromium reduction from Cr^{6+} to Cr^{3+}. Others degrade organic pollutants enzymatically.

4.4 Biomineralization and precipitation

Under certain pH and redox conditions, algae induce precipitation of metals as oxides, hydroxides, or carbonates.

4.5 Extracellular polymeric substances (EPSs)

EPSs secreted by algae form sticky matrices capable of trapping metals and organic compounds. This enhances stability and immobilization.

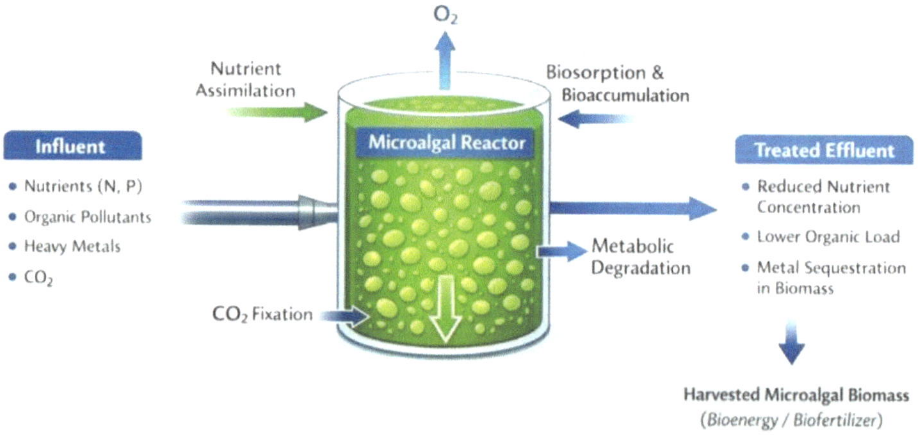

Figure 1: Conceptual pathway of pollutant removal in a microalgal reactor.

5 Environmental and operational factors affecting performance

Microalgal remediation efficiency is governed by a combination of physicochemical, biological, and operational parameters. Solution pH plays a critical role by influencing metal ion speciation, solubility, and binding affinity at the cell surface, while temperature directly affects enzymatic activity and overall metabolic rates. The concentration of metals in the influent is also a determining factor, as toxicity thresholds can inhibit growth and reduce uptake efficiency beyond certain levels [9]. Light intensity and photoperiod regulate photosynthetic performance and biomass productivity, whereas nutrient balance controls cellular growth and the synthesis of metal-binding biomolecules. In addition, culture density influences light penetration and mass transfer, and hydraulic retention time determines the contact duration between microalgae and contaminants [4]. Finally, strain selection is crucial, as different microalgal species exhibit distinct tolerances and uptake capacities for specific pollutants. Consequently, careful optimization of these interacting parameters is essential under field conditions, where pollutant loads and environmental variables often fluctuate [1].

6 Microalgal cultivation and reactor systems

A range of cultivation technologies exist:

6.1 Open raceway ponds

Advantages are slow capital cost simple operation; however, disadvantages are contamination risk, evaporation, and environmental variability.

6.2 Closed photobioreactors (PBRs)

Advantages are controlled environment; higher productivity improved biosafety.
Disadvantages are higher investment operational complexity.

6.3 Biofilm reactors

Biofilm systems reduce harvesting costs and improve resilience to toxins.

Figure 2: Comparison of open pond versus photobioreactor systems.

7 Representative strains and reported efficiencies

Microalgae strain	Target metals	Dominant mechanism	Reported Removal efficiency
Chlorella vulgaris	Pb and Cd	Biosorption + accumulation	80–95%
Scenedesmus obliquus	Cr^{6+}	Reduction + sorption	70–92%
Spirulina platensis	Cu and Zn	Cell wall binding	65–88%
Chlamydomonas reinhardtii	Hg	Intracellular sequestration	60–85%

Values reflect literature ranges under optimized laboratory conditions.

8 Biomass valorization and circularity

One of the defining advantages of microalgal remediation lies in the ability to valorize pollutant-laden biomass into a range of useful products. Depending on the nature of the contaminants and the post-treatment strategy, harvested biomass can be converted into biofertilizers, incorporated into polymer composites, or utilized for bioenergy generation where safety and regulatory conditions permit. In addition, microalgal biomass can serve as a precursor for catalysts as well as pigments and other specialty chemicals with commercial value. This integrated approach effectively links environmental remediation with revenue generation and waste minimization, thereby strengthening the economic viability and circularity of microalgae-based treatment systems.

9 Techno-economic dimensions of microalgal remediation

While microalgae-mediated remediation demonstrates strong scientific feasibility, large-scale implementation is governed by techno-economic realities. These realities determine whether the technology can move from laboratory environments to industrial application.

9.1 Capital and operational cost structure

The major cost components of microalgae-based remediation systems span both capital and operational expenditures. Capital costs are primarily associated with cultivation infrastructure, such as raceway ponds or photobioreactors, along with land acquisition and site preparation. Operational costs are dominated by energy inputs required for mixing, aeration, and pumping, as well as expenses related to nutrient supplementation to sustain biomass productivity [2]. Additional costs arise from biomass harvesting and dewatering processes, the deployment of monitoring and control systems to ensure stable performance, and ongoing labor and maintenance requirements [3]. Together, these factors largely determine the economic feasibility and scalability of microalgal remediation technologies. Photobioreactors, while efficient and controllable, involve higher capital expenditures compared with open ponds. However, in environments with high contamination or land scarcity, their precision and safety may justify the additional investment. Energy consumption varies considerably depending on design choices, such as whether passive illumination or artificial light supplementation is employed.

9.2 Industrial integration potential

Economic viability improves significantly when microalgal systems are integrated into existing industrial or municipal wastewater treatment infrastructure [3, 9]. In such configurations, microalgae can replace or complement conventional tertiary treatment processes, thereby allowing shared capital investment, reduced operational costs, and lower energy demand compared with standalone systems [4, 12]. When carbon dioxide derived from industrial flue gas is supplied to microalgal reactors, additional benefits are realized through enhanced biomass productivity and simultaneous carbon capture, contributing to climate mitigation objectives [3, 9].

9.3 Revenue streams and circular value

Cost recovery in microalgae-based remediation systems can be achieved through multiple value streams rather than a single revenue source. These include the sale of biomass-derived products, such as biofertilizers, bio-based materials, or specialty chemicals, as well as credits associated with nutrient recovery and carbon sequestration [5]. Additional economic benefits may arise from avoided environmental penalties and compliance costs, improved public health outcomes resulting from reduced pollutant exposure, and long-term ecological restoration that enhances ecosystem services [3]. This multi-value structure clearly differentiates microalgal remediation from conventional "end-of-pipe" treatment models, which typically involve continuous operational and disposal expenses without generating compensatory economic or societal value.

9.4 Economic barriers

However, several key barriers continue to limit the large-scale deployment of microalgal remediation systems. These include uncertainty surrounding the market value of biomass-derived products, volatility in metal and commodity prices that affects revenue predictability, and the absence of standardized performance benchmarks for evaluating technical and environmental outcomes. In addition, limited investor familiarity with biological remediation technologies often constrains access to finance. As a result, successful implementation typically depends on risk-sharing frameworks, supportive policy incentives, and well-designed demonstration pilots that can validate performance, reduce perceived risk, and build technical and financial confidence among stakeholders [11].

10 Hybrid bioremediation approaches

Microalgae rarely exist in isolation in natural ecosystems. Their interaction with bacteria, fungi, and other microorganisms has inspired hybrid remediation systems designed to exploit synergistic pollutant removal pathways [9].

10.1 Algae-bacteria consortia

In mixed cultures, bacteria mineralize organic matter and release CO_2, which microalgae use for photosynthesis. Oxygen generated by algae supports bacterial metabolism. This symbiotic exchange enhances pollutant tolerance, stabilizes community dynamics, and improves overall wastewater treatment efficiency. Such consortia are particularly effective for nutrient-rich, metal-contaminated effluents characteristic of informal e-waste processing zones.

10.2 Algae-fungi biopellets

Filamentous fungi form structural frameworks onto which microalgae adhere, creating dense pellets. These pellets settle rapidly, are easy to harvest, and maintain high pollutant-binding capacity. They are useful in systems where mechanical harvesting would otherwise dominate costs.

10.3 Integration with physicochemical processes

Microalgae can operate as either a primary or polishing remediation stage. For instance, chemical precipitation removes bulk metals, microalgae capture residual contaminants, and biological degradation addresses organics. This reduces chemical consumption while achieving discharge compliance.

11 Life cycle sustainability assessment

A comprehensive evaluation of microalgal remediation must consider life cycle impacts, ensuring that environmental benefits outweigh energy and material inputs.

11.1 Key life cycle metrics

Important indicators for assessing the sustainability of microalgal remediation systems encompass cumulative energy demand, greenhouse gas emissions, nutrient recovery rates, water footprint, toxicity potential, and land requirements. Together, these metrics provide a comprehensive life cycle perspective on environmental performance and resource efficiency. Empirical studies consistently show that integrating microalgal remediation with wastewater streams and renewable energy inputs can dramatically improve overall life cycle performance by reducing external energy and nutrient inputs while enhancing system efficiency and environmental benefits.

11.2 Environmental trade-offs

Potential trade-offs include electricity use during harvesting, embedded carbon in reactor materials, and nutrient reliance if not sourced from waste streams. However, when designed within a circular resource framework, the overall environmental balance remains favorable.

12 Social, ethical, and governance dimensions

E-waste management is fundamentally intertwined with equity, livelihoods, and environmental justice.

12.1 Informal sector realities

In this context, the deployment of microalgal remediation technologies must be designed to avoid the displacement of livelihoods or the reinforcement of existing social and economic inequalities. Rather than imposing centralized or externally controlled solutions, such systems should be locally managed and embedded within community structures, with an emphasis on cooperative-based ownership and operation. A strong focus on capacity building is essential to ensure that local stakeholders acquire the technical skills and institutional knowledge required for long-term system management. Public investment and policy support can further enable inclusive adoption by reducing financial barriers, strengthening governance frameworks, and ensuring that the social benefits of remediation are equitably distributed.

12.2 Regulatory support

In this context, the deployment of microalgal remediation technologies must be designed to avoid the displacement of livelihoods or the reinforcement of existing social and economic inequalities. Rather than imposing centralized or externally controlled solutions, such systems should be locally managed and embedded within community structures, with an emphasis on cooperative-based ownership and operation. A strong focus on capacity building is essential to ensure that local stakeholders acquire the technical skills and institutional knowledge required for long-term system management. Public investment and policy support can further enable inclusive adoption by reducing financial barriers, strengthening governance frameworks, and ensuring that the social benefits of remediation are equitably distributed.

12.3 Ethics of biotechnology deployment

Where genetically modified strains are proposed, transparent risk assessment, containment strategies, and stakeholder engagement are essential.

13 Operational risks and safeguards

Microalgal remediation, while environmentally friendly, requires careful management.

13.1 Metal-laden biomass handling

If microalgal biomass contains elevated concentrations of metals, stringent handling and management protocols are essential to prevent secondary contamination and human exposure. Such protocols should include controlled storage to avoid leaching or accidental release, the use of appropriate personal protective equipment during handling and processing, and the application of thermal treatment or targeted extraction methods to remove or recover accumulated metals. Where recovery is not feasible, stabilization of the biomass prior to disposal is necessary to immobilize contaminants and ensure compliance with environmental and occupational safety regulations.

13.2 Ecological containment

Closed systems minimize accidental release. Monitoring ensures that strains do not disrupt existing ecosystems.

13.3 Occupational safety

Workers must be trained in biosafety, waste handling, and emergency response – reducing the occupational health risks typical in informal recycling.

14 Comparative assessment with other remediation technologies

When compared with conventional chemical and physical treatment processes, microalgal remediation demonstrates clear advantages in terms of sustainability, lower energy intensity, and strong compatibility with circular economy frameworks, owing to its ability to recover nutrients and assimilate CO_2 while treating contaminated streams. These co-benefits enhance overall resource efficiency and environmental performance beyond simple pollutant removal. However, microalgal systems may lag behind conventional methods in terms of treatment speed under very high pollutant loads, resilience to sudden contamination spikes, and the operational complexity associated with biological monitoring and control. Consequently, hybrid treatment strategies that combine microalgal processes with physicochemical methods often represent the most practical and robust compromise, leveraging the strengths of each approach while mitigating their respective limitations.

15 Case illustrations and pilot experiences

A number of experimental and pilot-scale initiatives demonstrate the technical feasibility of microalgae-based remediation systems [3, 7]. Raceway pond systems treating metal-contaminated runoff have reported removal efficiencies exceeding 80–90% for selected heavy metals under optimized operational conditions, including controlled pH, hydraulic retention time, and biomass density [2, 6]. Biofilm-based microalgal reactors deployed near informal e-waste recycling zones have shown strong resilience to fluctuating pollutant loads while maintaining stable biomass productivity and metal uptake capacity [9]. In several regions, microalgal systems have also been piloted in conjunction with conventional municipal wastewater treatment plants to enhance effluent quality prior to environmental discharge, particularly with respect to residual metals and nutrients [3, 4]. Although commercial systems remain limited, each demonstration strengthens the evidence base needed for policy adoption and investment.

16 Future outlook and innovation trajectories

Key innovation pathways in microalgal remediation increasingly reflect advances at the intersection of biotechnology, digital systems, and circular economy design. High-throughput strain discovery platforms enable the rapid identification of microalgal species with enhanced metal uptake, stress tolerance, and biomass productivity [1, 2].

These capabilities are further extended through synthetic biology approaches that enhance tolerance to high metal concentrations and fluctuating environmental conditions by modifying metabolic and regulatory pathways [3, 4]. At the system level, the deployment of autonomous sensing, real-time analytics, and digital twin models is improving reactor monitoring and control, enabling adaptive operation under variable pollutant loads [5, 6]. Integration of microalgal remediation facilities with renewable micro-grids can substantially reduce energy footprints and improve life cycle performance [7]. In parallel, the development of metal-specific bioremediation consortia – combining microalgae with bacteria or fungi – offers improved selectivity and robustness for complex waste streams [8]. Finally, biorefinery-based biomass utilization frameworks allow pollutant-laden biomass to be fractionated into fuels, materials, and specialty chemicals, enhancing economic viability while minimizing waste [9, 10]. Together, the convergence of biotechnology, data science, and circular economy principles positions microalgae as a potentially transformative environmental technology for a low-carbon and resource-efficient future [4, 9].

18 Circular economy context and systemic integration

18.1 From linear waste to regenerative flows

Historically, electronic products have followed a linear model – extraction → manufacturing → consumption → disposal. This model externalizes waste and pollution, particularly in low- and middle-income regions where informal recycling dominates. A circular economy, by contrast, seeks to retain material value through reuse, repair, remanufacturing, and recycling while minimizing environmental leakage.

Microalgal remediation contributes to this transition in three fundamental ways:

18.1.1 Mitigating pollution leakage from dismantling and processing operations

18.1.2 Recovering nutrients and potentially metals, thereby reducing virgin material extraction

18.1.3 Generating biomass that supports bio-based economies

In other words, microalgae transform pollution into biological value, acting as both environmental guardians and resource processors.

18.2 Reinventing wastewater as a resource

Wastewater from e-waste recycling contains dissolved metals, surfactants, acids, alkalis, and organic residues. Rather than being viewed solely as a liability, this effluent becomes a feedstock for microalgal growth. The nutrients, dissolved inorganic carbon, and trace elements in contaminated water may support biomass formation – while metals are simultaneously sequestered. This dual role – pollution control + biomass cultivation – represents a powerful ecological innovation pathway.

18.3 Industrial symbiosis and eco-parks

A particularly promising model is the development of eco-industrial parks in which multiple firms exchange resources, e-waste recyclers supply wastewater, algae cultivation systems remediate pollutants, biomass processing units create fertilizers, bioproducts, or catalysts, and recovered metals or nutrients return to manufacturers. Such networks embody *industrial symbiosis*, minimizing waste while sustaining local economic activity.

18.4 Urban mining and bio-based recovery

Microalgal systems can also function as an effective pre-concentration step for metals present in dilute waste streams. Once metals are bound to algal biomass through surface complexation and intracellular uptake, they can be selectively recovered using mild chemical extraction, thermal treatment, or controlled pyrolysis, thereby supporting urban mining strategies [18]. Although this approach is still technologically emergent, studies indicate strong potential for integrating biological metal capture with downstream metallurgical recovery processes [19]. In the longer term, such systems could contribute to stabilizing supply chains for critical elements, including cobalt and rare earth elements, while simultaneously reducing dependence on environmentally destructive primary mining activities and associated geopolitical risks [20].

19 Socioeconomic and developmental implications

19.1 Green job creation

Electronic waste pollution has intensified due to rapid technological obsolescence, shortened product life cycles, and inadequate end-of-life management systems, particularly in low- and middle-income countries [13]. The environmental burden of discarded electronics extends beyond material loss, contributing to complex contamination pathways and long-term ecological and human health risks [17]. In response, biologically based remediation strategies are gaining prominence because of their low energy requirements, adaptability, and environmental compatibility compared with conventional physicochemical treatments [14]. Microalgae have attracted particular attention due to their strong affinity for metal ions, which arises from functional groups such as carboxyl, hydroxyl, and phosphate moieties on their cell surfaces [15]. These biological characteristics enable efficient pollutant removal while allowing remediation processes to be embedded within broader circular economy frameworks that emphasize resource recovery and value regeneration [16]. Together, these advances highlight a shift toward integrated, systems-level approaches that align pollution control with sustainable material management and long-term industrial resilience [13–17].

19.2 Opportunities for developing economies

Many developing countries currently function as destinations for e-waste flows due to lower labor and regulatory costs. Microalgal remediation offers such countries a technology-intensive alternative that adds value locally, retains human capital, attracts climate-aligned investment, and supports industrial upgrading. Thus, the technology is not merely environmental – it is developmental.

19.3 Gender and inclusion dimensions

Women frequently participate in low-wage, high-risk segments of informal recycling. Introducing safer remediation technologies – accompanied by training programs, cooperative models, and access to finance – may enhance gender equity. Social planning must therefore be integrated into technology deployment strategies.

20 Extended producer responsibility and policy levers

20.1 Role of extended producer responsibility (EPR)

Under EPR frameworks, manufacturers remain financially or logistically responsible for end-of-life products. When properly enforced, EPR encourages eco-design for disassembly, formal recycling networks, and funding for environmentally safe waste management technologies. Microalgal remediation systems may be financed partly through EPR revenues or producer-supported innovation funds.

20.2 Regulatory pathways

Enabling policy actions include establishing treatment performance standards, recognizing biological remediation in environmental legislation, approving safe biomass utilization routes, financing pilot projects, and integrating microalgae into national wastewater strategies.

20.3 International governance

Because e-waste flows cross borders, international coordination – such as under Basel Convention frameworks – is important for preventing "pollution offshoring" and ensuring universal adoption of safer treatment technologies.

21 Technical limitations and research challenges

Despite its promise, microalgal remediation still faces technical constraints requiring sustained research.

21.1 Biological variability

Algal performance varies by strain, pollutant load, climatic conditions, and geographical region. This variability complicates standardization and requires site-specific optimization.

21.2 Metal toxicity thresholds

At extremely high pollutant loads, metals may inhibit photosynthesis, respiration, and protein synthesis. Engineering reactor conditions that maintain performance without compromising biomass viability is therefore essential.

21.3 Harvesting complexity

Separating microalgal biomass from treated water remains a major cost driver. Techniques include centrifugation, filtration, flocculation, and biofilm harvesting. Each involves trade-offs in cost, energy use, and efficiency.

21.4 Scaling from laboratory to field

Laboratory success does not always predict field reliability. Real-world wastewater contains variable compositions, suspended solids, microbes, temperature fluctuations, and hydraulic disturbances. Reactors must therefore be designed for robust performance under uncertainty.

22 Environmental ethics and responsibility

Microalgal remediation operates at the intersection of technology, ecology, and social responsibility. Ethical considerations include ensuring fair benefit distribution, avoiding displacement of vulnerable workers, protecting ecosystems from biological releases, promoting community engagement, and respecting indigenous and local knowledge. Sustainable deployment means technology must serve people and ecosystems – not the reverse.

23 Alignment with the sustainable development goals

Microalgal remediation directly contributes to multiple sustainable development goals (SDGs). They support SDG 3 (Good Health and Well-Being) by reducing human exposure to toxic metals and hazardous substances and SDG 6 (Clean Water and Sanitation) through improvements in wastewater quality and reduced pollutant discharge. Economic benefits align with SDG 8 (Decent Work and Economic Growth) by enabling the creation of green jobs across recycling, recovery, and advanced materials sectors. Technological advancement and system-level innovation contribute to SDG 9 (Industry, Innovation, and Infrastructure) by fostering novel approaches in environmental biotechnology and circular industrial processes. Furthermore, enhanced material recovery and reuse advance SDG 12 (Responsible Consumption and Production) by strengthening material circularity and resource efficiency. Climate-related benefits address SDG 13 (Climate Action) through potential reductions in greenhouse gas emissions and carbon sequestration pathways, while the prevention of toxic leakage into ecosystems supports SDGs 14 and 15 (Life Below Water and Life on Land) by mitigating ecological contamination and preserving biodiversity. Thus, the technology reinforces integrated sustainability outcomes rather than addressing pollution in isolation.

24 Future outlook: the next generation of microalgal systems

24.1 Digitalization and smart monitoring

The integration of sensors, automation, and AI-based analytics allows real-time optimization of growth conditions, pollutant loads, and biomass yields. Predictive maintenance and digital twins further enhance reliability.

24.2 Advanced biorefinery integration

Future systems will increasingly operate as biorefineries, extracting multiple value streams from biomass – pigments, lipids, proteins, polymers, catalysts – thereby maximizing economic viability.

24.3 Metal-specific biological design

Selective metal-binding peptides and engineered metabolic pathways may allow highly targeted remediation, transforming algae into bioselective extraction tools for critical elements.

24.4 Climate-resilient deployment

Microalgal systems perform well in arid and warm climates, making them suitable for regions facing climate stress. Integration with renewable energy further enhances resilience.

25 Synthesis: why microalgae matter in the digital age

The rise of the digital economy is inseparable from the rise of e-waste. Microalgae offer a nature-inspired counterbalance – transforming toxic legacies into regenerative ecological processes. Key strengths include low-chemical operation, compatibility with wastewater treatment, biomass valorization potential, climate and biodiversity co-benefits, and alignment with circular economic principles. Microalgae therefore represent both a technological innovation and an ecological philosophy – one grounded in cooperation with living systems.

26 Conclusion

Electronic waste pollution represents one of the defining environmental and social challenges of the twenty-first century. The mishandling of discarded electronics has led to widespread contamination of water, soil, and air, with impacts falling disproportionately on vulnerable and marginalized communities involved in informal recycling or living near disposal and dumping sites. Within this context, microalgae-mediated remediation has emerged as a scientifically credible, environmentally regenerative, and socially transformative approach to addressing pollution derived from electronic waste. Microalgal systems remove contaminants through mechanisms such as biosorption, bioaccumulation, biotransformation, and biomineralization, which are governed by cell wall chemistry, metabolic activity, and the production of extracellular polymeric substances. These organisms can be cultivated in a wide range of reactor configurations, from open raceway ponds to advanced closed photobioreactors, and can be integrated into hybrid treatment systems. Such integration enhances operational stability, improves pollutant removal efficiency, and increases resilience to fluctuating contaminant loads. Beyond remediation, harvested algal biomass can be further valorized into bio-based products or used as a medium for metal recovery, embedding pollution control within a broader circular resource economy. The economic viability of microalgal remediation improves significantly when systems are integrated with existing municipal wastewater treatment infrastructure or industrial operations. This integration enables cost sharing, lowers energy demand, and generates additional co-benefits, including the utilization of carbon dioxide. From a social standpoint, the deployment of microalgal technologies offers safer and more formal employment pathways compared with informal recycling practices, while also supporting sustainable development objectives when aligned with inclusive governance structures and extended producer responsibility frameworks. Despite these advantages, several challenges persist, including biological variability among algal strains, costs associated with biomass harvesting and downstream processing, regulatory uncertainty, and the complexity of scaling laboratory and pilot systems to full field implementation. Nevertheless, ongoing advances in strain selection and genetic improvement, digital process monitoring, reactor engineering, and circular value-chain development are steadily strengthening technical robustness and improving economic feasibility.

Ultimately, microalgal remediation is more than a technical solution for pollution control. It represents a systems-level rethinking of how societies interact with waste, ecosystems, and industrial production. By linking environmental stewardship with technological progress, microalgae place biology at the core of sustainable, circular, and equitable strategies for managing electronic waste in the digital era.

References

[1] Forti V, Baldé CP, Kuehr R, Bel G. The Global e-waste Monitor 2020: Quantities, Flows and the Circular Economy Potential. Bonn, Germany: United Nations University; 2020, pp. 1–120.

[2] Bhattacharya P, Yadav A, Mishra S. Environmental and health impacts of electronic waste: A global perspective. Environmental Research. 2020;185:109386–109397.

[3] Priyadarshini M, Jain S, Sharma S. Circular economy strategies for e-waste valorization and sustainable resource recovery. Journal of Cleaner Production. 2022;334:130256–130268.

[4] Ali H, Khan E, Ilahi I. Environmental chemistry and ecotoxicology of hazardous heavy metals: Sources, impacts and remediation. Chemosphere. 2013;91:869–881.

[5] Gaur N, Flora G, Yadav M, Tiwari A. Biosorption of metals by algal biomass: Mechanisms and applications. Journal of Hazardous Materials. 2018;356–357:181–198.

[6] Rai M, Gade A, Yadav A. Bioremediation technologies for heavy metal removal: Principles and applications. Singapore: Springer; 2020, pp. 1–412.

[7] Mishra A, Medhi K, Malaviya P, Thakur IS. Microalgae as a sustainable technology for wastewater treatment: A review. Journal of Environmental Management. 2019;238:346–359.

[8] Wang J, Chen C. Biosorption of heavy metals by microbial biomass: A review of mechanisms and applications. Bioresource Technology. 2016;160:1–9.

[9] Zeraatkar A, Ahmadzadeh H, Talebi AF, Moheimani NR, McHenry MP. Microalgae in wastewater treatment: A review of operational parameters and sustainability. Renewable Energy. 2016;94:388–399.

[10] Rawat I, Ranjith Kumar R, Mutanda T, Bux F. Algal biofilm-based wastewater treatment: A review. RSC Advances. 2018;8:34149–34163.

[11] Sudhakar K, Mamat R, Samykano M, Azmi WH, Ishak WFW, Yusaf T. Microalgal technologies for environmental remediation and bioenergy production: A review. Renewable and Sustainable Energy Reviews. 2019;101:65–79.

[12] Singh R, Parihar P, Singh M, Bajguz A, Kumar J, Singh S, Vp S, Prasad SM. Algal systems for heavy metal removal and detoxification. Environmental Science and Pollution Research. 2021;28:4568–4590.

Tusheema Dutta*, Nishita Mathur, Himanshi Ohri,
Shubhankar Mahata, Bhavya Sachdeva, Aman Upadhyay

From hazard to resource: valorization of e-waste into value-added products for a sustainable future

Abstract: Electronic waste (e-waste) is among the fastest growing and most complex waste streams worldwide. While conventionally treated as hazardous solid waste, e-waste contains high concentrations of valuable materials such as precious metals, base metals, rare earth elements, engineered polymers, and glass. Valorization seeks to transform e-waste from an environmental liability into a regenerative resource through advanced material recovery, green processing technologies, and circular economic integration. This chapter critically examines strategies for converting e-waste into value-added products including recovered metals, nanomaterials, functional composites, catalysts, and energy materials. Green chemical and biotechnological approaches are emphasized for their ability to reduce secondary pollution. Business models, policy mechanisms, social implications, and technological barriers are also discussed, positioning e-waste valorization as a strategic pillar for sustainable industrial development aligned with SDG 9 and SDG 12.

Keywords: E-waste, valorization, circular economy, resource recovery, nanomaterials

1 Introduction

Modern digital life depends on an expanding ecosystem of electrical and electronic devices – smartphones, laptops, photovoltaic panels, electric vehicles, home appliances, industrial control systems, and telecommunication networks [1]. These devices fuel productivity, social connectivity, and technological progress. However, their eventual disposal produces electronic waste (e-waste), a complex mixture of metals, polymers, glass, ceramics, and additives that poses environmental, economic, and social challenges [2, 3]. Global e-waste generation has risen sharply over the past two decades and continues to accelerate. A large proportion remains uncollected, informally recycled, dumped, or stockpiled in households, despite containing substantial

***Corresponding author: Tusheema Dutta**, Center for Research and Innovation in Frugal Technology Management, FORE School of Management, New Delhi 110016. e-mail: 183113@fsm.ac.in
Nishita Mathur, Himanshi Ohri, Shubhankar Mahata, Bhavya Sachdeva, Aman Upadhyay, Center for Research and Innovation in Frugal Technology Management, FORE School of Management, New Delhi 110016

 | https://doi.org/10.1515/9783112225509-017

residual value [2, 4]. Printed circuit boards, connectors, wiring, and batteries hold concentrated quantities of copper, aluminum, gold, silver, palladium, and rare earth elements – often far richer in grade than natural ores [2, 5]. Plastics and fiberglass composites represent additional feedstocks for polymer processing and composite development [6, 7]. Historically, e-waste has been managed primarily through incineration, landfilling, or crude recycling techniques. These pathways result in toxic emissions [8], uncontrolled leaching of metals and flame retardants, occupational exposure risks [8], and the loss of valuable material resources. In contrast, valorization reframes e-waste as a high-potential secondary resource, to be reprocessed into new materials, components, and industrial inputs [9]. Rather than treating waste as an environmental burden, valorization embeds it into circular economic loops that retain material value, reduce the need for primary mining, and stimulate innovation ecosystems [9, 10]. This chapter adopts an integrated perspective on e-waste valorization – spanning materials science, environmental engineering, industrial ecology, business strategy, and policy [10]. It explores the technological foundations of material recovery and product development, analyzes enabling and constraining economic factors, and evaluates sustainability outcomes. Special emphasis is placed on green chemistry, bio-based processing, and low-carbon pathways, which minimize secondary environmental damage [11, 12].

2 Composition and resource potential of e-waste

2.1 A High-value urban ore

E-waste is often described as "urban ore" because the concentration of valuable materials it contains frequently exceeds that found in many natural mineral deposits [13]. Printed circuit boards, for example, may contain 10–30 times more gold per ton than conventional gold ore [14, 15], while copper concentrations in certain electronic components can exceed 20% by weight [16]. Lithium-ion batteries represent an important secondary source of critical metals such as cobalt, lithium, and nickel [17, 18], and rare earth elements are commonly embedded in components including permanent magnets and display systems [19, 20]. In addition to metals, e-waste contains a wide range of non-metallic yet valuable materials, including high-grade engineering plastics such as ABS, HIPS, and polycarbonate [18], fiberglass epoxy laminates [16], silicon wafers [3], and glass [6], along with various flame retardants and chemical additives [8, 10]. Consequently, e-waste represents both a significant environmental hazard and a strategic secondary resource base, particularly as global demand for critical materials continues to increase with the expansion of renewable energy technologies, electric mobility, and digital manufacturing systems [4, 5].

2.2 Environmental burden of conventional disposal

Landfilling and open dumping allow metals and additives to migrate into soils and groundwater. Incineration releases dioxins, furans, and heavy metal fumes. Informal acid leaching creates contaminated wastewater, while thermal stripping generates toxic particulates [7]. These practices not only damage ecosystems and human health but also destroy material value that could otherwise sustain economic development.

3 Concept of valorization

Valorization differs from basic recycling in that it aims not merely to recover materials, but to upgrade them into higher-value products through technological processing, redesign, and innovation [2]. The key objectives of advanced e-waste valorization and remediation strategies focus on the recovery of critical and precious metals to reduce reliance on primary resource extraction, alongside the transformation of recovered plastics into functional composites for secondary applications. These approaches also aim to convert residual fractions into value-added products such as catalysts and energy carriers, thereby improving overall resource efficiency. By strengthening domestic manufacturing through a stable supply of secondary raw materials, such systems help reduce import dependence while minimizing environmental leakage of hazardous substances. Collectively, these objectives emphasize the integration of diverse waste streams into circular value chains, reinforcing sustainability, resilience, and long-term material security. Valorization therefore bridges waste management, materials science, and sustainable industrial strategy.

4 Major valorization pathways

4.1 Metal recovery and refining

Metal recovery is a central pillar of e-waste valorization and is typically achieved through a sequence of complementary processing routes. Initial mechanical preprocessing steps – such as shredding, followed by magnetic and eddy-current separation – are used to liberate and concentrate metal-containing fractions. These are commonly followed by hydrometallurgical processes, which involve acid leaching and solvent extraction to selectively dissolve and recover target metals. In parallel, pyrometallurgical routes based on smelting and refining are employed for bulk metal recovery, particularly where complex or mixed material streams are involved. More recently, biohydrometallurgical approaches using microbial or algal leaching have emerged as lower-energy and potentially more environmentally benign alternatives. The recovered metals are reintegrated directly into high-value sectors, including electronics manufacturing, renewable energy systems, automobiles, aerospace components, and medical equipment, thereby closing material loops and reinforcing circular economy objectives. Metal recovery also reduces dependence on primary mining, which is often ecologically destructive and socially contentious.

4.2 Nanomaterials and advanced functional materials

Metals such as gold, silver, and copper recovered from e-waste can be further upgraded into high-value advanced materials rather than being limited to bulk reuse. Through appropriate refining and downstream processing, these metals can be converted into nanoparticles and nanowires with controlled size and morphology, as well as thin-film catalysts for chemical and electrochemical applications. In addition, recovered metals can be formulated into conductive inks for printed electronics and flexible devices, or engineered into sensing materials for environmental, biomedical, and industrial monitoring applications. Such upcycling pathways significantly enhance the economic value of recovered metals while supporting innovation in advanced manufacturing and nanotechnology sectors. These nanomaterials have high demand in electronics, catalysis, medicine, and environmental engineering, enabling both economic and technological upgrading.

4.3 Polymer and composite valorization

Plastics recovered from e-waste – particularly flame-retardant ABS, HIPS, and polycarbonates – represent a valuable secondary resource that can be upcycled rather than incinerated or down-cycled. Following appropriate purification and decontamination, these polymers can be directly reused or blended with natural fibers to improve sustainability profiles. They may also be reinforced with recovered metal particles or fillers, cured into construction-grade materials, or converted into lightweight composite panels with enhanced performance characteristics. Such composites often exhibit high mechanical strength, improved flame resistance, and long-term durability, making them suitable for applications in furniture, building panels, transport components, and a wide range of consumer goods. By enabling repeated material use at comparable or higher value, these strategies contribute to the development of circular polymer economies and reduce reliance on virgin plastic production.

4.4 Energy materials and fuel production

E-waste plastics and other polymeric residues can also be valorized through thermochemical conversion pathways when material recycling is not technically or economically feasible. Processes such as pyrolysis and gasification enable the transformation of these fractions into syngas, pyro-oils, and carbon-rich char, which can be utilized as energy carriers, chemical feedstocks, or solid fuels. In parallel, metals recovered from e-waste can be repurposed as catalysts in renewable energy systems, further enhancing overall resource efficiency. While energy valorization should not substitute for higher-value material recovery routes, it provides an important complementary option for managing otherwise unrecyclable fractions, thereby minimizing waste and improving the completeness of circular e-waste management systems.

5 Green and biotechnological valorization approaches

A growing research frontier in e-waste valorization focuses on green chemistry and biotechnology approaches that aim to minimize the use of toxic reagents and reduce secondary emissions. (Figure 1) These include the application of bioleaching microbes or microalgae capable of selectively solubilizing metals, as well as the replacement of strong mineral acids with ionic liquids and deep eutectic solvents that offer lower volatility and improved selectivity. Additional innovations involve supercritical CO_2 extraction, low-temperature selective dissolution processes, and enzymatic treatments for polymer depolymerization and modification. Collectively, these techniques significantly reduce secondary pollution, enhance process safety, and align more closely with circular sustainability frameworks by enabling cleaner recovery, reuse, and reintegration of materials into value chains.

6 Industrial and real-world applications

Recovered materials from e-waste are already being reintegrated into a wide range of high-value industrial and technological applications. Refined metals and purified materials support semiconductor and printed circuit board manufacturing, while precious metals are reused in jewelry production. In advanced manufacturing, recovered elements contribute to the development of high-efficiency catalysts, electric vehicle battery components, and smart sensing devices. Additional applications include antimicrobial coatings, conductive textiles for wearable electronics, and structural composites used in construction and transport sectors. These diverse end uses demonstrate the practical feasibility of e-waste valorization and highlight its role in closing material loops across multiple strategic industries. This indicates that valorization is not theoretical – it is emerging industrial practice (Table 1).

Table 1: Examples of value-added products from e-waste.

Recovered resource	Processing method	Value-added product	Application
Gold and silver	Hydrometallurgy/ bioleaching	Nanoparticles and thin films	Catalysis, electronics, and medical devices
Copper	Mechanical and electro-refining	Conductive inks and wiring	Electronics and motors
Lithium and cobalt	Battery recycling	New cathode materials	EV and energy storage
Plastics	Compounding and molding	Structural composites	Construction and furniture
Glass and silica	Re-melting and reformulation	Insulation and specialty glass	Buildings and electronics

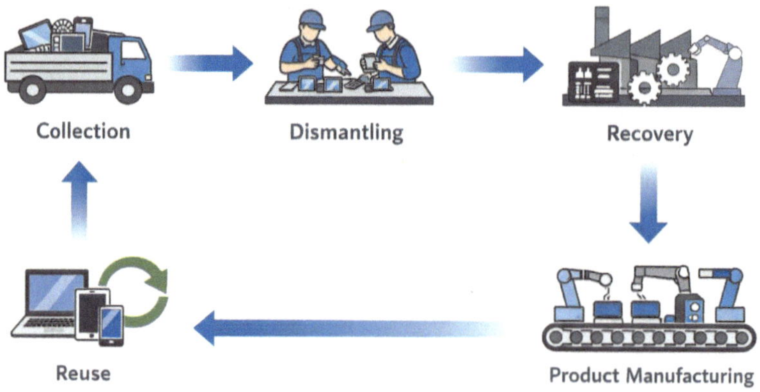

Figure 1: Circular flow of e-waste valorization.

7 Environmental performance and sustainability

E-waste valorization delivers significant environmental benefits by substantially reducing the volume of waste sent to landfills and limiting the release of toxic substances into surrounding soil and water systems. It also contributes to lower greenhouse gas emissions by decreasing energy demand associated with primary material production and waste disposal. By offsetting the need for virgin material extraction, valorization reduces ecosystem disturbance, habitat loss, and other environmental impacts linked to mining activities. Collectively, these outcomes highlight the role of e-waste valorization as a key strategy for mitigating environmental degradation while advancing resource efficiency and circular economy objectives. Life cycle assessments generally show lower energy and carbon footprints compared with primary production – especially for precious and specialty metals.

8 Role of business models and markets

The success of e-waste valorization depends on a combination of technical, economic, and institutional factors, including a stable and sufficient feedstock supply, sustained market demand for recovered materials, investor confidence, technological readiness, and alignment with regulatory frameworks. In response to these requirements, a range of implementation models has emerged, such as specialized urban mining firms, social enterprises focused on inclusive recycling, producer-financed recycling networks under extended producer responsibility schemes, and innovation-driven start-ups developing advanced recovery technologies. Countries that succeed in establishing robust domestic valorization capacity not only improve environmental outcomes but also enhance strategic autonomy over critical resources, reducing dependence on volatile global supply chains.

8.1 Challenges and risks

Despite its strong potential, e-waste valorization continues to face several persistent barriers that hinder large-scale and consistent implementation. The heterogeneous composition of e-waste and the presence of contaminants significantly increase separation and processing complexity, while high initial capital requirements pose challenges for new entrants and technology deployment. In many regions, limited formal collection systems constrain feedstock availability and quality, and competition with informal recycling sectors can undermine material flows into regulated facilities. These challenges are further compounded by price volatility in global metal markets, which affects revenue stability, and by policy inconsistency or weak enforcement that creates uncertainty for investors and operators. Together, these factors highlight the need for coordinated technical, financial, and regulatory interventions to unlock the full potential of e-waste valorization. (Figure 2) Technical, economic, and social innovation must therefore evolve in parallel.

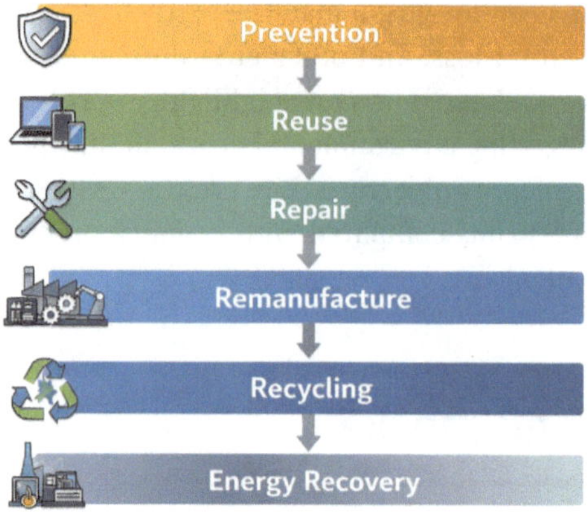

Figure 2: Hierarchy of e-waste valorization strategies.

8.2 Ethical and social considerations

Millions of informal workers rely on recycling for subsistence. Transition strategies must improve safety, formalize labor rights, provide training, avoid displacement, and enhance income stability. Valorization should support human dignity and inclusion, not only technological progress.

9 Detailed processing techniques for e-waste valorization

Although the concept of valorization is straightforward – recovering and upgrading materials – the underlying processing pathways are technically sophisticated [1]. They must be designed not only for efficiency but also for environmental safety and worker protection [12].

9.1 Mechanical preprocessing

Mechanical preprocessing is the front-end stage of most valorization chains. It includes manual dismantling, shredding and granulation, magnetic separation, eddy-current separation, density, and optical sorting. The objective is to separate materials into purer fractions, reducing contamination before advanced recovery steps. Skilled manual dismantling remains crucial, as it enables selective extraction of high-value components – printed circuit boards, CPUs, RAM chips, wiring, and batteries – while minimizing cross-contamination.

However, mechanization is increasing as safety and productivity standards evolve. Dust control systems and enclosed processing environments are important to prevent inhalation of fine particulates containing metals and flame retardants.

9.2 Hydrometallurgical processing

Hydrometallurgy involves leaching metals into solution using aqueous reagents. It is often preferred over pyrometallurgy for lower energy consumption and higher metal selectivity. Typical steps include leaching with acids or alternative solvents, purification through precipitation or solvent extraction, and electrowinning or crystallization. Gold, silver, copper, and rare metals can be recovered at high purity. Recent advances emphasize eco-friendly solvents, such as deep eutectic solvents and ionic liquids, which reduce environmental risk and improve recyclability of reagents.

9.3 Pyrometallurgical processing

Pyrometallurgy uses high-temperature furnaces to melt and separate metals. It is effective for heavily mixed or contaminated fractions and enables efficient recovery of copper, precious metals, nickel, and cobalt. However, drawbacks include high energy demand, emission control requirements, potential for dioxin formation, and infrastructure cost. Therefore, modern systems incorporate advanced emission scrubbing, thermal efficiency optimization, and continuous monitoring.

9.4 Bio-hydrometallurgy (bioleaching)

Bio-hydrometallurgy employs microorganisms – bacteria, fungi, or algae – that naturally solubilize metals [7, 11]. The organisms oxidize or complex metal ions, allowing their extraction at lower energy intensity compared with conventional metallurgical processes [7]. Advantages include mild operating conditions [11], compatibility with low-grade streams [7, 11], and reduced chemical consumption [7]. Challenges include slower reaction rates and sensitivity to environmental fluctuations [11]. Nevertheless, as biological processing technologies mature, bio-hydrometallurgy is emerging as a key pillar of green valorization within sustainable e-waste management systems [5, 11].

9.5 Polymer valorization and composite engineering

Plastics in e-waste are often fire-retarded, blended, and contaminated with metals – complicating recycling. However, advanced polymer science allows selective dissolution and purification, melt blending, compatibilization, fiber reinforcement, and incorporation into hybrid composites. These methods enable the production of structural materials, insulation boards, furniture panels, and automotive components. When combined with natural fibers such as jute, hemp, or bamboo, they support biocomposite innovation, linking industrial ecology with sustainable agriculture.

10 Case studies of e-waste valorization in practice

10.1 Industrial metal recovery facilities

Several countries operate large-scale plants where printed circuit boards are processed into high-value metal concentrates. Copper acts as a collector during smelting, absorbing precious metals, which are later refined into bullion. These facilities demonstrate that urban mining is already an industrial reality, not merely a research concept.

10.2 Start-ups and innovation ecosystems

An increasing number of start-ups specialize in environmentally safe PCB recycling, battery material recovery, polymer composite production, and e-waste-derived nanomaterials. These firms often partner with universities and public agencies, contributing to technology transfer and workforce development.

10.3 Social enterprises and community-based recycling

In several regions, cooperatives organize informal recyclers into safer, better-regulated networks. By combining protective equipment, training, fair payment, and access to technology, they generate both social and environmental value. Valorization becomes not only a technical strategy but a vehicle for social justice and inclusive development.

11 Circular economy integration

Valorization exists within a broader circular economy framework designed to eliminate waste while retaining material value.

11.1 Designing for disassembly and re-manufacture

One of the most powerful interventions occurs at the product design stage. Modular, repairable, and easily separable components dramatically increase recyclability and reduce processing complexity. Manufacturers adopting design-for-circularity principles can reduce end-of-life treatment costs, increase material recovery rates, and improve regulatory compliance.

11.2 Reverse logistics and collection systems

For valorization to function, e-waste must reach processing facilities. Efficient collection requires consumer awareness, deposit–refund systems, retailer take-back programs, municipal collection networks, and digital tracking platforms. Without consistent feedstock supply, even the most advanced plants operate below capacity – reducing economic sustainability.

11.3 Industrial symbiosis and eco-innovation clusters

Colocating dismantling centers, metal recovery plants, plastic reprocessors, composite manufacturers, and research laboratories creates eco-innovation clusters where by-products from one facility serve as inputs for another. This reduces logistics costs and strengthens local value chains.

12 Life cycle assessment and environmental performance

12.1 Reducing environmental burden

Compared with primary metal production, urban mining generally requires significantly less energy, fewer greenhouse gas emissions, lower land disturbance, and minimal water contamination. Life cycle studies consistently show that recycling gold, copper, and rare metals from e-waste has a substantially smaller environmental footprint than mining virgin ore.

12.2 Avoiding burden shifting

Sustainable valorization must avoid toxic solvent release, dioxin emissions, uncontrolled residues, and worker exposure. Therefore, stringent process control and occupational health standards are essential.

12.3 Business strategy, markets, and finance

Valorization does not occur in a vacuum; it is shaped by economic signals and investment dynamics.

12.4 Revenue streams

Income may derive from sale of recovered metals, composite product manufacturing, catalyst or nanomaterial production, service contracts for safe recycling, and carbon and resource recovery credits. Profitability depends on metal prices, regulatory policies, and operational efficiency.

12.5 Investment barriers

Barriers include high initial capital costs, market price volatility, uncertainty in feedstock availability, competition from informal recyclers, and limited access to affordable finance. Risk-sharing mechanisms, public–private partnerships, and green financing instruments can mitigate these constraints.

12.6 Role of innovation policy

Government support through:
- Tax incentives
- R&D grants
- Public procurement
- Technology incubators

helps accelerate industrial maturity and commercialization.

12.7 Social equity and development

Valorization must align with social sustainability, particularly in regions where informal recycling dominates.

12.8 Worker safety and health

Valorization shifts hazardous open-air practices toward regulated, enclosed facilities with ventilation, protective equipment, medical monitoring, and formal employment protections. This substantially reduces exposure to toxic metals and fumes.

12.9 Inclusive transition strategies

Policies should ensure that informal workers receive training, gain stable employment, participate in cooperatives, and share economic benefits. A just transition prevents marginalization and economic displacement.

12.10 Gender and youth participation

Women and young people frequently participate in low-paid dismantling labor. Valorization initiatives that integrate skills training and enterprise support can create **pathways to higher-value technical employment.**

13 Policy and regulatory frameworks

13.1 Extended producer responsibility (EPR)

Under EPR, manufacturers retain responsibility for post-consumer product management. This encourages collection networks, formal recycling partnerships, eco-design innovation, and financing mechanisms for safe processing. EPR is now adopted – in varying forms – across many jurisdictions.

13.2 Standards and certification

Certification frameworks ensure that facilities meet environmental and safety standards. Third-party monitoring increases transparency and market credibility.

13.3 International governance

Global e-waste movement is shaped by transboundary agreements that restrict dumping and encourage responsible recycling.

13.4 Risks and ethical considerations

Valorization, if poorly managed, may externalize pollution risk to poorer communities, promote greenwashing, and generate hazardous residues. Ethical implementation requires accountability, transparency, and community participation.

13.5 Alignment with the sustainable development goals

E-waste valorization contributes to:
- SDG 8 – Decent Work and Economic Growth
- SDG 9 – Industry, Innovation, and Infrastructure
- SDG 11 – Sustainable Cities
- SDG 12 – Responsible Consumption and Production
- SDG 13 – Climate Action

while indirectly supporting health and biodiversity goals.

13.6 Environmental justice and global inequality dimensions

13.6.1 Unequal distribution of risk

Although e-waste is generated globally, its environmental and health burdens fall disproportionately upon low- and middle-income regions. Informal recycling hubs often emerge in peri-urban settlements where housing is insecure, sanitation infrastructure is weak, and occupational protections are minimal. Communities already facing poverty and social exclusion then experience heightened toxic exposure, reinforcing cycles of vulnerability. Children play and attend school near burning sites; women frequently work in sorting and dismantling; and groundwater used for drinking and irrigation may become contaminated. Exposure occurs not only through occupational contact but also via dust inhalation, food chain bioaccumulation, and indoor contamination. Valorization – when inclusive, regulated, and community-oriented – offers an opportunity to reverse this pattern. Rather than exporting pollution to the margins, society can invest in safer, technologically advanced resource recovery systems that distribute benefits more equitably.

13.6.2 Responsible innovation

Responsible innovation places strong emphasis on the anticipation of unintended consequences, continuous reflexive evaluation of technological choices, inclusive stakeholder participation, and responsiveness to community concerns. When applied to e-waste valorization, this approach requires transparency across the entire value chain, including process selection, emission control measures, waste transport logistics, and the safety of final products derived from recovered materials. Meaningful community involvement in planning, implementation, and monitoring is essential to ensure that environmental and social safeguards are substantive and enforceable, rather than merely symbolic, thereby strengthening public trust and long-term sustainability outcomes.

13.6.3 Avoiding eco-colonialism

A key ethical question in the global e-waste economy concerns whether high-income countries externalize environmental and technological risks by exporting e-waste to regions with weaker regulatory oversight. Sustainable e-waste valorization explicitly rejects such practices and instead emphasizes the development of domestic processing capacity wherever feasible, complemented by strong international governance mechanisms. It also promotes equitable technology transfer partnerships that build local technical capability and institutional strength, alongside fair pricing mechanisms for recovered resources. Collectively, these principles aim to ensure that the benefits of e-waste valorization are distributed equitably, advancing shared prosperity rather than reproducing patterns of unequal extraction and environmental injustice.

14 Risk management and safety protocols

14.1 Occupational safety

Even in technologically advanced e-waste processing facilities, occupational risks persist, including chemical exposure, electrical hazards, injuries from sharp components, and the inhalation of fine particulates. To mitigate these risks, facilities must integrate comprehensive safety measures such as effective ventilation and dust control systems, appropriate personal protective equipment, and ergonomic work design to reduce physical strain and injury. Regular medical surveillance, ongoing safety training, and well-defined emergency response systems are equally essential to ensure preparedness and early detection of health impacts. Ultimately, the protection of worker dignity and health constitutes a non-negotiable pillar of sustainable e-waste valorization and responsible industrial practice.

14.2 Environmental controls

Environmental control strategies are essential to ensure that e-waste valorization does not generate secondary pollution or unintended environmental harm. Key measures include the implementation of zero-liquid-discharge wastewater systems to prevent effluent release, high-efficiency particulate filtration to capture fine airborne contaminants, and continuous monitoring of dioxins and furans associated with thermal processes. Secure stabilization and containment of residual wastes are necessary to prevent leaching, while continuous emissions assessment enables real-time compliance and rapid corrective action. In addition, groundwater protection systems play a critical role in safeguarding surrounding ecosystems and drinking water sources. Together, these measures reflect a precautionary approach that ensures valorization activities enhance environmental performance rather than creating new pollution pathways.

14.3 Product safety and standardization

Recovered materials entering consumer or industrial markets must comply with purity specifications, contaminant limits, and durability standards. Certification builds trust and encourages adoption of recycled-content products.

15 Innovation futures and emerging directions

The future of e-waste valorization will be shaped by converging technological, economic, and social trends.

15.1 Digitalization and traceability

Digital product passports, blockchain tracking, and AI-driven sorting systems will improve collection efficiency, authenticate recovered material, enhance transparency, optimize logistics and reduce fraud and leakage. This ushers in a data-enabled circular economy, where materials remain visible throughout their life cycle.

15.2 Next-generation battery recycling

The rapid expansion of electric mobility is expected to generate vast quantities of end-of-life lithium-ion batteries, making their sustainable management a critical priority within the energy transition. Key innovation areas include the development of low-energy hydrometallurgical processes that reduce environmental footprints, along with safe discharge techniques to minimize fire and explosion risks during handling and processing. Advances in selective solvent systems are enabling more efficient separation and recovery of valuable metals, while emerging approaches increasingly target the recovery of anode materials and electrolytes rather than focusing solely on cathode metals. Crucially, these recovered components can be reintegrated into the manufacture of new battery cells, enabling closed-loop material cycles. As a result, circular battery systems are becoming a central pillar of energy transition strategies, enhancing resource security, reducing dependence on primary mining, and improving the overall sustainability of electric mobility.

15.3 Advanced polymer valorization

Polymer valorization is increasingly shifting toward advanced recovery and upgrading pathways that enable higher-value reuse of plastic fractions. Emerging approaches include chemical depolymerization to break polymers back into reusable monomers, the use of green solvent systems that improve selectivity while reducing toxicity, and catalytic cracking processes that convert complex polymers into valuable chemical intermediates. In parallel, blending recovered polymers with bioplastics is gaining traction as a means of enhancing material performance and sustainability. Together, these strategies reduce dependence on petroleum-based feedstocks while enabling the production of high-performance recycled materials suitable for demanding industrial applications.

15.4 E-waste as a platform for nanotechnology

Recovered metals may form the basis for low-cost nanomaterial synthesis, expanding applications in catalysis, sensing and diagnostics, environmental remediation, energy storage, and smart materials. In this way, e-waste becomes a source of technological enrichment rather than environmental degradation.

15.5 Role of SMEs, start-ups, and social enterprises

Small and medium enterprises (SMEs) are particularly well suited to operate within e-waste valorization ecosystems due to their agility, innovation-driven culture, and strong embedding within local communities. Their flexibility enables them to collaborate effectively across sectors, bridging gaps between technology developers, recyclers, manufacturers, and public institutions. Within this ecosystem, start-ups often drive breakthrough processes and novel recovery technologies, while social enterprises play a crucial role in integrating informal workers into fair, regulated labor structures. SMEs, in turn, help establish localized value chains that complement large-scale industrial facilities by adding processing, customization, and service capacity. Targeted capacity-building programs – such as incubation hubs, grants, soft loans, and technical training – can significantly accelerate entrepreneurial participation and strengthen the resilience and inclusivity of e-waste valorization systems.

15.6 Education, skills, and workforce development

A skilled workforce underpins successful valorization systems. Key disciplines include materials science, chemical engineering, biotechnology, environmental management, occupational health and safety, and logistics and supply chain management. Universities, technical institutes, and vocational schools must update curricula to reflect the needs of a circular industrial economy. Public awareness campaigns can also reshape consumer behavior – increasing formal returns and acceptance of recycled-content products.

16 Financing the transition to circular e-waste economies

16.1 Green finance instruments

Financing tools supporting valorization include sustainability-linked loans, green bonds, blended finance, climate-aligned investment funds, and development finance institution support. These instruments help reduce capital barriers and attract institutional investors.

16.2 Public–private partnerships

Partnerships between municipalities, producers, recyclers, and technology firms spread risk while building infrastructure capacity. Such arrangements can reinforce compliance with extended producer responsibility and environmental obligations.

16.3 Market assurance mechanisms

Governments may use procurement policies to guarantee demand for recycled-content materials – stabilizing markets during early development phases.

17 National and regional strategy development

Countries seeking to build sustainable e-waste valorization capacity should adopt a coordinated policy and institutional approach that spans the entire material life cycle, from product design and consumption to end-of-life recovery and secondary material use [18]. Establishing clear and enforceable collection mandates is essential to ensure a reliable feedstock supply, while national circular economy roadmaps help align e-waste management with broader industrial development, resource security, and sustainability objectives [19]. Incentivizing domestic processing and refining capacity can retain material value within national economies, generate skilled employment, and reduce exposure to volatile global secondary material markets. Equally important are robust measures to discourage illegal dumping and transboundary movement of hazardous electronic waste, supported by strengthened customs enforcement and international cooperation to curb illicit trade flows [18]. In parallel, sustained public investment in research, pilot, and demonstration facilities plays a critical role in accelerating technology maturation, reducing uncertainty, and de-risking private sector participation in emerging valorization pathways [19]. For smaller or emerging markets, regional collaboration offers additional advantages by enabling economies of scale, shared infrastructure, knowledge exchange, and harmonized technical and regulatory standards, thereby improving the overall viability, efficiency, and environmental effectiveness of e-waste valorization initiatives [19, 20].

18 Monitoring, evaluation, and continuous improvement

Effective e-waste valorization depends on continuous and systematic performance monitoring across environmental, social, and economic dimensions to ensure long-term sustainability and accountability [13]. Core technical indicators such as material recovery rates and resource efficiency metrics are essential for evaluating process performance, while monitoring emissions, effluents, and residual waste streams ensures compliance with environmental regulations and safeguards surrounding ecosystems [17]. Equally important are occupational health and safety indicators, which help protect workers involved in collection, dismantling, and processing activities, particularly in contexts transitioning from informal to formalized systems [14]. Beyond technical and environmental performance, social inclusion indicators are critical for assessing the integration of informal-sector workers, fairness in benefit sharing, and the broader social impacts of valorization initiatives on local communities [13]. Economic indicators – including cost recovery, profitability, and market stability of secondary materials – determine whether valorization systems can remain viable without long-term dependence on subsidies [17]. In addition, consumer acceptance of products derived from recovered materials serves as an important signal of market readiness, trust, and the perceived quality of circular products [16]. Together, a data-driven monitoring and evaluation framework enables adaptive management, enhances transparency, and supports alignment between operational practices and wider sustainability and circular economy objectives [13–17].

A profound shift is underway in how societies conceptualize waste. Whereas e-waste was once viewed primarily as a symbol of environmental degradation and technological discard, it is increasingly recognized as a strategic material resource, a catalyst for innovation, and a driver of sustainability. Through valorization, discarded materials are reintroduced into industrial supply chains, environmental impacts are reduced, and security of critical minerals is strengthened. At the same time, new economic sectors and employment opportunities emerge, and broader social development objectives are supported. This transformation is not merely technological but fundamentally philosophical, as it redefines how value is created, preserved, and regenerated in an increasingly resource-constrained world.

19 Conclusion

Electronic waste valorization stands at the intersection of materials science, environmental stewardship, and industrial innovation. Rather than a burden to be disposed of, e-waste is increasingly understood as a regenerative input to the global economy, supporting sustainable growth while simultaneously mitigating ecological harm. Key valorization pathways include the recovery and refining of precious, base, and rare earth metals; polymer recycling and composite manufacturing; nanomaterial and catalyst production; energy material development; and the creation of eco-industrial symbiosis systems. Together, these pathways enable discarded electronic materials to reenter productive value chains and reduce dependence on virgin resource extraction. Green chemistry and biotechnology play a critical role in minimizing secondary pollution and preventing environmental leakage during valorization processes. At the same time, business models, policy frameworks, and social enterprise initiatives shape how benefits and risks are distributed across stakeholders, influencing whether valorization systems promote inclusivity, resilience, and long-term sustainability. Significant challenges remain, including variability in e-waste feedstocks, high technology and infrastructure costs, reliance on informal recycling sectors, and volatility in secondary materials markets. Despite these constraints, the broader trajectory is clear: urban mining and circular resource recovery are becoming central pillars of sustainable industrial development. Ultimately, e-waste valorization extends far beyond conventional recycling. It represents a reimagining of the material foundations of modern civilization – one in which waste is transformed into possibility, resources circulate rather than deplete, and innovation is directed toward the well-being of both people and the planet.

References

[1] Forti V, Baldé CP, Kuehr R. The global e-waste monitor 2020: Quantities, flows and the circular economy potential. Bonn, Germany: United Nations University; 2020. pp. 1–120.

[2] Baldé CP, Wang F, Kuehr R, Huisman J. The global e-waste monitor – 2014: Quantities, flows, and resources. Renewable and Sustainable Energy Reviews. 2015;52:1586–1596.

[3] Widmer R, Oswald-Krapf H, Sinha-Khetriwal D, Schnellmann M, Böni H. Global perspectives on e-waste. Environmental Impact Assessment Review. 2005;25:436–458.

[4] Lepawsky J, McNabb C. Mapping international flows of electronic waste. Geographical Journal. 2010;176:1–12.

[5] Cucchiella F, D'Adamo I, Koh S. Urban mining and e-waste management: A review of strategies and challenges. Renewable and Sustainable Energy Reviews. 2015;51:263–272.

[6] Tanskanen P. Management and recycling of electronic waste. Acta Materialia. 2013;61:1001–1011.

[7] Zeng X, Li J. Innovations in e-waste recycling technologies: From materials recovery to circular economy. Journal of Cleaner Production. 2017;144:328–339.

[8] Akcil A, Erust C, Gahan C, Ozgun M, Sahin M. Precious metal recovery from waste printed circuit boards using hydrometallurgical processes. Waste Management. 2015;45:258–271.

[9] Sun Z, Xiao Y. Hydrometallurgical recovery of metals from electronic waste: Process development and challenges. Metals. 2016;6:145–158.

[10] Isernia C, Passaro R, d'Angelo AM. Bioleaching technologies for metal recovery from electronic waste: Current status and future prospects. Journal of Hazardous Materials. 2020;387:121689–121701.

[11] Li J, Lopez N. Recycling of e-waste plastics: Material recovery and environmental implications. Waste and Biomass Valorization. 2018;9:1583–1596.

[12] Zuo X, Li J, Chen H, Liu L. Circular economy approaches for electronic waste management: A systems perspective. Resources, Conservation & Recycling. 2018;135:297–305.
[13] Geissdoerfer M, Savaget P, Bocken NMP, Hultink EJ. The circular economy: A new sustainability paradigm. Journal of Cleaner Production. 2017;143:757–768.
[14] Kirchherr J, Reike D, Hekkert M. Conceptualizing the circular economy: An analysis of 114 definitions. Resources, Conservation & Recycling. 2017;127:221–232.
[15] European Commission. Circular economy action plan: Electronics and batteries directive implementation report. Brussels, Belgium: European Commission; 2020. pp. 1–98.
[16] Sinha-Khetriwal D, Kraeuchi P, Schwaninger M. A comparison of electronic waste recycling in Switzerland and India: Social dimensions and policy implications. Environmental Impact Assessment Review. 2015;45:1–12.
[17] Bakhiyi B, Gravel S, Ceballos D, Flynn MA, Zayed J. Has the question of e-waste opened a Pandora's box? An overview of occupational and environmental health risks. Environment International. 2018;110:173–192.
[18] Gadd GM. Biosorption: Critical review of scientific rationale, environmental importance and significance for pollution treatment. Journal of Chemical Technology and Biotechnology. 2009;84:13–28.
[19] Monteiro CM, Castro PM, Malcata FX. Metal uptake by microalgae: Underlying mechanisms and practical applications. Biotechnology Progress. 2012;28:299–311.
[20] Mazzucato M. Mission-oriented innovation policies: Challenges and opportunities. Industrial and Corporate Change. 2018;27:803–815.

Index

 | https://doi.org/10.1515/9783112225509-018